# MATERIALS
# PRINCIPLES AND
# PRACTICE

# MATERIALS PRINCIPLES AND PRACTICE

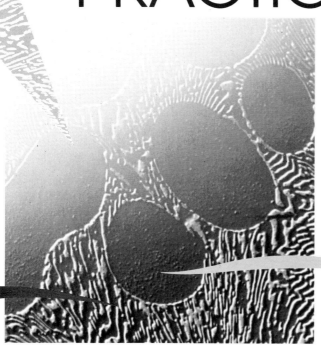

EDITED BY CHARLES NEWEY AND
GRAHAM WEAVER

The Open
University

MATERIALS DEPARTMENT
OPEN UNIVERSITY, MILTON KEYNES, ENGLAND

**Butterworths**
LONDON  BOSTON  SINGAPORE
SYDNEY  TORONTO  WELLINGTON

**British Library Cataloguing in Publication Data**
Materials principles and practice.
 1. Materials
 I. Newey, Charles    II. Weaver, Graham    III. Series
 620.1′1
 ISBN 0-408-02730-4

**Library of Congress Cataloging-in-Publication Data**
Materials principles and practice/edited by Charles Newey and Graham Weaver.
   p.   cm.–(Materials in action series)
 This text forms part of an Open University Materials Department course.
 Bibliography: p.
 Includes index.
 ISBN 0-408-02730-4
  1. Materials.   I. Newey, Charles (Charles W. A.)    II. Weaver, Graham
(Graham H.)    III. Open University. Materials Dept.
IV. Series.
TA403.M3455 1990

Butterworth Scientific Ltd

Part of Reed International P.L.C.

Designed by the Graphic Design Group of the Open University

Typeset and printed by Alden Press (London & Northampton) Ltd, London, England

This text forms part of an Open University course. Further information on Open University courses may be obtained from the Admissions Office, The Open University, PO Box 48, Walton Hall, Milton Keynes, MK7 6AB.

ISBN 0408 02730 4

# Series Preface

The four volumes in this series are part of a set of courses presented by the Materials Department of the Open University. The books are nevertheless self-contained. They assume that you are just starting to study materials, and that you are already competent in pre-university mathematics and physical science.

Unlike many introductory texts on the subject, this series covers materials science in the technological context of making and using materials. This approach is founded on a belief that the behaviour of materials should be studied in a comparative way, and a conviction that intelligent use of materials requires a sound appreciation of the strong links between product design, manufacturing processes and materials properties.

The interconnected nature of the subject is embodied in these books by the use of two sorts of text. The main theme (or story line) of each chapter is in larger, black type. Linked to this are other aspects, such as theoretical derivations, practical techniques, applications and so on, which are printed in red. The links are flagged in the main text by a reference such as ▼Assessing hardness▲, and the linked text, under this heading, appears nearby. Both sorts of text are important, but this format should enable you to decide your own study route through them.

The books encourage you to 'learn by doing' by providing exercises and self-assessment questions (SAQs). Answers are given at the end of each chapter, together with a set of objectives. The objectives are statements of what you should be able to do after studying the chapter. They are matched to the self-assessment questions.

This series, and the Open University courses it is part of, are the result of many people's labours. Their names are listed after the prefaces. I should particularly like to thank Professor Michael Ashby of Cambridge University for reading and commenting on drafts of all the books, and the group of student 'guinea pigs' who worked through early drafts. Finally, thanks to my colleagues on the course team and our consultants. Without them this project would not have been possible.

Further information on Open University courses may be had from the address on the back of the title page.

**Charles Newey**
Open University
January 1990

# Preface

To succeed in the market, products must — at least — use the right materials and be made by efficient processes. Getting both the materials and the processes right, from both the service and the manufacturing points of view, is what we mean by the series title MATERIALS IN ACTION. This book introduces that series.

We start in the context of production. We see that different materials provide different 'portfolios' of properties. These determine what the materials are suitable for, and how they may be processed. But materials' properties can be modified, and understanding how requires an appreciation of some basic science of materials. This is the nub of the middle chapters, where the state of matter is modelled as a balance between the tendencies of atoms to stick together (by chemical bonding) or rattle apart (by thermal agitation). The last three chapters are about property modification by control of microstructure. The agencies for change here are thermal, mechanical and chemical.

All the chapters of the book have benefitted greatly from the comments, criticisms and suggestions of our colleagues on the course team; we are particularly grateful to Andrew Greasley in this respect. Nick Braithwaite was a co-author of Chapter 3, and we should like to thank Nigel Mills and Peter Lewis for their important contributions to Chapters 5 and 7 respectively.

We are especially indebted to our editor, Allan Jones, for his forbearance and unstinting quest for clarity, to Andy Harding, our course manager, for his indispensable help in making it all happen, and to Phil Thompson, who took a large share of the burden of 'de-bugging' each chapter. We should also like to thank our secretaries Lesley Phelps, Lisa Emmington and Tracy Bartlett for working on so many (messy!) drafts, and Naomi Williams and Richard Black for many of the micrographs we have used.

**Charles Newey and Graham Weaver**

# Open University Materials in Action course team

MATERIALS ACADEMICS
Dr Nicholas Braithwaite (Module chair)
Dr Lyndon Edwards (Module chair)
Mark Endean (Module chair)
Dr Andrew Greasley
Dr Peter Lewis
Professor Charles Newey (Course and module chair)
Professor Nick Reid
Ken Reynolds
Graham Weaver
Dr George Weidmann (Module chair)

PRODUCTION
Phil Ashby (Producer, BBC)
Gerald Copp (Editor)
Debbie Crouch (Designer)
Alison George (Illustrator)
Andy Harding (Course manager)
Allan Jones (Editor)
Carol Russell (Editor)
Ernie Taylor (Course manager)
Pam Taylor (Producer, BBC)

TECHNICAL STAFF
Richard Black
Naomi Williams

SECRETARIES
Tracy Bartlett
Lisa Emmington
Angelina Palmiero
Lesley Phelps
Anita Sargent

CONSULTANTS FOR THIS BOOK
Dr Charles May (City of London Polytechnic)
Dr Nigel Mills (Birmingham University)
Dr Jerome Way (Bristol Polytechnic)

EXTERNAL ASSESSOR
Professor Michael Ashby (Cambridge University)

# Contents

# Chapter 5  Controlling the mix

# Chapter 6  Mechanical properties for processing and use

# Chapter 7  Chemical properties for processing and use

# Appendix 1  Periodic table

# Appendix 2  Names, abbreviations and descriptions of common polymers

# Chapter 1 Products, properties, processes and principles

In this chapter we shall explore the properties of different kinds of material, the ways in which materials can be shaped, and the uses to which they can be put. In the course of this exploration we shall see something of the relationship between the internal structure of materials and their properties. Knowledge of this relationship, and of the processes by which materials can be shaped, is fundamental to the design and manufacture of products, which we shall also consider. Later chapters, and other books in this series, will develop these themes more extensively.

## 1.1 The role of materials

The materials we are concerned with begin as physical resources in or on the Earth. They occur as gases, as liquids (especially water and oil), as inorganic solids (minerals) and as organic solids both dead (coal for example) and alive (wood for example). Thousands of these raw materials are used in making the myriad products to meet our needs (and wants) of food, shelter, communication, medicine, entertainment and so on. We are concerned in this book with the materials from which these artefacts are made. They are usually called **engineering materials**, that is materials designed, made and used for a practical purpose — carrying mechanical loads, electric currents, providing thermal insulation, resisting corrosion or whatever. *? vague*

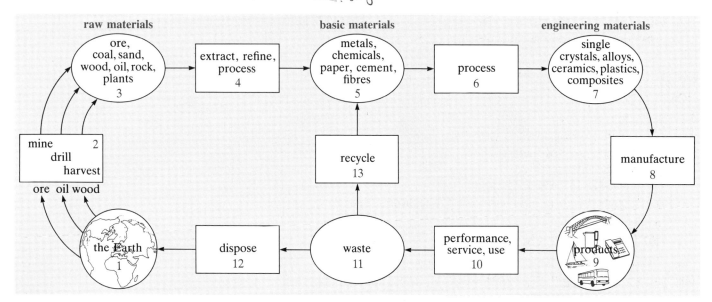

Figure 1.1 'Life-cycle' of engineering materials

Figure 1.1 puts a global perspective on the life-cycle of engineering materials: a short manufacturing cycle links into a very long geological one within the Earth. The symbols used are conventional in diagrams of this and related types. The ovals (or circles) represent states, in this diagram states of a material. The rectangles are processes, and arrows indicate flows. The diagram is in a highly generalized form and the details of processes and states differ from one material to another. For instance, notice that the recycling of waste products offers a short-cut in the cycle. Recycling is used extensively for some materials — about 60% of lead is recycled for example and up to 30% of a glass bottle comes from old broken glass — but, at present, there is very little recycling of plastics and none at all for some other materials such as concrete. (See ▼Recycling and re-using▲.)

# ▼Recycling and re-using▲

The term **recycling** is often used loosely. Strictly, it means feeding waste material (scrap) back into the *same* processing cycle — the product is of the same quality as the original. Making glass bottles from old ones and using old car bodies in steelmaking are examples. This is different from **re-using** scrap that is degraded to make some other, less demanding, product; for example, re-using old newspapers for cardboard and old concrete for road foundations.

Two sorts of scrap may be recycled or re-used: products at the end of their useful life (called **old scrap**) and waste material, such as rejects and swarf from machining, produced during manufacture (**new scrap**). Figure 1.2 illustrates the distinction between old and new scrap in the context of aluminium drinks cans.

The cost of materials to a manufacturer is usually the largest element in the overall cost of a product. This is true of products as varied as aeroengines, concrete building blocks, and shoes and garments (where traditionally labour has been the greatest cost). There is thus a strong financial incentive to recycle and re-use materials.

Why isn't recycling of old scrap more prevalent?

Cost and technical feasibility. Scrap usually consists of blended or combined materials from which the original ingredients cannot easily be recovered. For a manufacturer, recycling is only justified if the cost of collecting scrap and separating out the required material is less than the cost of new starting materials.

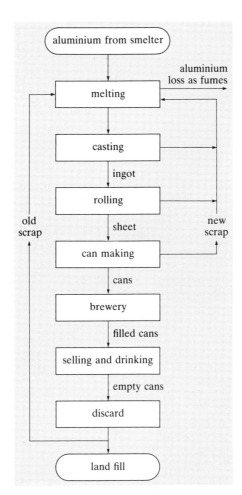

Figure 1.2 Recycling of scrap in manufacture of aluminium cans

In the early stages of human existence, technological, activity was limited to the naturally occurring materials to hand: wood, bone, hide, natural fibres, stone, flint and so on. Even then, the raw materials were usually manipulated in some way to make them more suitable for particular uses, albeit by simple processes such as drying and chipping. Figure 1.3 shows some examples. Some of the rudiments of shaping were obviously appreciated, especially the importance of the relative hardness of materials for chipping and grinding — a very hard material would be needed to form the boring tool. (See ▼**Assessing hardness**▲.)

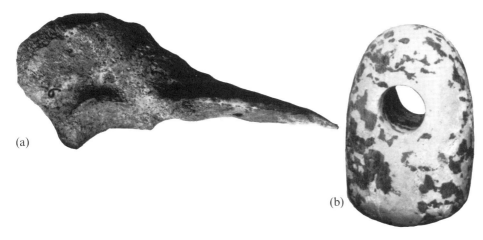

(a)

(c)

(b)

Figure 1.3 Neolithic artefacts, about 3000BC. (a) Flint boring tool. (b) Ceremonial stone hammer head. (c) Bone needle (approx. 10 cm long). Courtesy of the National Museums of Scotland

# ▼Assessing hardness▲

A **hard** material is difficult to scratch, wear away by abrasion or to indent. Hardness is not a fundamental property of a material: for each method of measuring it, it is some combination of elastic, plastic, and (in some cases) fracture properties. Hardness can be measured only by comparison with a material used as a scratcher or indenter and has objective meaning only in terms of a specific type of test. For example, glass will scratch steel but fractures more readily under indentation; nylon has a high resistance to wear but not to indentation.

The first systematic hardness scale was proposed by Mohs in 1822. Ten standard minerals, ranging in hardness from talc to diamond, were used as the reference scale. The hardness of a material under test was determined by which of the reference minerals could scratch the material. The scale was not very sensitive to different degrees of hardness.

Nowadays hardness is usually measured in terms of resistance to indentation, and the most common test is the **Vickers hardness test**. A diamond indenter, in the form of a square pyramid with an included face angle of 136°, is pushed with constant force into the surface of a sample (Figure 1.4a). The resulting 'square' impression (Figure 1.4b) is viewed in a measuring microscope and the two diagonals measured.

The hardness is given a number $H_V$ which is calculated as the load (in kg) on the indenter divided by the area of the faces of the indentation. Most hardness testing machines have a set of tables from which, having determined the average of the two diagonals of the 'indent', the user can read off the Vickers hardness number.

The Vickers test is more common than the earlier Brinell test (BH), which uses a steel ball as the indenter. Another test, the Rockwell, uses different loads with a diamond indenter for hard materials and a range of steel balls for softer materials.

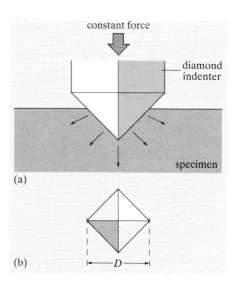

constant force

diamond indenter

specimen

(a)

(b) ←— $D$ —→

Figure 1.4 Vickers hardness test

Nowadays most of the materials we use are artificial, and many more processes are available for manipulating and shaping them. Of course, some naturally occurring materials are still used in relatively unchanged form — timber in furniture and stone in buildings for instance — but really only for simple functions. In their original state, these materials are not able to meet other, more demanding functions. On the other hand, the engineering materials used in the artefacts that provide our vastly improved living conditions are purpose made. For artefacts as simple as paperclips, pens, and window panes, let alone those as complex as computers, satellites and nuclear power stations, raw materials are refined, combined and processed into engineering materials with utterly different, or vastly improved, characteristics. Polyethylene kitchen utensils don't look a bit like the crude oil from whence they came, and the highly pure and perfect single crystals used in silicon chips (Figure 1.5) have come a long way from grains of sand. ▼ **Silicon single crystals** ▲ gives a brief summary of how silicon electronic devices are made.

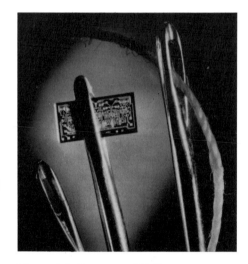

Figure 1.5 Courtesy of Philips Components Ltd

# ▼Silicon single crystals▲

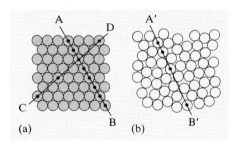

Figure 1.6 (a) Crystal. (b) Glass (or amorphous) solid

In simple terms, solids can be **crystalline**, **glassy** or a mixture of the two. Figure 1.6 illustrates the difference. The glass in Figure 1.6(b) lacks the long-range order of the crystal in Figure 1.6(a) — compare the line AB passing through the centres of some atoms in the crystal with A'B' for the glass. Notice that the glass has short-range order in that it consists of similar small clusters — in this case three-atom clusters.

Silicon can be made in either a crystalline or an amorphous form. Crystalline silicon is the basis of most electronic devices such as transistors, diodes and integrated circuits. Such devices put stringent requirements on the material from which they are made. In particular, they usually have to be **single crystals**, that is, a crystalline solid in which the long-range order extends throughout its volume.

The starting material for a device is a **wafer** — a flat disc cut from a single crystal rather as slices are cut from a salami sausage (Figure 1.7). An array of devices is built onto and into the surface of the wafer by a combination of oxidation, masking, etching and deposition of other elements. The wafer is then cut into a number of chips (typically 250).

Silicon single crystals begin as silica

Figure 1.7 Polished silicon wafer, courtesy of Monsanto Electronic Materials Company. For electronic devices, the wafer must be of the highest possible quality

($SiO_2$), which is reduced to silicon rather as iron oxide is reduced to iron. The product contains about 98% silicon. Further elaborate purification procedures are necessary before it is adequate for electronic use.

A common way of producing a single crystal is the Czochralski technique, shown in Figure 1.8. A small single crystal of silicon is used as a 'seed' on which to grow a large crystal (up to 25 cm in diameter). It is held in a water-cooled chuck and lowered into the molten silicon. When the temperature is just right for the seed to grow, it is slowly rotated and withdrawn.

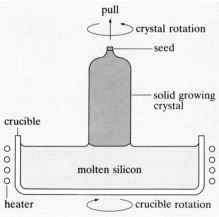

Figure 1.8 Growing a single crystal of silicon

The photograph in Figure 1.5 was popular in text books of the 1960s. I've included it because of its counterpoint with Figure 1.3. The eye in the Neolithic needle is much larger. The microcircuit is fabricated on a silicon chip and contains 120 electronic components (resistors, transistors and so on). Today, a chip this size could contain 10 000 components!

The vast range of engineering materials in use today is the result of deliberately searching for ways of making new and improved ones and invention continues apace, especially the invention of new plastics and of materials for new electronic devices. Over the past 100 years or so, the development of new materials has been one of the instrumental factors determining our 'quality of life' and from time to time particular materials and processes have had sudden and dramatic consequences. Especially important examples are the mass production of steel (about 1860) — see ▼Conventional steelmaking▲ — plastics such as polyethylene and PVC (from the 1930s), the separation of nuclear fissile material, uranium 235 (early 1940s) and silicon, the cornerstone of the electronics industry (early 1950s).

The factors and circumstances that have led to such significant advances are well beyond the scope of this book but in essence they centre on economic 'ends' and scientific 'means'. We shall be concerned primarily with scientific means. But, as you will see, the economic ends cannot be ignored. As you know, in many 'developed' countries economic ends stem from an economic/political system that relies on the production of artefacts (goods in economic parlance) for profit. It is a system of supply and demand, which generates competition. One important consequence of this is a drive to develop new materials that can provide new, improved or cheaper products.

# ▼Conventional steelmaking▲

Steel is not one material but hundreds. All are alloys of iron containing less than 2% by weight of carbon (alloys with more carbon are called **cast irons**). Even at such low concentrations, varying the carbon content and the way it is distributed can provide materials as diverse as those used, for instance, in car body panels (fairly strong yet dentable), bolts (strong), piano wire (very strong), cutting tools (very hard but brittle). Alloying with other elements in addition to carbon provides the different steels used in boiler tubes (high strength and oxidation resistance at high temperature), stainless kitchen utensils and surgical instruments (corrosion resistance), tools (hard and strong), transformer cores (magnetically 'soft'), permanent magnets (magnetically 'hard').

Figure 1.9 shows the process of conventional steelmaking. The chemical information is much simplified.

The pig iron from the blast furnace is brittle and useless for engineering products because it contains too much carbon (about 4%) and other impurities such as sulphur. The carbon content is reduced by controlled oxidation in a converter. Bessemer's original converter (*c*.1860), which provided the means of mass producing steel, relied on the oxygen in a blast of air. Modern versions use nearly pure oxygen. Notice that a substantial part of the metal input to the converter is scrap — both new and old — such as car bodies.

The molten steel is poured into a large mould and solidifies as an ingot.

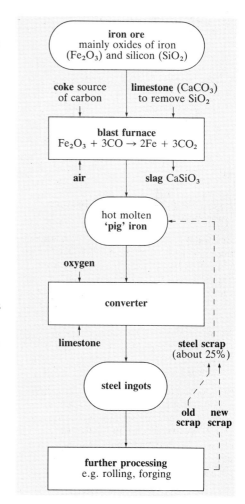

Figure 1.9 Steelmaking

Prominent amongst the means to these ends is an increasingly well developed scientific understanding of why materials behave in the way they do: an appreciation of how their properties relate directly to their internal structure (the types of atom or molecule they contain and how these are bonded together and spatially arranged) which, in turn, is strongly influenced by their thermal and mechanical history, that is, how the material has been processed from its ingredients. Elucidation of the fundamental structure–property–process relationships that have been established and their technological applications is a major theme of this book. These scientific concepts and principles have led to a better idea of where to go searching for new and better materials. However, it is important to appreciate that the development, production and use of materials is by no means a completely science-based industry. Most materials are complex and far from fully understood.

There are all sorts of materials, made up of different types and numbers of atoms or molecules held together by different forces and spatially arranged in different ways. A striking example is that combinations of carbon, hydrogen and oxygen alone give rise to very many compounds as diverse as alcohols, cellulose, fats, and poly(methyl methacrylate) (Perspex). Clearly the details of a material's make-up are crucial. As a consequence of its individual make-up, each type of material behaves in a unique way: it has a particular set of properties.

Now, the design of a product depends critically on the properties of the material(s) from which it is made. Two aspects of the properties are important: those that determine how the product performs in service, and those that determine the processes by which the material is transformed to the required shape. For instance, a functional requirement of window panes is that they must be transparent, and glasses can provide this property (together with brittleness). But it is the viscosity of glasses and the way it falls with increasing temperature that determines the conditions for making glass into flat sheets.

EXERCISE 1.1 Consider the Neolithic 'borer' in Figure 1.3. What interdependencies can you see between the design requirements of the artefact, the properties of the material, and the way it is made?

Interdependencies such as these apply to all products. Since we shall be referring to them frequently, I'm going to abbreviate them to PPP: P for product (design), P for properties (of materials) and P for processes.

Now PPP is really to do with the technical performance of a product — whether it functions properly and uses the right materials and manufacturing processes to meet its technical design specification. These aspects of product design and manufacture are of direct concern to engineers. Unfortunately, many engineers think that this is all they need be concerned about. You may, for instance, have heard or read quotes like: 'We can make products that perform well but we can't sell them.' Clearly it is unwise to concentrate solely on getting the technical bits right and to ignore the possibility of a product failing for other reasons. Let's look at the other reasons a bit more closely.

# 1.2 Products and materials

To be successful, a product must do more than meet its technical specification: it must sell. Designing a product to sell means taking into account a wide variety of factors: cost, marketing, reliability, competition, and so on. We shall call these **external factors** because they can be considered as external to the technical (PPP) factors. The two are intimately connected as you will see.

Selling is about supply and demand. Put simply, in a competitive market the **price** of a product depends on the **demand** for it. The materials, equipment, administration and so on needed to manufacture the product determine its **cost**. Usually, to stay in business a manufacturer has to make a profit. The cost must be less than the price.

Figure 1.10 is my attempt to illustrate the interconnected nature of the factors that influence product cost and price. This is an example of a **pattern diagram**, and if it appears messy, don't be put off! It will become much clearer and its usefulness apparent, when we have followed a few connections. The technical (PPP) factors are highlighted at the centre of the diagram; in here are included all the technical decisions about the design of the product and the choice of material and manufacturing process. The cost elements are to the right and the price elements to the left. Follow the links in the diagram.

Figure 1.10

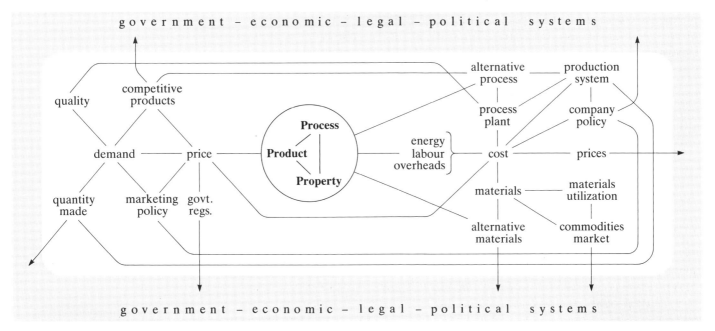

17

The cost of a product depends on the prices the manufacturer has to pay for materials, equipment, labour and other elements of overheads such as administration, research and development, and marketing. Some of these are themselves subject to the vagaries of supply and demand, particularly ▼The price of materials▲. The price put on the finished product is strongly influenced by the demand for the product; and both price and demand depend on the price of competitive products, quality of the product, the marketing policy adopted, government regulations and so on. So, the further away the links get from PPP, the more unquantifiable and uncontrollable the dependences become. Quite quickly they can get involved with government policy, the economic system and international politics.

EXERCISE 1.2   Using Figure 1.10 or your own modified version of it, suggest possible ways of avoiding increases in the cost of a product in

(a) the short term
(b) the long term.

The difficulty with this sort of complexity is thinking about it in an organized way. One useful way is the **systems approach**, that is, looking at a complex problem as a system, being clear about why you are looking at it, and then making judgements about the factors to include in an analysis. For consistency, we need a few basic definitions and these are given in ▼The systems approach▲

SAQ1.1   (Objective 1.1)

(a) The manufacturer of a particular product is concerned only with using the materials and the manufacturing processes that give the best technical performance of the product. Specify the factors in Figure 1.10 you would include (i) inside your system boundary, and (ii) in the environment of your system.

(b) Evaluation of the properties of candidate materials is obviously a crucial part of selecting the one from which to make a product. In such an analysis, it is common practice to include the cost of materials to the manufacturer as a 'property'. Using Figure 1.10 comment on (i) the advantages and (ii) limitations of this tactic.

In Section 1.8 we consider PPP further and look at a way of drawing up a design specification for a product which encompasses a wide variety of external factors. Before doing this, we need to consider P for properties and P for processes in more detail.

# ▼The price of materials▲

The price to a manufacturer of different materials varies enormously. Table 1.1 is a list of prices at the time of writing of a range of engineering materials. Some prices are for the material in its basic form — bags of cement, plastics granules, metal ingots and so on; other prices are for simple fabricated shapes or formulations — bars, beams, composites and so on.

The difference between the two reflects the costs involved in processing the basic material into a more directly usable (and therefore valuable) form. At all stages of manufacture, the extra cost from further processing is called the **added value**. In our context it is more appropriate to think of it as the **processing cost**. You will notice from Table 1.1 that the processing cost is significant and sometimes startlingly high.

Essentially, the price of basic materials is determined by two factors:

(a) Supply and demand. These can create short-term fluctuations of price and, for some metals in particular, are subject to speculation in the commodity markets, and to factors such as strikes and political actions that affect the output of mines.

(b) The actual cost of producing the material. This is the source of long-term changes in price. The real cost is that of extracting, transporting and refining the material into its basic form (this is often called primary processing). Inflation and energy costs also drive the price up, as they do all processing stages. The energy cost can be especially important because with increasing consumption of physical resources, metals in particular have to be gleaned from poorer and poorer ores.

Looking at the prices in Table 1.1, it is not surprising that concrete, wood and mild steel are so popular.

Table 1.1   Prices in 1989

| Material | UK£/tonne | US$/tonne |
|---|---|---|
| diamonds (industrial high quality) | 320 000 000 | 500 000 000 |
| platinum | 10 600 000 | 16 500 000 |
| gold | 9 300 000 | 14 500 000 |
| silicon (single crystal wafers) | 6 400 000 | 10 000 000 |
| silicon (as extracted) | 2 000 | 1 300 |
| palladium | 2 600 000 | 4 050 000 |
| silver | 1 380 000 | 2 150 000 |
| CFRP (special rod) | 110 000 | 170 000 |
| GFRP (special rod) | 25 000 | 39 000 |
| glass fibres | 1 000 | 1 500 |
| carbon fibres | 30 000 | 45 000 |
| epoxy resin | 4 000 | 6 000 |
| polyether (etherketone) (PEEK, a heat-resistant polymer) | 20 000 | 30 000 |
| silicon carbide (for engineering ceramics) | 18 000 | 27 000 |
| silicon carbide (for abrasives) | 900 | 1 400 |
| silicon carbide (for refractory bricks) | 500 | 750 |
| tungsten | 13 000 | 19 500 |
| cobalt | 11 000 | 17 000 |
| titanium | 5 300 | 8 300 |
| polycarbonate (pellets) | 3 500 | 5 300 |
| brass (60/40 as sheets) | 2 500 | 3 750 |
| brass (ingots) | 1 100 | 1 650 |
| aluminium | 1 600 | 2 400 |
| steel (stainless) | 1 800 | 2 700 |
| steel (high speed) | 800 | 1 200 |
| cast iron (grey) | 500 | 830 |
| steel (low alloy) | 500 | 750 |
| steel (mild — as angles) | 230 | 350 |
| hard wood (teak veneer) | 1 100 | 1 650 |
| hard woods (structural) | 350 | 530 |
| soft woods (structural) | 230 | 350 |
| synthetic rubber | 900 | 1 400 |
| natural rubber (commercial) | 580 | 870 |
| polystyrene | 850 | 1 300 |
| lead (sheet) | 780 | 1 200 |
| lead (ingot) | 380 | 570 |
| polyethylene | 700 | 1 100 |
| poly(vinyl chloride) | 650 | 1 000 |
| glass | 500 | 750 |
| alumina | 230 | 350 |
| magnesia | 220 | 330 |
| concrete (reinforced beams) | 220 | 330 |
| cement | 45 | 70 |

# ▼The systems approach▲

The first step is to have a purpose for studying an issue or problem, and also to be clear about the point from which you are viewing it. This is of fundamental importance because everything follows from it. For example, given a common purpose — assessing why a particular manufacturing company went bankrupt — the viewpoint of a manager or a machine operator within the company would be different from one another and from that of a customer, a supplier or a competitor.

We need some basic definitions:

**System** An assembly of components that relate to each other in an organized way, and which act together to do something. Hence, legal system, hi-fi system, an educational assessment system, and the aluminium can system shown in Figure 1.2.

**Systems boundary** The imaginary line drawn round everything that is to be included in the system. It is often difficult to decide what to put in and what to leave out. But the line has to be drawn somewhere! Deciding where the system boundary is can be crucial to the outcome — arguments often arise because people take different things into account and come to opposing conclusions.

Being clear about the purpose of your study makes it easier to decide where to put the system boundary. For example if you wanted to know the yearly wastage in an aluminium drinks can factory you could use the system shown in Figure 1.11 with numbers for the annual input of aluminium and the annual output of cans.

If, on the other hand, you wanted to know the sources of the waste at each stage of can making, you would have to 'open up the box' of Figure 1.11. This is done in Figure 1.12.

Figure 1.11 System representation of production of aluminium cans

Figure 1.12 The can-making system

**Subsystem** A part of a system that is looked at separately for a particular purpose. If you wanted to know, say, the scrap produced at each stage in the can-making plant, Figure 1.12 would be the system for study. It is a subsystem of the aluminium can manufacturing system.

**System environment** It would be artificial and, in practice, disastrous to ignore everything that goes on outside the system boundary. This is where the idea of the system environment comes in. It is the region outside the boundary that contains those aspects you think might have an important influence on the behaviour of the system. So, as well as direct influences through the supply and demand subsystems, it might include a whole range of external factors such as tax law, government grants and regulations on pollution control.

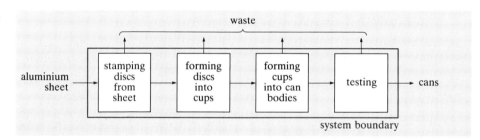

## Summary

A successful product design must take into account:
(a) the technical factors concerned with its design and the materials and processes used in its manufacture,
(b) a wide variety of external factors that determine the cost to make it and its price in the market.

These technical and external factors are interrelated in complex ways and materials are involved in both.

# 1.3 Materials properties

With so many materials in use, it is convenient to classify them into types. Traditionally materials are divided into **metals**, **ceramics**, **plastics** and **natural materials**. Like all classification systems, this one is imperfect, but it's still useful as a way of learning about materials, so we shall use it.

The first three types, metals, ceramics and plastics, are artificial and they were developed independently of each other — presumably because they are chemically very different. Essentially, metals are based on metallic chemical elements, plastics on large organic molecules and ceramics on inorganic compounds. Their production industries are thus quite distinct from one another as ▼**Working clay**▲,  ▼**Producing polyethylene**▲ and 'Conventional steelmaking' (Section 1.2) demonstrate.

Ceramics were originally clay-based materials — 'ceramic' comes from the Greek *keramos*, meaning burnt earth — but they are now often considered to include glass and stone (already you can see the classifications are being bent to accommodate exceptions). They are quite different from natural materials, which we define as consisting of large natural organic molecules — plastics, remember, are based on artificial organic molecules.

## ▼Working clay▲

Clays are produced by the weathering of igneous rocks such as granite. They consist of various aluminium silicates, such as kaolinite [$Al_2O_3.2SiO_2.2H_2O$], together with small amounts of other mineral such as limonite $Fe_2O_3 \cdot H_2O$, which gives most clays their reddy-brown colour, and quartz ($SiO_2$). When mixed with the right amount of water, clay becomes mouldable (plastic). It becomes hard if it is then baked.

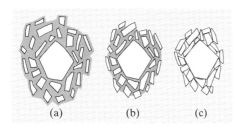

(a)          (b)          (c)

Figure 1.13  Clay artefact drying

In wet clay (Figure 1.13a) the clay particles are separated by a film of water — some 25% of the volume is water — with only surface tension forces holding the whole lot together. During baking (typically at about 1200 K) the water evaporates and the mixture shrinks until the particles are touching, Figure 1.13(b). This first stage of drying must be done very slowly because the object will crack if it dries non-uniformly. Coarse-grained material, such as flint, added to the clay helps to prevent this mishap by improving the strength of the mix. Further drying removes water from between the particles leaving behind spaces filled with air, Figure 1.13(c). Artefacts made in this way, for example bricks, and roof tiles, are called **earthenware**. They are porous and brittle.

If the clay mix is fired at a higher temperature, some ingredients melt together to form a glass that spreads around the solid particles and, on cooling, binds the whole together. The result, called **stoneware** or **vitrous ceramic**, is stronger than earthenware but usually still porous. The surface can be sealed by glazing. A glaze is a low-melting temperature glass applied to the artefact as a powder after firing. When the artefact is re-fired, the glaze flows over the surface and enters the pores by capillary action. Glazing allows decorative colours and textures to be applied; hence our familiar cups, plates and pots.

Figure 1.14 Chinese Guan ware vase, 13th C. Courtesy of the Percival David Foundation of Chinese Art

In the pot in Figure 1.14 the glaze has been deliberately made to crack to provide a decorative feature — a clear indication that the Chinese had a sound awareness of materials properties and accurate process control.

# ▼Producing polyethylene▲

More polyethylene is used than any other plastics material. It is ubiquitous in containers of all sorts, packaging and electrical insulation. Polyethylene usually begins as crude oil. Crude oil is a viscous liquid containing a complex mixture of chemical compounds of carbon and hydrogen (called hydrocarbons).

The remarkable feature of the petrochemical industry is the strategy of taking the feedstock (oil or natural gas) and breaking it down into a basic set of small molecules or **intermediates** that are then used as building blocks for a wide variety of products; importantly, the mix of products can be changed according to demand. First, in a refinery crude oil is distilled into major fractions — gasolene (petrol), naphtha, kerosine, bitumen etc. The naphtha fraction is then broken down (cracked) into intermediates.

Ethylene, the precursor for polyethylene, is an intermediate. Its composition is $C_2H_4$ and its structural formula is

$$
\begin{array}{cc}
H & H \\
| & | \\
C & = C \\
| & | \\
H & H
\end{array}
$$

Note the double bond between the carbon atoms. Being a simple molecule, it can be obtained from sources other than crude oil — mainly natural gas in the USA and coal in South Africa.

Polyethylene is a type of polymer. Polymers are very large organic molecules made by the successive linking together of small molecules. The small molecules are called **monomers** and provide the repeat units from which the poly (many) 'mers' are built. Converting monomers into a polymer is called **polymerization**.

Ethylene is the monomer for the polymer polyethylene. In essence, during polymerization the double bond in each monomer is opened up, thus allowing them to join up. This process can be described as

$$
\begin{array}{c}
H \quad H \quad\; H \quad H \quad\quad\; H \quad H \quad H \quad H \\
| \quad\; | \quad\quad | \quad\; | \quad\quad\;\; | \quad\; | \quad\; | \quad\; | \\
C = C \; + \; C = C \; \rightarrow \; - \; C-C-C-C- \dots \text{etc.} \\
| \quad\; | \quad\quad | \quad\; | \quad\quad\;\; | \quad\; | \quad\; | \quad\; | \\
H \quad H \quad\; H \quad H \quad\quad\; H \quad H \quad H \quad H
\end{array}
$$

(Actually, this polymerization is a single-ended chain building reaction; it is considered in detail in Chapter 7.) The length and linearity of the molecular chains in commercial polyethlene depend on the polymerizing conditions. There are two types of process route. The first uses low temperatures (about 350 K) and complex catalysts and yields linear molecules very much like those you would expect from the configuration shown. The molecules pack closely together and the product is called **high density polyethylene** (HDPE). The other process uses simple catalysts at higher temperatures and pressures and produces molecules which are not linear; they have numerous short side branches attached to the backbone of the molecule, as Figure 1.15 illustrates. Clearly, such molecules cannot pack together so closely. This is **low density polyethylene** (LDPE).

Polyethylene made from molecular chains containing fewer than about 300 repeat units is very soft and has no strength — the short molecules slide apart very easily and it is rather like candlewax. If the molecules are too long, more than about 3000 repeat units, the polymer becomes too viscous to be moulded to shape. Commercial polyethylenes have molecular lengths between 500 and 2000 repeat units.

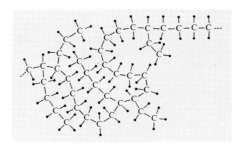

Figure 1.15 Branched LDPE molecule

We identify materials and distinguish between them by their characteristic qualities rather than by, for instance, their content or indeed their origin. Thus we say a material is hard, floppy, magnetic; it rusts, bends, breaks easily; it is clear, shiny and so on.

In order to compare materials objectively we need to translate these qualities into properties we can measure. How do such properties relate to the materials types in our classification? Exercise 1.3 should help you to answer this. (You may need to look at ▼Review of basic properties▲ first.)

# ▼Review of basic properties▲

Exercise 1.3 assumes that you are familiar with thermal and electrical conductivity and with a range of mechanical properties that can be determined from the stress–strain curve of a material. If you want to check these basic ideas, try SAQ 1.2 before you try Exercise 1.3.

SAQ 1.2   (Revision)

(a) Define (i) thermal conductivity, (ii) electrical conductivity and (iii) electrical resistivity, and state the SI the units in which they are measured.

Figure 1.16 shows the stress–strain curve for a range of hypothetical materials, numbered 1 to 5.

(b) If material 1 is unloaded at point F, by how much will it be permanently extended?

(c) Which material has the highest
 (i) Young's modulus
 (ii) yield stress
(iii) tensile strength
(iv) ductility?
(d) Define the term 'work hardening'.

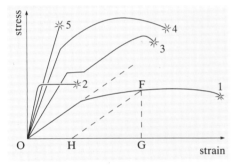

Figure 1.16

EXERCISE 1.3   Try completing Table 1.2 overleaf. You should:
(a) Give four common examples of each type of material.
(b) Give an assessment of the properties of each type. Use any scheme you like, such as high, medium, low. (Toughness, a property not covered in SAQ 1.2, is a measure of robustness; a material that breaks easily isn't tough. More about toughness in Chapter 2.)
(c) Give four examples of artefacts (or their components) in which such materials are used.
(d) Comment on the properties that could influence how each type is shaped into an artefact; its ductility for instance.
(e) Give examples of materials which do not fit readily into this classification.

For this exercise you may find it helpful to think about the articles around your home — their shapes and the materials used.

Table 1.2  Your characterization of metals, plastics, ceramics and natural materials

| | Plastics | Metals | Ceramics | Natural Materials |
|---|---|---|---|---|
| **Common examples (four for each)** | | | | |
| Young's modulus | | | | |
| Toughness | | | | |
| Tensile strength | | | | |
| Compressive strength | | | | |
| Electrical conductivity | | | | |
| Thermal conductivity | | | | |
| Density | | | | |
| Melting or softening temperature | | | | |
| Ductility | | | | |
| Hardness or wear resistance | | | | |
| Optical characteristics (e.g. transparency) | | | | |
| Resistance to deterioration in different chemical and biological environments | | | | |
| **Examples of artefacts (four for each)** | | | | |
| Comments on properties that influence processing to shape | | | | |
| Materials that do not fit this classification | | | | |

What conclusions did you reach whilst working through Exercise 1.3? My main ones were:

(a) The classification is rather crude. Within each group there is a large variation in behaviour and some particular materials are exceptional. For instance, traditional cast iron has a toughness and tensile strength typical of a ceramic; unlike other ceramics, glass flows and melts at reasonable temperatures; rubber is very floppy. So, in terms of properties, there are no clear boundaries between the groups.

(b) As indicated by toughness, strength, stiffness and wear resistance, there seem to be two distinct types of plastic. In fact there are; they are called ▼**Thermoplastics and thermosets**▲

(c) Ductility and softening or melting temperature are important in determining how easily materials can be shaped into products. More about this later.

What about the exceptions?

There are two important kinds of exception. You will probably have spotted one of them because it is typified by silicon. I will call them **non-metallic elements**. Other important examples are germanium (once

# ▼Thermoplastics and thermosets▲

Before I define these two types of plastics it is useful to distinguish between **plastics** and **polymers**. As pointed out in 'Producing polyethylene', polymers like polyethylene are very large molecules formed by the successive linking of one or more types of small molecules (monomers). Natural rubber, fats, proteins, nucleic acid and so on are natural polymers; wood is essentially fibres of one polymer (cellulose) embedded in another (lignin). The adjective 'plastic' is generally applied to materials that undergo a permanent change of shape under an applied force; it applies to metals, clay, putty and many others. The noun 'plastics' has been adopted to describe the class of engineering materials based on synthetic polymers. Commercial plastics are tailored products in which compounds may be added to basic polymers. For example, plasticizers may be added to enhance flexibility; fillers, such as wood, flour, glass, and chalk powder, to stiffen and reduce cost; antioxidants to inhibit degradation in service; flame retardants to reduce flammability; and pigments for colour.

There are two types of plastics: **thermoplastics** and **thermosets** (more formally **thermosetting plastics**). Thermoplastics become fluid at higher temperatures; as their name implies, they can be moulded at these temperatures under pressure, the new shape being retained on cooling. An important characteristic of thermoplastics is that the process of softening by heating is reversible and repeatable. Examples are polyethylene, poly(methyl methacrylate) (that is, Perspex or Plexiglass), and polyamides (for example, Nylon).

Thermosets, again as the name implies, can be moulded by heat, but only once. The process is irreversible. Because of the chemical reactions that take place during polymerization, the material sets into a rigid mass. Examples of thermosets are epoxy resins (Araldite), phenol–formaldehyde (Bakelite) and polyester resin used in glass reinforced plastics.

Many polymers are known by trade names: Perspex, Nylon, Terylene, Melamine, Araldite and so on. These names are not particularly useful because each commercial plastic has a different formulation of polymer(s) and additives. It is now standard practice to use abbreviations for the common engineering polymers. PE is polyethylene, PMMA is poly(methyl methacrylate), PVC is poly(vinyl chloride), and so on. A list of the common polymers and their abbreviations is given in Appendix 2.

widely used in the electronics industry) and carbon, as both diamond and graphite. In many of their applications these materials are used in the form of single crystals. As 'Single silicon crystals' showed, the production of single crystals requires very special techniques and is, therefore, very costly compared with the normal polycrystalline form. So the reasons for using them have to be fairly compelling. Many electronic devices need to be made from structurally perfect materials, so single crystals are used. See ▼**Silicon devices: polycrystal or single crystal?**▲

The second exception in the materials classification is **composites**. These are physical mixtures of two materials (usually from different groups in the classification). In a successful composite, the combination of materials has better properties than the individual components. The most well known example is glass fibre reinforced plastics (GFRP). Here, the glass fibres stiffen and toughen the thermoset host material (usually polyester or epoxy). GFRP is used, for instance, in some car bodies, yacht hulls, the nose-cones of aeroplanes and fast trains. But composites aren't a new idea. In ancient Egypt it was appreciated that fewer bricks broke during manufacture if straw was mixed with the clay; in mediaeval times wall plaster was improved by incorporating horse hair, and Victorian papier mâché (paper and paste) was stronger and tougher than you might think (Figure 1.17).

Wood, bone and cartilage are also composite materials. In all of these examples fibres are incorporated, but many composites use particulate materials — stones in concrete, chalk, wood or glass powder in plastics, tungsten carbide in cobalt alloys for machine cutting tools, and so on. These few examples demonstrate how the idea of a composite cuts right across the classification we have used.

Figure 1.17 Nineteenth-century papier-mâché chair

# ▼Silicon devices: polycrystal or single crystal?▲

Silicon electronic devices are usually made from single crystals; but most solids are produced by the cooling of a liquid, which generally produces a polycrystalline material. Figure 1.18 illustrates, schematically, what happens.

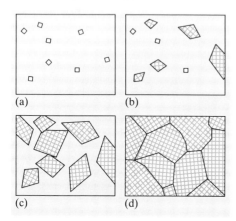

(a)   (b)   (c)   (d)

Figure 1.18 Grain boundaries forming during solidification

In (a) solid particles in the liquid act as nuclei for the forming solid; the crystals grow, (b) and (c), until solidification is complete, (d). The grid of lines in each crystal (or grain) represents the repeating pattern within the crystal. Notice that the interfaces between the crystals — called **grain boundaries** — are irregular in shape because the growth of an individual grain is restricted by the growth of its neighbours.

Figure 1.19(a) shows the surface of a piece of polycrystalline silicon produced by cooling from the liquid. The dark lines are the grain boundaries. This is compared directly in Figure 1.19(b) with a schematic drawing to the same scale of the plan view of a typical repeat unit used in some integrated circuits.

Silicon is so useful and versatile because its electrical resistivity can be controlled by 'doping' the silicon with atoms of another element. These atoms are persuaded to migrate into the crystal, by a process called diffusion. The dopant can be placed in the crystal where it is required. (More about this in Chapters 3 and 4.) The various areas in the integrated circuit in Figure 1.19(b) require different concentrations of dopant.

It is clear from Figure 1.19(a) that the grains of silicon would form a 'crazy paving' structure within the active area of the device. If the device was made from such a polycrystal, it would not have the sharp demarcations of dopant defined in Figure 1.19(b). Doping the silicon would be like pouring water onto crazy paving. Water would go quickly down the cracks but tend to sit on the surface on the actual paving stones. Diffusion of the dopant atoms occurs in a similar manner. Because a grain boundary is a region of disorder between grains — a sort of amorphous region between the regularity of neighbouring crystals — dopant atoms

would move quickly down between the grains causing havoc to controlled doping. (Diffusion along grain boundaries is about a hundred times faster than through the bulk of each grain.) The obvious solution is to remove the grain boundaries by using a single crystal.

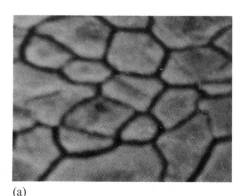

(a)

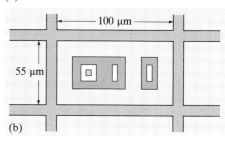

(b)

Figure 1.19 (a) Polycrystalline silicon revealed by a scanning electron microscope. (b) Part of an integrated circuit, to same scale as (a)

I hope that Exercise 1.3 has helped you to appreciate that the traditional classification of materials into metals, ceramics and plastics is very restricting. It tends to emphasize the differences between materials and not their similarities. In the search for a material from which to make a product, the required set of properties should be compared with those of all materials. Such an approach is less convenient because it requires a knowledge and understanding of a broad range of materials and their processing technologies.

Now let's consider how the properties of materials determine how they can be processed.

# 1.4 Processing of materials

Each engineering material undergoes a series of processes in its transformation from raw material to artefact. In general, the processing achieves one or more of the following:

(a) Making the basic (starting) materials. The production of silicon, steel and polyethylene, described earlier, illustrate the diversity of the processes used.

(b) Producing the shape required for a product. Steel as a girder or a bolt, silicon as a chip, polyethylene as a bag or a cable sheath, and so on. This is **processing for shape**.

(c) Creating the structure needed *in* the material, that is the structure which gives the required properties. As I said earlier, two sorts of property are involved. First, those such as ductility, viscosity and melting temperature, which determine the possible routes for (b). Second, the properties required in the product. The structure which produces the latter is achieved by what I will call **processing for performance**; the growth of silicon single crystals is an example.

The distinction between the last two purposes of processing is important because the properties required for an artefact in service may be very different from those needed to facilitate its shaping. For instance, solid metals are often shaped in a soft, ductile condition and then strengthened for performance by a 'heat treatment' process. You will meet a range of examples of this in later chapters. For the moment we will concentrate on processing for shape.

Starting with a given mass of material, a pile of thermoplastic granules, an ingot of steel, a lump of clay or whatever, the basic process routes for manipulating it into a specified shape are very limited. Essentially, they are pouring, squeezing, cutting and joining. However, with the wide range of engineering materials available, there are many variations on each of these themes. Here we will consider only the common processes. Again, it is useful to classify them, this time in terms of the type of materials flow involved. There are three types of process:

Mass conserving   This includes all those processes in which there is an emphasis on achieving a shape change without losing any material; waste means inefficiency. It covers those processes that involve pouring, squeezing (and pulling).

Mass diverging   This is a rather grand name for shaping processes that inevitably produce waste because material is pared away. Various ways of removing material can be employed — machining, grinding and dissolving with chemicals for example.

Mass converging   Here, as the name implies, a product is formed by joining different pieces together; examples are using adhesives and riveting.

When choosing a process for shaping we see the interdependence of product design, properties of materials and process route in action again. The choice of process depends, at least, on the answers to these questions:

• Can the process produce the desired shape? More intricate shapes can be made by casting or gluing pieces together for instance than by, say, rolling.

• Do the properties of the material suit the shaping process? For example, some materials can be melted and shaped as liquids and some are sufficiently ductile to be squeezed in the solid state; those which lack ductility and have very high melting points can be manipulated as a paste of particles in a suitable liquid.

Sections 1.5 to 1.7 look briefly at the main mass-conserving processes. We begin with processes for molten material. See ▼**When does a solid become a liquid?**▲

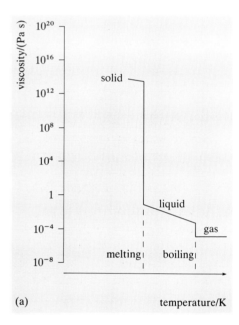

(a)

# ▼When does a solid become a liquid?▲

Distinguishing between the solid and liquid forms of some materials is straightforward: the solid is rigid and, on heating, melts to a 'runny' liquid — ice and electrical solder for instance. But some liquids are much more viscous than others. A slab of toffee is rigid but if you keep it in your pocket for a while it will come out a different shape. Is it now a liquid? Many engineering materials, especially thermoplastics and glasses, behave like toffee. One way of deciding when such solids become liquids is in terms of their viscosity.

I assume that you are familiar with the term viscosity $\eta$ as a measure of how readily a liquid flows and know that it is defined by

$$\tau = \eta\frac{d\gamma}{dt}$$

where $d\gamma/dt$ is the shear strain rate produced by a shear stress $\tau$ acting on the liquid. It has units of pascal seconds, Pa s ($N\,m^{-2}\,s$).

Assuming that treacle and water are Newtonian liquids, that is their viscosities obey the above relationship, which has the higher value of $\eta$ at room temperature?

Treacle. If you think of pouring them out

of a container, $\tau$ (provided by gravity) can be assumed to be about the same in both cases. The term $d\gamma/dt$ will be much less for treacle and, therefore, will have a higher value of $\eta$.

Many engineering materials, particularly thermoplastics and glasses, do not obey this simple relationship — they are non-Newtonian liquids. For the present we will ignore this complication.

Compare the two curves in Figures 1.20.

Figure 1.20(a) confirms the dramatically clear-cut distinction between a solid and a liquid for a material with a sharply defined melting temperature; the viscosity drops discontinuously by about 13 orders of magnitude! But what about the glass in (b)? As it is heated, its viscosity decreases slowly and progressively. At about 1600 K its viscosity is similar to that for a liquid metal. So, clearly, in such cases the demarcation between a solid and a liquid is somewhat arbitrary. In practice, glass can be blown to shape at a viscosity of about $10^7$ Pa s. This is the viscosity of which materials (including toffee) will flow (slowly) under their own weight, which is a useful way of defining a liquid — in this condition they have to be kept in containers.

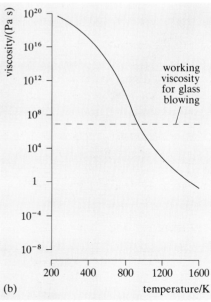

(b)

Figure 1.20 Viscosity against temperature for (a) a typical metal, (b) ordinary glass

# 1.5 From the liquid

In the processes described here the material is heated to melt it or, at least, to soften it sufficiently for it to flow into a mould and thereby 'copy' the mould shape.

## 1.5.1 Casting

This is a popular way of moulding metals and their alloys because many of them have relatively low melting points and when molten their viscosities are very low (similar to that of water). There are many variations on the casting theme but the main ones are sand casting, gravity die casting, pressure die casting, and investment casting.

**Sand casting** is illustrated in Figure 1.21. The mould is made of sand and is destroyed when the solid casting is removed. A solid replica (usually wood or plastic) of the required object is made by machining; this is called a pattern. Sand is rammed around the pattern in a 'moulding box' and when the pattern is removed it leaves a shaped cavity behind. The runners and risers act as reservoirs of liquid to top up the casting as the metal contracts on solidification; they are usually substantial and the residual metal in them is scrap. As Figure 1.21 illustrates, hollow castings can be produced using cores.

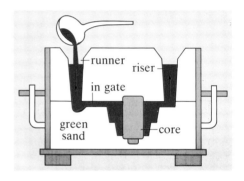

Figure 1.21 Sandcasting

**Gravity die casting** is like sand casting except that the mould is machined from solid metal, usually cast iron. Being of metal the mould can be machined accurately and, with a good thermal conductivity, allows the casting to cool quickly. Engine cylinder heads and pistons are typical products.

**Pressure die casting** is a development of gravity die casting in which the molten metal is squirted into a steel mould under pressure; it is the metal equivalent of injection moulding (see below). This quickens the process. It is used for carburettors, door handles, electric-iron base plates and so on.

**Investment casting** This process is really only suitable for smallish precision castings with a good surface finish, or for one-off artefacts made by jewellers and sculptors. A pattern made of wax is coated with a ceramic slurry using the slip-casting process described later. During firing of the slip, the wax burns out to leave a preheated ceramic mould — the investment — which is broken to get the casting out. Figure 1.22 shows a turbine blade for a jet engine made by investment casting.

Figure 1.22 Investment-cast turbine blade

## 1.5.2 Extrusion and injection moulding

Thermoplastics when heated do not have as low a viscosity as molten metals, so they cannot be shaped by gravity-fed casting methods; they have to be forced into a shape. Extrusion and injection moulding (and also blow moulding and vacuum forming, see below) have been devised for the processing of such viscous materials.

The process of extrusion is similar to squeezing toothpaste from a tube. An extruder is sketched in Figure 1.23. A screw pushes thermoplastic granules through a heated cylinder and then through an opening — usually called a die. The granules mix and soften as they are moved along the barrel. The polymer emerges through the die with a constant area of cross-section (extrusion always produces a shape with a constant section).

In this basic form, extrusion is only suitable for making long thin objects with a constant cross-section. Nevertheless it is very versatile; rods, sheet, film, tube and irregular cross sections are possible — curtain rails and window frames for example. A form of extrusion is also used for shaping metals and clay products such as bricks and pipes.

Three-dimensional objects can be made by, in effect, replacing the open die of an extruder with a closed die. The die is in the form of a split-mould, the two parts of which interlock very closely. Figure 1.24 illustrates the process, which is called **injection moulding.**

Here, when sufficient molten thermoplastic has accumulated ahead of the screw, the screw moves as a ram and injects the plastic into the closed mould.

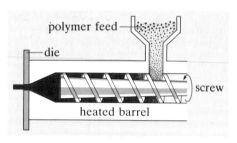

Figure 1.23 An extruder

Figure 1.24 Injection moulding

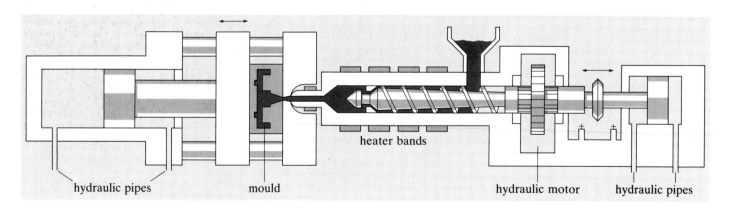

There is a 'dwell time' during which the injection pressure is maintained in order to feed the moulding with more material while it cools and contracts. There is usually very little flash on the moulding. **Flash** is a term used commonly in processing parlance; it is so-called because it 'flashes' out between the two halves of the mould. However, there is a **sprue** — the plastic that remains in the channel through the die which is the equivalent of the runner and riser in a casting. This is broken off and leaves a tell-tale mark on the moulding.

It is an extremely versatile process; you will be familiar with many examples of it from spoons and washing-up bowls to telephones and dustbins. It produces high precision mouldings with good surface finish, but it is not used for objects with closed re-entrant cavities — such as bottles — because a 'core' would be needed which would be difficult to remove from the moulding.

Injection moulding is also used for moulding thermoset products, especially those in the form of a foam (such as polyurethane shoe soles) and those reinforced with short glass fibres (such as the reflector housing of car headlamps).

## 1.5.3 Blow moulding

This process was invented in Roman times. It is used to make hollow objects from relatively viscous materials such as molten glasses and thermoplastics. These materials can be heated to a temperature at which their viscosity becomes low enough for the material to be blown to shape, but not so low that the material runs away. This temperature is a property of the particular material being used.

You have probably already seen pictures of glass blowers twirling a 'gob' of red hot glass on the end of a blowpipe (Figure 1.25). Nowadays the process is usually automated (Figure 1.26). In the first stage, a partially blown bottle — a **parison** — is blown. The distribution of glass in the parison is such that when it is blown out to full size (the second stage), the bottle has a uniform wall thickness. Thermoplastic bottles and other containers can be produced by a similar process.

The most widely used glasses in blow moulding are based on the ubiquitous silica ($SiO_2$). Molten silica is very viscous and has to be

Figure 1.25 Glass blower

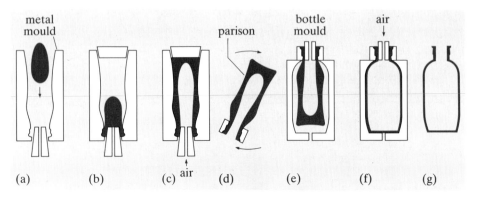

Figure 1.26 Automatic production of glass bottles

heated to about 2300 K to make a workable glass. However, the viscosity can be lowered by the addition of other oxides. Soda–lime glass is silica with additions of $Na_2O$ and $CaO$ and is used for 90% of all glass products; it is the glass in window panes, bottles and light bulbs.

## 1.5.4 Compression moulding

This is a common process for shaping thermosets. Once formed, thermosets cannot be made to flow, so in any shaping process that involves flow, polymerization must proceed during processing. Therefore, the starting materials for compression moulding are the mix of the molecular species that react to form the thermoset.

Basically the process is a simple one. See Figure 1.27.

The starting materials, as powder or pellets, are placed in the bottom half of a preheated mould. The material is made to melt and flow by the heat and pressure applied when the mould is closed. After it has set, the moulding is ejected. The excess material, the 'flash', is then removed.

Thermosets made this way can also include other materials such as wood-flour or chalk for cheapness, or short glass fibres or beads for reinforcement. Such materials are called **fillers**.

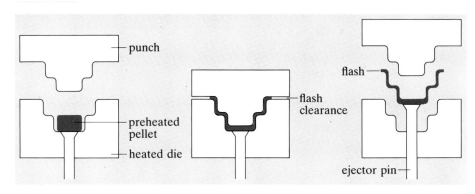

Figure 1.27 Compression moulding

# 1.6 From the solid

There are two routes to a shape here:

- Forcing a solid lump into a required shape. This is dependent on the plastic properties, particularly the ductility, of the material and is thus largely confined to the shaping of metals.
- Consolidating materials which start as powders. It is used for plastics, ceramics and metals. Let's look at them in turn.

## 1.6.1 The shaping of solid metals

The processes here are called **mechanical working** and are what blacksmiths, goldsmiths and silversmiths have been doing for centuries.

At ordinary temperatures, steel is very strong and difficult to deform, but on heating to red-heat it becomes soft enough to shape by hammering — hence the blacksmith's art. This is an example of **hot working**. It is important to appreciate that the temperature required for hot working is peculiar to each metal.

The stress–strain curves of materials 1, 3 and 4 in Figure 1.16 exhibit a period of work-hardening — the metal gets progressively stronger as it is plastically deformed. When work-hardening occurs, the deformation processes are called **cold working**. It is what happens when, for instance, commercial aluminium is worked at room temperature. If however aluminium is deformed at about 500 K, it shows little or no work-hardening; beyond the yield stress its stress–strain curve is flat, rather like that of material 2 in Figure 1.16. In addition, its yield stress is lower at the higher temperature. Deformation at such temperatures is what is meant by hot working.

Copper becomes soft at roughly 700 K and mild steel at about 900 K. In general, metals begin to soften significantly at temperatures above about $0.5T_m$, where $T_m$ is the melting temperature in kelvins. (This rule of thumb depends on many factors, such as the amount of alloying — more about this in later chapters.) So lead, with a melting temperature of 600 K, is hot-worked at ambient temperature, whilst tungsten, with a melting temperature of 3680 K, is cold-worked when deformed at 1000 °C (1273 K)!

The big advantages of hot working are that because the metals are soft and not strengthened by work hardening, large changes of shape can be made with relatively small forces (compared with cold working). However, the process cannot provide the increased strength, dimensional accuracy and good surface finish of which cold-working is capable — witness aluminium cooking foil and copper flex.

Now for a quick review of the common shaping processes.

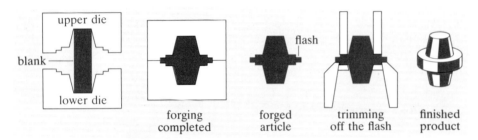

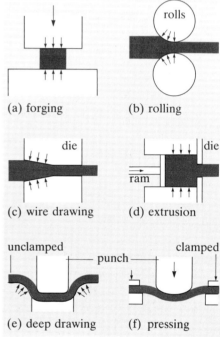

Figure 1.28 Closed-die forging

Shaping a metal by squeezing it between a pair of dies is called **forging** and is exemplified by the blacksmith's anvil and hammer. A variant of this widely used in manufacturing industry is **closed-die forging**, Figure 1.28. It is usually a hot-working process.

Forging is sometimes used to prepare a starting ingot produced from primary processes (such as steelmaking) for other mechanical working processes, in particular rolling, extrusion and wire-drawing. Very simple sketches of these are shown in Figure 1.29. (The arrows indicate the major forces acting on the material.)

**Rolling** (b) is really like a continuous forging operation and works on the same principle as a mangle. **Extrusion** (d) you have already met for thermoplastics. Here, because the metal is solid, a ram is used rather than a screw. Extrusion is usually a hot working process and all sorts of cross-sections are possible. **Wire drawing** (c) is similar to extrusion but the material is pulled through the die rather than pushed.

**Deep drawing and pressing**, (e) and (f) respectively in Figure 1.29, are two important ways of deforming sheets into a variety of shapes. In pressing, the material is clamped and is stretched to shape by a punch. In deep drawing the unclamped material is drawn down into the die. It is similar to the vacuum forming of plastics, which is our next process.

Figure 1.29 Common mechanical working methods

## 1.6.2 Vacuum forming

This is a way of sucking, as opposed to blowing, thermoplastics to shape. It is used for a wide range of products, from small trays and disposable cartons to domestic baths. The starting material is in sheet form (Figure 1.30). The sheet, softened by radiant heaters, is forced against the mould when the air below is pumped out.

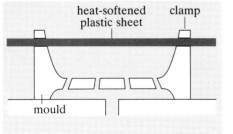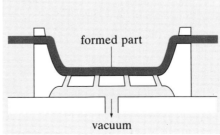

Figure 1.30 Vacuum forming

## 1.6.3 Pressing of glass

This is very similar to compression moulding. Compare Figure 1.31 with Figure 1.27.

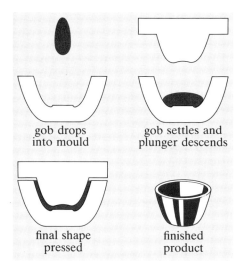

gob drops
into mould

gob settles and
plunger descends

final shape
pressed

finished
product

Figure 1.31 Glass pressing

At the right temperature for flow, the correct amount of glass — a gob — is dropped into the mould and the plunger presses it to shape. Pressing is used to make open-topped containers (which allow the plunger to be withdrawn) such as tumblers and ovenware. It is very different from the process of the same name used on metal sheet (Figure 1.29f) and on powders (below).

## 1.6.4 Pressing and sintering

This process starts with the material in powder form and is used for a wide variety of materials, especially those with properties which preclude shaping by melting and casting, or by plastically deforming large lumps. Examples are ceramics such as alumina and silicon carbide; brittle metals with high melting temperatures (over 2300 K), such as tungsten; and the polymer PTFE, poly(tetrafluorethylene), which has a viscosity too high to suit other moulding methods. Some composite materials can only be produced this way — tungsten carbide particles in cobalt for masonry and metal-cutting tools for instance.

The simplest form of the pressing stage is illustrated in Figure 1.32.

Powder, mixed with a lubricant/binder, is moulded to shape by compressing it in a die to form what is called a **green compact**. Although

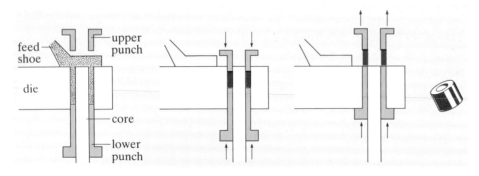

Figure 1.32 Pressing powder to the shape of a simple bearing

this is highly porous, it has enough rigidity to support its own weight and to permit gentle handling. The compact is then **sintered**, that is, heated at a high temperature for a prolonged time. During sintering the powder particles coalesce and grow to fill the pore space; the compact shrinks correspondingly. The mechanisms of microscopic mass transport (called diffusion) that occur during sintering are considered in Chapter 4.

In this form, pressing and sintering is generally only used for rudimentary shapes such as rods, hollow cylinders and simple gear wheels. Even so it has attractions, in addition to providing a solution for otherwise intractable materials, because this route may be more economic for some metal products than casting, forging or machining.

Can you see why? (Compare Figures 1.32, 1.21 and 1.28; also look again at Figure 1.2.)

This process route is efficient in its use of materials, which can be a significant element in the cost of a product. This is because very little waste is created by the pressing operation compared with the runners and risers from castings, the flash from forgings, and the swarf from machining. It is a good mass conserving process.

Obviously, whether this route is actually chosen depends on other factors, such as the volume of production and the materials properties required in the product.

# 1.7 From a paste

Mixing a powdered solid in a suitable liquid makes a paste which can be manipulated to shape. This route is used for a variety of materials and products. The wet clay used by potters is the obvious example; and water added to plaster of Paris, to concrete and to cake mixes allows them to be shaped, besides contributing to their chemical reactions. Many new ceramic materials and refractory metals such as tungsten and molybdenum are shaped as pastes, then dried and sintered (Chapter 4) into continuous solids.

A very runny paste lends itself to the process called **slip casting**. The process is illustrated in Figure 1.33. The powder and water mix is called a **slip**. The mould is made from a material that readily absorbs moisture from the slip (such as plaster of Paris), so leaving a thin damp shell which, on removal of the mould, is a green compact (see above). This is allowed to dry further and then fired in a kiln to promote sintering.

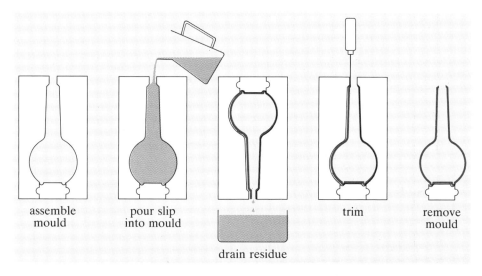

assemble mould     pour slip into mould     drain residue     trim     remove mould

Figure 1.33 Slip casting

## Summary

There are three stages of processing: obtaining the basic material, manipulating its structure to provide the properties required of the product in service (processing for performance), and producing the appropriate shape (processing for shape). These stages may not be separate.

Shaping processes can be classified into mass conserving, mass diverging and mass converging.

Mass conserving processes can start with material in the form of a liquid, a solid or a slurry, and the variety of methods described emphasizes how the choice of process depends on the geometry of the shape required (product design) and on the properties of the material involved.

It is convenient to say that solid materials that do not have sharp melting temperatures become liquid at temperatures at which they begin to flow slowly under their own weight. This means at a viscosity of about $10^7 \, \text{Pa s}$.

SAQ 1.3  (Objective 1.2)
List the properties of materials that determine the processes that can be used to shape them. For each property, comment on the process(es) that can be used and the possible implications for the product.

SAQ 1.4  (Objective 1.3)
Which of the processes described in the preceding sections would you choose to make the following products? Make notes on the reasons for your choice. Consider only the processes described in this section and the conventional form of the product — we are not concerned with novel designs, such as carbon fibre church bells, even if they are feasible.

- glass beer mug
- glass ashtray
- steel table knife
- metal end cap of a lamp bulb
- church bell
- plastics casing for a television set
- casing for an electric plug
- the investment for an investment casting
- thin-walled disposable plastic beaker.

# 1.8 Exploring the problem of choice

I want to introduce now a number of important aspects of product design, and of properties and processes, and I shall use two very different products, an ordinary electric light bulb and a drinking vessel, as examples.

## 1.8.1 An electric light bulb

Nowadays, a huge variety of engineering materials is used, each chosen for particular reasons determined by the design of the product, the required performance and processing properties of the material and, of course, cost. The humble domestic electric light bulb is an illuminating example. Its invention had very far reaching effects as ▼Some consequences of Thomas Edison's lamp▲ indicates. You might like to reflect on the system of supply and demand we considered earlier.

A typical bulb together with a connector and supply lead is illustrated schematically in Figure 1.35 with the component parts numbered and named. Table 1.3 describes the materials and the main manufacturing processes used for each part.

There are *twenty-three* different components using at least *seventeen* different materials and *twelve* processing methods! The process technology is very impressive; from the blow moulding of the glass bulbs at up to 2000 a minute, to the production of the coiled-coil filament of tungsten wire with a diameter of about 0.05 mm (about half the thickness of a human hair, see Figure 1.34). And the whole lot (bulb plus holder) is made and sold to a retailer for less than £1.

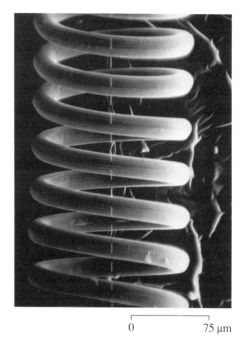

0       75 μm

Figure 1.34 Enlargement of bulb filament.

# ▼Some consequences of Thomas Edison's lamp▲

(The following is part of a paper by J. W. Guyon of GEC, presented at the First International Tungsten Symposium, Stockholm 1979).

On October 21, 1879, Thomas Edison's famous lamp attained an incredible life of 40 hours.

The entire electrical industry has as its foundation this single invention — the incandescent lamp. At the time the lamp was invented, it was of little use because homes were not wired for electricity nor was sufficient electrical power being generated, nor were there electrical distribution systems in place. Thus, not only was lamp manufacturing needed, but such things as electrical generating equipment, wiring, insulation, fixtures, switches, and protective devices had to be

built and installed before Edison's invention could be used.

Then, after everyone had electrical power — and homes, offices and factories were nicely lighted — the industry looked for other devices which could use electricity. Things like electrical heaters, flat irons, refrigerators, washing machines, clothes dryers, radios, television sets, kitchen ranges, vacuum cleaners, bread toasters, fans, air conditioning, humidifiers, dehumidifiers, electric drills, saws, hedge clippers, and even devices of more questionable utility, such as can openers, electric knives, and toothbrushes were developed and manufactured. And, of course, the electrification of the world's production facilities required yet a different set of electrical devices to operate and control the tools of industry.

That was not all. This electrical equipment, including the lamp, required materials with specific properties — not always available, particularly in the quantities and forms that were needed. Wires to conduct the electrical current and insulating materials to keep it contained are simple examples that come to mind. Plastic and ceramic businesses became natural spin-offs of the electrical industry. In the case of the incandescent lamp, the glass required to enclose the hot filament put our company into the glass business; the gases ultimately used to surround the filament put us — at least for a while — into the gas business and, of course, the filament itself put us into the tungsten business. All of this activity was generated by that single invention — the incandescent lamp.

It is clear that, with so many different materials being used, each must be selected for a specific set of reasons — it would be a lot quicker and cheaper to make all of the wire bits out of the same metal! Very importantly, notice that the comments in the final column of Table 1.3 clearly indicate that the reasons for selection are in terms of both properties for performance and properties for processing.

The filament of an incandescent lamp is a good example of PPP. Edison's original lamp did not contain tungsten. He used a carbon filament made from some Coats Thread Company cord 29 taken from Mrs. Edison's sewing basket. It didn't last very long.

For about 30 years there was intensive competition to find a filament material and bulb design which could run at a high enough temperature with a reasonably long life. Processing the competing materials to shape was often the stumbling block. Various filament materials were used, including mixtures of magnesium and calcium oxides, tantalum and osmium, before tungsten became the best choice. But again there were processing problems, because bulk tungsten is brittle. Tungsten has not

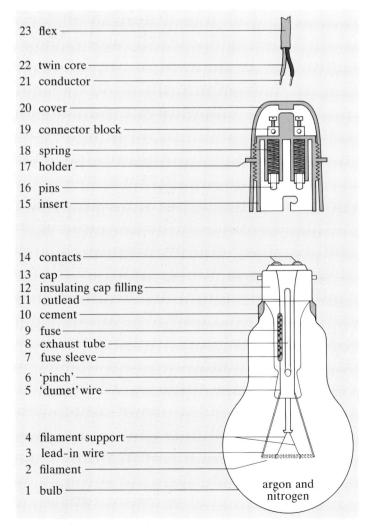

23 flex
22 twin core
21 conductor
20 cover
19 connector block
18 spring
17 holder
16 pins
15 insert

14 contacts
13 cap
12 insulating cap filling
11 outlead
10 cement
 9 fuse
 8 exhaust tube
 7 fuse sleeve
 6 'pinch'
 5 'dumet' wire
 4 filament support
 3 lead-in wire
 2 filament
 1 bulb

argon and nitrogen

Figure 1.35 Construction of bulb and holder

looked back since Coolidge produced wire from powder by pressing, sintering and then forging. We shall be using this example later. Now for another product that is also remarkable, but for quite different reasons.

Table 1.3  Parts of an electric bulb and holder. Part numbers refer to Figure 1.35

| Part no. | Part name | Material | Main manufacturing processes | Comments on material selection |
|---|---|---|---|---|
| 1 | bulb | soda–lime glass | blow-moulding | low cost, transparency, viscosity |
| 2 | filament | tungsten | hot-pressing + wire-drawing | high melting temp. (above 'glowing' temp.), electrical conductor, low vapour pressure, high strength and sufficient ductility |
| 3 | lead-in wire | nickel or nickel-plated | wire-drawing (nickel plating) | electrical conductor, oxidation resistant during processing, ductility |
| 4 | filament support | molybdenum | wire-drawing | high melting temp., ductility |
| 5 | 'dumet' wire | nickel-iron alloy | wire-drawing | coefficient of thermal expansion has to match that of glass in (6) |
| 6 | 'pinch' | lead glass ($SiO_2$ + 20–30 wt% PbO) | pressing | |
| 7 | fuse sleeve | soda–lime glass | drawing | as (1) |
| 8 | exhaust tube | lead–glass | drawing | as (6) |
| 9 | fuse | copper–nickel alloy | wire-drawing + welding | correct fusing characteristics; suited to automatic welding |
| 10 | cement | phenol–formaldehyde | | good adhesion at 400 K for life |
| 11 | outlead | copper or copper-clad steel | wire-drawing (electroplating) | electrical conductor, easy to solder |
| 12 | cap filling | opaque glass | casting | high-temp. insulation, melting temp. |
| 13 | cap | aluminium or brass | pressing | choice depends on cost of raw material |
| 14 | contacts | solder (Pb/Sn) | | melt temp. has to be higher than contact operating temp. |
| *holder* | | | | |
| 15 | insert | brass (70 Cu/30 Zn) | pressing | ductility |
| 16 | pins | brass (60 Cu/40 Zn) | extrusion | electrical conductivity, ductility |
| 17 | holder | phenol-formaldehyde (filled with wood/paper) | compression moulding | |
| 18 | spring | phosphor bronze (Cu/6Sn/0.2 P) | wire-drawing | electrical conductivity, strength and low Young's modulus |
| 19 | connector block | brass (60 Cu/40 Zn) | forging + machining | electrical conductivity, machinability |
| 20 | cover | urea-formaldehyde (filled with glass) | | can be coloured white with glass powder filler |
| 21 | conductor | copper | wire-drawing | electrical conductivity, ductility |
| 22 | twin core | } PVC (plasticized) | extrusion | electrical insulation, viscosity |
| 23 | flex | | extrusion | electrical insulation, viscosity |

## 1.8.2 A drinking vessel

Another way of appreciating the diversity of materials, and some important aspects of product design, is to contemplate the variety of products available to meet a particular demand. Consider a common example: a drinking vessel. The function seems very simple, namely to hold liquid, but we are unlikely to be satisfied by solutions such as jam jars or coconut shells. Clearly the function is more complex than this.

Looking round my home, I found one or more of each of the following drinking vessels:

- pottery cup
- pottery mugs, some glazed, all decorated
- enamelled steel mug
- Melmex (trade name for a melamine formaldehyde thermoset) mug
- Pyrex (trade name for glass resistant to thermal shock) beaker
- bone-china cup
- expanded polystyrene beaker
- polypropylene beakers (tapered and with half-handles)
- glasses of different shapes
- cut-glass wine glass
- glass beer mug
- pewter tankard
- Maxpax (trade name of company which merchandizes pre-packed drinks) plastic coffee cup, Figure 1.36.

What motives have led to so many different materials being chosen to meet this single function (and to my giving them house-room)?

Figure 1.36 'Maxpax' cup, cut to show section

EXERCISE 1.4   Study my list of drinking vessels (if you wish, add any others you have) and classify the articles against the headings below. Also make brief comments on how you think the manufacturers perceived their markets and how this influenced the design of the product and the choice of materials.

(a) Cheap and practical.
(b) Intentionally disposable.
(c) Unbreakable (relatively).
(d) Suitable for hot drinks.
(e) Prestigious to own.
(f) Aesthetic/curiosity/sentimental value.
(g) Specialized function met by design.
(h) Specialized function met by materials properties.

I think the most important points to come out of this exercise are:

• An ostensibly simple function — to hold a liquid — has a complex set of sub-functions. Attempting to specify a complete range of functions is an important aspect of product design because the various sub-functions can lead to conflicts between the materials properties required. One material may hold a liquid but not look attractive; one that holds liquid and looks attractive may not be robust and so on.

• A particular function can sometimes be met by a product design solution and sometimes by a materials solution, for example the two solutions to the problem of thermally insulating a beaker (Figures 1.36 and 1.37). Expanded polystyrene is the material route, the Maxpax cup is the product design route — a cup of hot liquid can be held (if you are careful) by gripping the flange, which is remote from the liquid.

SAQ 1.5  (Objective 1.4)
Make notes on the PPP links you perceive for

(a) the glass bulb of an incandescent lamp,
(b) the Maxpax disposable cup.

In both the light bulb and the drinking vessel many materials and processes are used — in the light bulb all in one product and in the drinking vessel in a variety of products. Without materials such as tungsten, copper and glass, lighting devices would be very different. On the other hand, almost any materials with any properties can be used for a cup. So, properties are important in different ways. We look at this more closely in the next section.

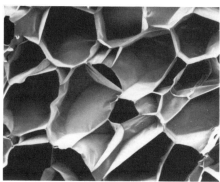

0      100 μm

Figure 1.37 (a) Expanded polystyrene beaker. (b) 'Pore' structure of beaker

## 1.8.3 Which properties matter?

First of all, when thinking about which material to use for a particular product, it is important to distinguish between the property of a *material* and the property of a *product*. To illustrate the point, imagine trying to bend a branch of a tree and then a twig on the branch — no problem with the twig! Now, the resistance of the material to bending — the Young's modulus — is about the same for the wood in the branch and in the twig. So the difference must be due to the shape of the piece, in this case its diameter. Similarly for other shapes; for example, a piece of copper central heating pipe is stiffer than a piece of copper electrical flex. In general, the two properties are related by the expression:

property of an artefact = materials property × geometrical factor.

SAQ 1.6  (Objective 1.5)
Use the above expression to relate electrical resistivity to electrical resistance.

Sometimes a material has a particular property which makes the material uniquely able to meet a particular need. Without that material there would be no product. For such a material, P for property overrides P for product and P for process. For example, were a material discovered that was superconducting at or near ambient temperatures, then the design of a generator using it would have to cope with whatever other properties the material had. Fortunately designers usually have a range of materials to make their choice from.

For many products more than one material property is important, as the comments column in Table 1.3 illustrates. In such cases it is often very useful to combine the properties in an approximate way as a criterion for materials selection. The criterion is then called a **merit index**.

SAQ 1.7 (Objective 1.6)

Which material makes the 'best' conductor in the following applications?

(a) An electrical machine which requires as many conducting elements as possible in a given space.

(b) An overhead power cable for which the weight of conductor is a major design consideration, but for which there is no space problem.

Assuming a conductor of given length and resistance, for each application,

i specify a property criterion for the selection of a material for the conductor,

ii using the property data in Table 1.4 rank the three best materials

Table 1.4

| Material | Density $\rho$ $\mathrm{Mg\,m^{-3}}$ | Electrical resistivity $\rho_e$ $10^{-8}\,\Omega\,\mathrm{m}$ |
|---|---|---|
| aluminium | 2.7 | 2.6 |
| copper | 8.9 | 1.7 |
| iron | 7.9 | 9.7 |
| silver | 10.5 | 1.6 |

For the power cable in SAQ 1.7, $\rho_e\rho$ is a merit index. You will meet other examples later.

A material with an attractive combination of certain properties may, however, turn out to be a poor choice when further properties are considered — see ▼Choosing the material for a dinghy▲. This leads to the important idea that a material has a materials property profile (MPP); that is, a suite of properties, such as tensile strength, electrical resistivity, oxidation resistance, machinability and so on. If one particular property is wanted, the others have to be accepted — and some way of coping with the undesirable properties must be found. This might be done by looking at other aspects of PPP. For example, because of tungsten's lack of ductility, early bulb filaments were made by sticking powder particles together with gum arabic, which was then burnt off. They were fragile, but worked — for a while. Old-style steel dustbins were galvanized to prevent rusting. They could cope with hot ashes. Plastics dustbins cannot, but this materials limitation can be overcome by product design — by having an internal metal base plate, for instance.

Choosing the material and the process route are two vital stages in designing a product. How does the MPP fit into the design activity?

Table 1.5

| Material | $E/\rho$ $\mathrm{MPa\,kg^{-1}m^3}$ | $E^{\frac{1}{3}}/\rho$ $\mathrm{MPa^{\frac{1}{3}}kg^{-1}m^3}$ |
|---|---|---|
| alumina | 97.4 | 18.5 |
| silicon nitride | 82.1 | 21.9 |
| soda–glass | 29.8 | 17.1 |
| mild steel | 26.5 | 7.6 |
| aluminium | 26.3 | 15.5 |
| GFRP (chopped strand mat) | 24.0 | 13.4 |
| oak (along grain) | 23.2 | 49.0 |
| concrete | 20.8 | |
| polypropylene | 16.5 | 12.7 |

# ▼Choosing a material for a dinghy▲

Suppose we want to choose a material for the hull of a small sailing dinghy, such as that illustrated in Figure 1.38.

Two important criteria are that the material should be rigid and lightweight. Of course, in practice a boat designer can change the overall stiffness of the hull by incorporating bulkheads, ribs, increasing the wall thickness and so on, but for this exercise we will assume that the design is fixed and that it is the stiffness of the material that matters. To compare possible materials, we need to see how much stiffness we get for a given mass. This can be done by calculating $E/\rho$ where $E$ is the Young's modulus and $\rho$ is the density. $E/\rho$ is another example of a merit index.

$E/\rho$ for a range of possible candidates is listed in Table 1.5. How is it that polypropylene, at the bottom of the list, is the material used?

On the criterion $E/\rho$ alone, the clear winners are alumina and silicon nitride. Maybe a way will be found of overcoming their shortcomings but, at present, along with concrete and glass, they are too brittle to be contenders. In addition, the conventional methods of processing ceramics cannot cope with artefacts this

large. Going down the list, aluminium and steel are better than GFRP and oak. But the rusting of steel is a major handicap. Interestingly, steel wins for ships; one of the main reasons being that it can be fabricated into large structures by welding plates together — a processing reason. Aluminium has been used for canoes (aluminium sheets are riveted together) but it also corrodes and is easily dented. The materials actually employed are wood (as plywood) and GFRP.

GFRP was used for dinghies like that in Figure 1.38 when they were hand made. However, there was thought to be a large market for these boats. Polypropylene, the worst candidate in terms of $E/\rho$ is well suited to mass production by injection moulding, which is why it is the material used.

To be fairer to polypropylene, I should point out that another merit index $E^{1/3}/\rho$ is also relevant to hull design. Under sail, a dinghy undergoes bending. The mast is well forward and the crew spend most of their time in the stern, so the hull tends to droop at the ends and rise in the middle. $E^{1/3}/\rho$ is the appropriate merit index for the bending of a beam. As you can see from Table 1.5, polypropylene does rather better on this index.

Figure 1.38 'Topper' single-person dinghy

## 1.8.4 Product design

Figure 1.39 is one way of describing the activity of product design. The arrows represent the flow of information.

The first step is to identify the 'need' — usually called the market need. As Exercise 1.4 demonstrated, there are many ways of performing a function. In identifying the need we define a set of requirements that is more precise than simply meeting a particular function.

In the conceptual design stage, ways of meeting the perceived market need are conceived. One of these is chosen to be designed in detail. If this is satisfactory, we move on to manufacture and sale; if not, it's back to the conceptual design. In practice, the process would be much more complicated, with lots of analysis, checking and rethinking in the light of developments.

A ▼Product design specification (PDS)▲ is one way of organizing the design activity from the initial identification of the market need through to a detailed design. If you have met design specifications before, beware. A PDS is not itself a design specification.

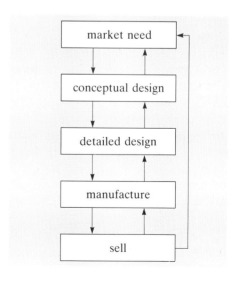

Figure 1.39 Product design

# ▼Product design specification▲

A product design specification (PDS) begins as a means for defining the market need for a product and evolves as the design proceeds. It ends up as a detailed statement of the requirements the product should meet. It is not a detailed design specification, which will consist of engineering drawings, process, materials and marketing specifications and so on.

A comprehensive PDS covers all the elements that should be involved in the design of a product to fit its defined market. Some important elements are shown in Figure 1.40. The idea is that each element gives rise to questions about the product which have to be answered. For instance, company constraints (4) might prompt the questions 'Can our current production plant cope? Will it affect personnel?' Standards (26) might suggest 'Is the product to be designed to national or international standards?' or 'Does this apply to the materials involved?'

Some of the elements have materials implications. At the detailed design stage these would be brought together to form the detailed materials specification.

EXERCISE 1.5  For each of the elements in Figure 1.40 numbered 3, 6, 10, 14, 22 and 24, make notes on the sorts of questions you might ask concerning the material(s) for a product.

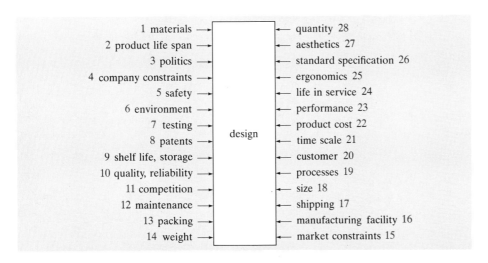

```
 1 materials  ──▶                  ◀──  quantity 28
 2 product life span ──▶           ◀──  aesthetics 27
 3 politics  ──▶                   ◀──  standard specification 26
 4 company constraints ──▶         ◀──  ergonomics 25
 5 safety  ──▶                     ◀──  life in service 24
 6 environment ──▶                 ◀──  performance 23
 7 testing ──▶         design      ◀──  product cost 22
 8 patents ──▶                     ◀──  time scale 21
 9 shelf life, storage ──▶         ◀──  customer 20
10 quality, reliability ──▶        ◀──  processes 19
11 competition ──▶                 ◀──  size 18
12 maintenance ──▶                 ◀──  shipping 17
13 packing ──▶                     ◀──  manufacturing facility 16
14 weight ──▶                      ◀──  market constraints 15
```

Figure 1.40 Elements for consideration in a PDS

Now, a complete and comprehensive detailed design should include a specification for every aspect of the product's function, manufacture and marketing. There will be specifications for the type of material(s) and processing method. The detailed design stage will create a set of requirements that the material has to meet — a certain strength, corrosion resistance, thermal conductivity and so on. In general, the material can be selected by comparing these requirements with the property profiles of likely candidates.

SAQ 1.8 is an elementary example of part of the above procedure.

SAQ 1.8 (Objectives 1.6 and 1.7)
A simple product specification for the envelope of an incandescent lamp is:

- transparency,
- impermeability to gases at the operating temperature,
- capable of being formed to the desired shape and not to distort or degrade in use,
- robust, for protection of the filament structure.

From your first-hand experience of the property profiles of (a) glass and (b) a thermoplastic, comment on their suitability for the envelope.

## Summary

A material may be selected for a product because it uniquely has a particular property (for example, a superconducting material) or because it has a combination of desirable properties. Merit indices are useful for comparing combinations of properties from material to material.

The materials property profile (MPP) summarizes all of a material's properties.

Undesirable properties have to be accommodated or accepted. Other aspects of PPP, such as processing or product design, may afford ways of overcoming them.

The product design specification (PDS) is a method of ensuring that all elements are considered in the detailed design specification.

# 1.9 Products, properties, processes and principles

In earlier sections of this chapter I have tried to show that the properties of a material, the way it is processed and the design of the product in which it is used are strongly interdependent and that the relationships (PPP) are very important. Finally in this chapter, I'm going to add another P: P for principles! By a **principle** I mean a fundamental or theoretical basis on which action can be taken. This takes us back to the scientific understanding and empirical laws I referred to at the beginning of the chapter. Thus, we have four interrelated Ps, which I have portrayed as a tetrahedral symbol in Figure 1.41(a).

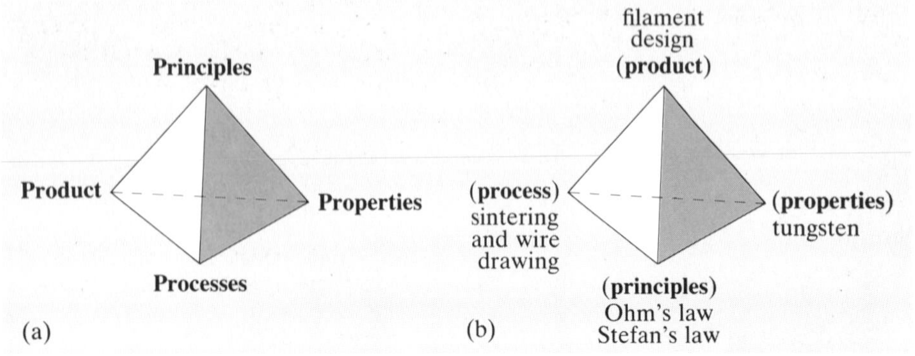

Figure 1.41 (a) The PPPP tetrahedron. (b) PPPP tetrahedron for a lamp filament

This tetrahedron is a useful device for thinking about materials and their uses. Notice that its edges join all possible *pairs* of interacting Ps and that its four faces present *trios* of interactions. You can begin investigating a material at any of the corners. For example

- What principles are needed to understand its properties?
- What can you make from it?
- How can it be processed?

For instance, let's start at P for product (design) and consider the design of the lamp filament in Figure 1.35. The length and diameter of the tungsten filament for a lamp can be calculated from the P for principles (in this case Ohm's law and Stefan's law) and P for properties (electrical resistivity) of the chosen material (tungsten). See ▼Thinking about a lamp filament▲. The P for processes is then concerned with the problem of producing long, very thin wire and is dependent upon the pertinent flow properties of tungsten such as tensile strength and ductility. In simple terms the tetrahedron would look like that in Figure 1.41(b). We consider other examples later.

# ▼Thinking about a lamp filament▲

What should be the dimensions of the filament in a 40 watt bulb for use on 240 volts? We shall take the operating temperature of the filament to be 3000 K, at which the resistivity of tungsten ($\rho$) is about $1.5 \times 10^{-6} \, \Omega \, m$.

Letting $P$ stand for power dissipated, $V$ for supply voltage and $R$ for resistance,

$$P = \frac{V^2}{R}$$

From SAQ 1.2, $R = \rho l / A$, where $l$ is length of conductor and $A$ is cross-sectional area. So

$$P = \frac{V^2}{R} = \frac{V^2 A}{\rho l}$$

hence

$$\frac{l}{A} = \frac{V^2}{P \rho}$$

$$= \frac{240^2}{40 \times 1.5 \times 10^{-6}} \, m^{-1}$$

$$\approx 1 \times 10^9 \, m^{-1}$$

What is the value of $l/A$ for (a) 314 km of 20 mm diameter wire and (b) 31.4 m of 0.2 mm diameter wire?

In each case $l/A$ equals nearly $1 \times 10^9$ $m^{-1}$ but (b) contains $10^8$ times less material than (a). Clearly neither 314 km

of 20 mm wire nor 31.4 m of 0.2 mm wire will fit inside the glass bulb of a lamp we are familiar with. Essentially, a smaller volume choice has to be made. But should it be of the order of 314 mm of 0.02 mm wire or, say, 31.4 mm of 0.0063 mm diameter?

If there is to be a unique solution to the design we need to extend our modelling of the processes within and around the filament. Let's first tackle the problem qualitatively. Forty watts of input power has to maintain the filament at its chosen operating temperature, and we shall assume that the energy loss by the filament occurs by radiation from the surface of the wire.

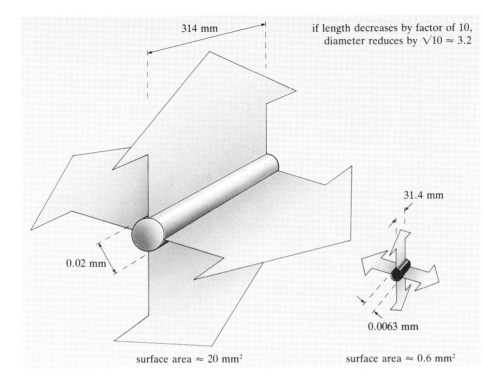

314 mm

if length decreases by factor of 10, diameter reduces by $\sqrt{10} \approx 3.2$

31.4 mm

0.02 mm

0.0063 mm

surface area $\approx 20$ mm$^2$

surface area $\approx 0.6$ mm$^2$

Figure 1.42 Radiative losses from two geometries of filament

A wire with a relatively large surface area loses energy more easily than one with a smaller surface area; so for a given power dissipation (40 W) the larger wire will not get as hot as the smaller (Figure 1.42).

Thus a wire longer than optimum (but with the correct value of $l/A$) will dissipate the input 40 watts without reaching the required temperature. A wire much shorter and thinner than optimum (but with the correct $l/A$) will not be able to dissipate the energy quickly enough before it melts.

Radiation losses can be assessed by Stefan's law, which states that the power $P$ radiated from a surface at temperature $T$ is given by

$$P = a\varepsilon\sigma \, (T^4 - T_{\text{surr}}^4),$$

where $a$ is the area of the surface, $\varepsilon$ is a quality called its emissivity and $T_{\text{surr}}$ is the temperature of the surroundings of the hot body. The term $\sigma$ is a measured constant and is about $5.7 \times 10^{-8}$ W m$^{-2}$ K$^{-4}$.

$T$ is expected to be very high (about 3000 K), $T_{\text{surr}}$ is around 300 K. So $T_{\text{surr}}^4$ is negligible compared with $T^4$. Also, the emissivity is a rather uncertain fraction, around 0.8 to 1 for dull surfaces. So the calculation is going to be a bit approximate. With $\varepsilon \approx 1$ we get

$$P \approx a\sigma T^4$$

$$a \approx \frac{P}{\sigma T^4}$$

Now if the filament is a wire of circular cross-section, its surface area is $2\pi rl$. So

$$2\pi rl = \frac{P}{\sigma T^4}$$

hence

$$rl = \frac{P}{2\pi\sigma T^4}$$

Putting in the given values,

$$rl = \frac{40}{2\pi \times 5.7 \times 10^{-8} \times (3000)^4} \text{ m}^2$$

$$\approx 1.4 \times 10^{-6} \text{ m}^2$$

So, we now have two geometrical requirements, one derived from the input power equation, the other from the output power.

Input power criterion,

$$\frac{l}{A} = \frac{l}{\pi r^2} = 1 \times 10^9 \text{ m}^{-1}.$$

Output power criterion,

$$rl = 1.4 \times 10^{-6} \text{ m}^2.$$

These are two equations which the design must simultaneously meet. The unique solution requires $l = 0.18$ m and $r = 7.7$ $\mu$m: that is, 180 mm of wire thinner than human hair made from a metal with the highest known melting temperature (3680 K). As I said in Section 1.8.1, it's very impressive!

## Objectives for Chapter 1

You should now be able to do the following.

1.1 Demonstrate, by means of examples, the use of the systems approach, especially as a way of handling the technical and external factors that affect product manufacture. (SAQ 1.1)

1.2 Using examples specify, and comment on, the factors that determine the choice of process for a material and a product. (SAQ 1.3)

1.3 Choose, giving your reasons, a process route for specified products. (SAQ 1.4)

1.4 By means of examples, demonstrate the interdependence of P for product design, P for processes and P for properties of materials. (SAQ 1.5)

1.5 Through the geometrical factors involved, relate the properties of a material to those of a product. (SAQ 1.6)

1.6 Describe using examples how (a) materials properties and (b) merit indices can be used as aids in the selection of a material for a product. (SAQ 1.7)

1.7 Explain, using examples, the usefulness of the materials property profile (MPP). (SAQ 1.8)

1.8 Describe the following methods of processing:
extrusion, wire-drawing, rolling, forging, injection moulding, blow moulding, vacuum forming, compression moulding, pressing, deep drawing, casting, slip casting, single crystal growth.

1.9 Define or distinguish between the following terms and concepts:
recycling, re-using
thermoset, thermoplastic
composite
crystal, glass, single crystal, polycrystal
system, boundary, environment, subsystem
merit indices
product design specication (PDS)
technical and external factors in product cost and price
materials property profile (MPP)
runner, riser
flash, sprue
mould, die
hot working, cold working.

# Answers to Exercises

EXERCISE 1.1 To meet its design function, the 'boring' end of the tool has to be hard — at least as hard as the materials in which it is intended to make holes, and preferably harder. It also needs to be of the size that allows it to make holes of the required size.

Along with 'modern' hard engineering materials, stone is not deformable as a solid — being hard and deformable is really a contradiction in terms. It also has a high melting point which makes it difficult (certainly in those days) to mould it to shape as a liquid. So it has to be shaped by chipping and/or grinding, using a material at least as hard.

EXERCISE 1.2
(a) Short term savings might be achieved by using (cheaper) alternative materials, reducing scrap, saving energy, cutting staff, reducing research and development, stockpiling raw materials, selling equipment.
(b) Long term savings could be sought, for instance, through the development of new materials or processes, recycling of materials, finding new markets, changing from one-off or batch production to mass-production, and redesigning the product to use less material.

EXERCISE 1.3 Table 1.7 (overleaf) shows my attempt at completing Table 1.2. I've included more examples than I asked you for but, even so, some of your answers may differ from mine.

EXERCISE 1.4 My answers are included in Table 1.6.

Table 1.6

| Type of vessel | Heading | Comments |
|---|---|---|
| pottery cups | (a), (d) | |
| pottery mugs | (a), (d), (f) | |
| enamelled steel mugs | (a), (c), (d) | |
| Melmex (thermoset) cups | (c), (d) | stylish plastic cup |
| Pyrex (glass) beakers | (a), (c), (d), (h) | special glass composition |
| bone-china cups | (d), (e), (f), (h) | composition designed to give translucence |
| expanded polystyrene beakers | (a), (b), (d), (g), (h) | tapering to allow stacking; foam structure gives thermal insulation |
| polypropylene beakers | (a), (c), (d), (g) | tapering and half-handles allow stacking |
| sundry glasses | (a), (g) | some of shapes associated with particular wines, sherry etc. |
| cut-glass wine glasses | (e), (f), (g) | |
| glass beer mug | (a), (g) | |
| pewter tankard | (e), (f), (h) | attractive because of unusual lustre of pewter |
| Maxpax cups | (a), (b), (d), (g) | double wall gives thermal insulation |

EXERCISE 1.5 Lots of points are possible, depending on the type of product. Here are some examples.

3 Politics. Is the supply of materials likely to be uncertain due to political uncertainty in particular countries, or the action of international cartels? Relying on a single supplier may be hazardous.

6 Environment. What are the likely operating temperatures, humidities, atmospheres, chemical environments? Will the materials involved be affected? What pollution hazards could be produced by materials processing?

10 Quality/reliability. What level of product reliability is necessary to ensure success of the product? Are failures due to material failures? What tests of materials should be carreid out?

14 Weight. Is product weight important? Can lighter materials/smaller components be used?

22 Product cost. What is target cost? Can performance be achieved with cheaper materials? Are less costly processes possible? Energy saving during processing? Reduce wastage of materials or recycle it?

24 Life in service. How long should the product last? Is a material property likely to cause premature failure? If so, how can it be combated?

Table 1.7

| | Plastics | Metals | Ceramics | Natural material |
|---|---|---|---|---|
| Common examples | polyethylene<br>polystyrene<br>nylon<br>PVC<br>rubber<br>polyurethane<br>Bakelite<br>(i.e. phenol<br>formaldehyde)<br>epoxy (e.g. Araldite) | steel<br>solder (Pb + Sn)<br>copper<br>cast iron<br>brass (Cu + Zn)<br>aluminium<br>chromium<br>silver | glass<br>concrete<br>alumina<br>pottery<br>brick<br>bone<br>porcelain | leather<br>wood<br>cotton<br>wool<br>bone<br>hair |
| Young's modulus | low or very high | high | high | some high, some low |
| Toughness | some good, some poor | good | poor | good |
| Tensile strength | some fair, some poor | high | low | fair to good |
| Compressive strength | fair | fairly high | high | fair |
| Electrical conductivity | very low | high | very low | very low |
| Thermal conductivity | low | high | low | low |
| Density | low | high | medium | low |
| Melting or softening temperature | low | fairly high to high | high | low |
| Ductility | some good, some nil | good | nil | fairly low |
| Hardness or wear resistance | some low, some moderate | moderate | high | low to fair |
| Optical characteristics | some transparent/translucent, some opaque | opaque, naturally coloured | some transparent/translucent, some opaque; can be artificially coloured | opaque and coloured |
| Deterioration in different chemical solutions | good | some good; some poor | good | poor |
| Examples of artefacts | cold water plumbing, containers and panels of all sorts, adhesives, packaging, kitchenware, clothing, electrical/thermal insulation | electrical wiring, window frames, 'whitegoods' (cooker, fridge, etc.), car parts, machinery, jewellery, bridges, railway lines | windows, lintels, cups, abrasives, bricks, ovenware, tableware | clothing<br>furnishings<br>jewellery<br>books<br>bags |
| Comments on properties that influence processing | some ductile and can be softened or melted at reasonable temperatures, hence can be shaped in solid form; others not ductile, cannot be shaped in solid state | many ductile, so can be shaped as solid lump e.g. rolled; can usually be melted and cast to shape from liquid | high melting point, so difficult to contain and process when molten; not ductile at ambient temperatures, so shaping as a solid lump usually not possible | Making to shape other than the natural ones is limited to cutting and reassembling in some way |
| Examples of materials that don't fit the classification | non-metallic elements, such as silicon, germanium, diamond<br>physical mixtures of materials (called composites, e.g. plastics reinforced with glass fibres, or containing powders for cheapness<br><br>tungsten carbide particles in cobalt for use in machine cutting tools<br>glasses, which are unlike other ceramics mainly because they can be melted and therefore readily shaped at relatively low temperatures | | | |

# Answers to self-assessment questions

## SAQ 1.1

(a) (i) The appropriate system boundary is probably that drawn round PPP in Figure 1.10. The purpose requires consideration only of the technical aspects I defined earlier.

   (ii) Since the interest is in the (technically) best materials and processes, I would certainly include in the environment: alternative materials and processes, the existing process plant (its performance would need to be monitored), and those aspects of overheads that cover product testing, research and development.

(b) (i) The cost of the material is a high proportion of the cost of most products. The advantage of including it as a 'property' is that it can be taken into account explicitly and quantitatively. You will meet examples later.

   (ii) There are two important limitations to the approach:

   Cost isn't a property of a material. This is important because unlike other properties it can change for reasons that have nothing to do with the structure of the material. Thus, materials costs can rise suddenly and for unexpected reasons: oil 'crises' in the Middle-East, political strife in cobalt ore mines, and attempts to 'corner the market' in silver are examples. The consequences can be serious.

   Costing materials in this way can give a false impression of overall cost because it emphasizes only one element. The cost of processing can vary a lot between materials. This can happen even when the same processes are used (different melting temperatures and mechanical force requirements, for instance) and would certainly be the case if different materials require different process routes. PPP again!

## SAQ 1.2

(a) (i) Thermal conductivity is an intrinsic materials property. It is a determinant of heat flow rate through a material and is usually represented by $\kappa$ in equations. The other determinants of heat flow rate are not material-specific. They are the temperature difference between one side of an object and the other, and the shape — heat flows more readily through a broad, thick sample than through a long narrow one. So using $A$ to represent cross-sectional area, $\Delta T$ for temperature difference, $l$ for thickness and $\dot{Q}$ for rate of heat flow

$$\dot{Q} = \frac{A \times \Delta T \times \kappa}{l}$$

Thus,

$$\kappa = \frac{l\dot{Q}}{A\Delta T}$$

Putting experimental values into this equation gives a numerical value for $\kappa$. The equation also tells us that the units of $\kappa$ are

$$\frac{\text{m} \times \text{W}}{\text{m}^2 \times \text{K}} = \text{W}\,\text{m}^{-1}\text{K}^{-1}$$

(ii) **Electrical conductivity** is the materials property which measures how readily electrons flow through a material. The current $I$ flowing through a sample of a material is determined by its cross-sectional area $A$, its length $l$, the applied e.m.f. $V$ and the intrinsic electrical conductivity $\sigma$ of the material. Thus

$$I = \frac{A \times V \times \sigma}{l}$$

so $\sigma = \dfrac{Il}{AV}$

Since $V = IR$ (Ohm's law)

$$\sigma = \frac{l}{RA},$$

the units of which are

$$\frac{\text{m}}{\Omega \times \text{m}^2} = \Omega^{-1}\text{m}^{-1}$$

(iii) **Electrical resistivity** $\rho_e$ is the reciprocal of electrical conductivity.

$$\rho_e = \frac{1}{\sigma} = \frac{RA}{l},$$

which has units of $\Omega$ m.

(b) The testpiece will be permanently extended by the strain OH; this is the amount of plastic strain that will remain in the testpiece if it is unloaded at the point F on the stress–strain curve. If you are unclear about this, you may not have appreciated that elastic strain still continues at stresses beyond the yield stress. When the load is removed, the testpiece relaxes elastically along the line FH. GH is the relaxed elastic strain.

(c) (i) Material 2 has the highest Young's modulus. The initial (linear) portion of the stress–strain curve is solely elastic strain. Elastic strain is reversible deformation. Here, stress is proportional to strain and the constant of proportionality is called the elastic modulus. For a tensile or compressive stress the modulus is called the Young's modulus. Since strain is dimensionless, the Young's modulus has the same unit as stress, the pascal (Pa). One pascal is a stress of 1 N m$^{-2}$, and sometimes stress (and Young's modulus) is given in N m$^{-2}$. Values of Young's modulus for ceramics and metals are high and you will often see them expressed as GPa (gigapascals, $10^9$ Pa) or GN m$^{-2}$. The material with the steepest slope in its elastic region on the stress–strain curve has the highest Young's modulus; in this case, material 2.

(ii) Material 4. The yield stress is the stress at which **plastic strain** commences. Plastic strain is irreversible (permanent) deformation. Material 5 doesn't show plastic deformation and therefore has no yield stress.

(iii) Material 4 also has the highest tensile strength. The tensile strength is the highest stress a material can sustain without breaking. This is so whether or not the material exhibits plastic deformation.

(iv) Material 1 also has the highest ductility. The plastic range of its stress–strain curve is more extensive than that for the other materials. Check for yourself by drawing in the elastic unloading line from the point of fracture for each one. Ductility is considered in more detail in Chapter 2.

(d) **Work-hardening** is the strengthening of a material by plastic deformation. So, a progressively increasing stress is needed to deform it. This is manifested as a positive (rising) slope of the stress–strain curve beyond the yield strength. Of the materials in Figure 1.16, all exhibit work-hardening except 2 and 5.

SAQ 1.3 The main 'processing' properties of materials covered in Section 1.3 are: viscosity, ductility, work-hardening and melting temperature. This is not a complete list: a number of other properties can have a significant influence on processes, for example thermal conductivity and oxidation resistance.

**Viscosity**
Materials with low viscosities, such as metals, can flow under gravity and so can be cast; those with high viscosities have to be forced into moulds, hence injection moulding and blow moulding, and vacuum forming of glasses and thermoplastics

**Ductility**
Good ductility allows shaping in the solid state, hence the various ways of mechanically working metals. Cold working, which produces work hardening, requires higher forces but provides better accuracy and surface finish; hot working (no work-hardening) can produce large changes in shape.

No ductility, that is, the material is brittle, means that flow in the solid state cannot be used. This applies to most ceramics and some metals, for example tungsten, when cold.

**Melting temperature**
This can be so high that processing from the liquid is prohibited. If the material is also brittle, processing using powders in another liquid is possible, for example slip casting, pressing and sintering.

SAQ 1.4
(a) Glass beer mug. Blow moulding. Handle is moulded at same time — you will find a mould line (the line along which the two halves of the mould meet) all round a mug. Pressing can't produce the handle.

(b) Glass ashtray. Pressing. Plunger is shaped to produce the inside surface, and a mould to produce the outside. Could be cast?

(c) Steel table knife. Forged. Flash removed by grinding. Much stronger and better surface finish than a casting.

(d) Metal end cap of light bulb. Pressing from sheet. Needs a fairly ductile metal.

(e) Church bell. Sand casting. It's a metal, so very low viscosity, surface finish not crucial. Large object, so metal mould would be very expensive.

(f) Plastics casing for television set. Injection moulding. Thermoplastic. Gives dimensional accuracy, good surface finish and enables moulding of integral ribs for stiffness. Case needs intricate internal structure for other components etc.

(g) Casing for an electric plug. Compression moulding, thermoset assumed.

(h) Investment. Slip casting. The way of producing a thin-walled ceramic object.

(i) Disposable plastic beaker. Vacuum formed. Quicker (and cheaper than injection moulding. Blow moulding couldn't produce the shape.

SAQ 1.5
(a) The glass envelope is designed to provide a protective surround for the filament. To achieve this it must (at least) have the following properties:

• transparency,
• sufficient strength and impermeability at the *surface temperature of the envelope*.

The surface temperature is determined by the product design — by the power rating of the filament and the distance between the filament and the envelope, for example. The surface temperature in turn sets the property requirements for strength and impermeability.

The surface temperature also links properties to processing. The envelope must not flow out of shape during service.

So, if the envelope is shaped by flow, it has to be processed at a temperature above the operating surface temperature.

Finally, the product design requires an envelope with an open end. For glass this indicates processing by blow moulding, which requires a particular viscosity.

(b) Important requirements for the Maxpax (thermoplastic) disposable cup are:

• For performance. Lightweight (low density) and sufficient stiffness to support the contents. The stiffness can be achieved by using a material of suitable Young's modulus or by using the appropriate wall thickness for the cup.
• For processing to shape. The product design requires a material with a viscosity that allows processing into a quite complex shape with a thin wall.

Note that we can identify these links because we know the materials that are used and that they are, by and large, successful. If the materials were not suitable, the design of lamp bulbs and cups would be quite different.

SAQ 1.6
You know from SAQ 1.2 that the resistance $R$ of a wire is given by

$$R = \rho_e \times \frac{l}{A}$$

where $\rho_e$ is the electrical resistivity of the material (the material property), $A$ is the cross-sectional area of the wire and $l$ is its length. $R$ is a property of the product — it varies with the dimensions of the wire. So,

$R$ (property of product)

$\quad = \rho_e$ (material property)

$\quad \times \frac{l}{A}$ (geometrical factor)

SAQ 1.7
(a) Packing as many elements as possible into a given space requires each to have the smallest possible cross-sectional area. Since

$$R = \frac{\rho_e l}{A},$$

where $R$ is resistance, $\rho_e$ is resistivity, $l$ is length and $A$ is cross-sectional area, and $l$ and $R$ are fixed, $A$ is lowest for the material with the smallest electrical resistivity. So the criterion is simply 'minimum electrical resistivity'. From Table 1.4, silver is first choice, followed by copper and then aluminium. In practice copper is usually used; silver is scarce and expensive and its margin of gain over copper is small.

(b) Here the cross-sectional area of the conductor is of less concern than its mass. The requirement is a cable of fixed $R$ and $l$, and minimum mass. Now

$$\text{density } (\rho) = \frac{\text{mass } (m)}{\text{volume } (V)} \qquad (1.1)$$

and, using the same symbols as in part (a),

$$R = \frac{\rho_e l}{A} \qquad (1.2)$$

From Equation (1.1), the mass of the cable is given by $m = \rho V = \rho Al$, so $A = m/rl$. Substituting for $A$ in Equation (1.2) gives

$$R = \frac{\rho_e \rho l^2}{m}$$

or

$$m = \frac{\rho_e \rho l^2}{R}$$

The criterion for minimum mass of the cable is a minimum value of $\rho_e \rho$. Values of $\rho_e \rho$ for silver, copper and aluminium are shown in Table 1.8. The ranking is the reverse of what it was for the machine.

SAQ 1.8

(a) At this simple level, glass seems to meet requirements for transparency, gas-tightness and viscosity for blow moulding. It fails on robustness as we are often reminded, but this failing is tolerated. The ideal bulb material has not yet been developed.

(b) A number of thermoplastics are transparent; for example poly(ethylene teraphthalate) (PETP) found in large drinks bottles; polycarbonate (PC) used as a substitute for glass in greenhouses; poly(methylmethacrylate) (PMMA), used for windows in aircraft and boats.

Thermoplastics are tougher than glass. Their most severe limitations are their permeability to gases and their softening and distorting when hot; after all, they are usually moulded at temperatures lower than the surface temperature of a lamp bulb.

Table 1.8

| Material | Density $\rho$ | Electrical resistivity | $\rho_e \rho$ | Ranking |
|---|---|---|---|---|
| | $\mathrm{Mg\,m^{-3}}$ | $10^{-8}\,\Omega\,\mathrm{m}$ | $\mathrm{Mg\,m^{-2}\,\Omega}$ | |
| aluminium | 2.7 | 2.7 | 7.0 | 1 |
| copper | 8.9 | 1.9 | 15.1 | 2 |
| silver | 10.5 | 1.6 | 16.8 | 3 |

# Chapter 2 The nature of materials

## 2.1 Scale and mixture

One of the important ideas developed in Chapter 1 was that each material has a unique set of properties, a property profile, and that whilst some of these properties have a crucial bearing on whether or not it is selected for service in an artefact, others determine if and how it can be made in the required shape. We saw, for instance, that tungsten's melting temperature and electrical resistivity make it a good material for light bulb filaments, but that its lack of ductility precluded its production in the required filament form until a special process route from powder was developed.

In Chapter 1 we also established that properties depend on the internal structure of the material. It follows that an important aim of materials development is to create materials with the internal structures that provide useful *combinations* of properties, both for service and for manufacture. Notice the emphasis on combinations. Creating an internal structure that improves one property often has a deleterious effect on another desirable property. For instance, as you will see in this chapter, increasing the yield stress of metals decreases their ductility and hence their ease of being, say, drawn into wire. Interestingly, separating properties into those for service and those for manufacture highlights the possibility of manipulating a material's structure in stages to suit both purposes.

## ▼A hacksaw blade▲

The blade (Figure 2.2) is held in the frame by two pins. To prevent buckling and twisting it is pre-tensioned by the frame. The material round the pinholes must be tough enough not to fracture, and the teeth must be hard and non-deformable.

One way of achieving this is to use a steel which can be hardened locally; Figure 2.1 shows the processing steps. Cast ingots (a) of initially soft alloy are hot rolled to final thickness, then guillotined to blade-sized blanks. The teeth are machined and set (b) by passing the blade through a series of 'gear wheels'. The alloy is still soft at this stage. The toothed edge passes through a gas flame (c) and is then quenched in water to produce hard, brittle teeth. Heating at 450 K (d) reduces the brittleness while retaining useful hardness.

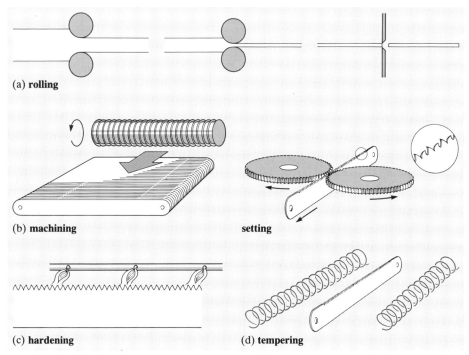

(a) **rolling**

(b) **machining**     **setting**

(c) **hardening**     (d) **tempering**

Figure 2.1 Processing a hacksaw blade

Figure 2.2 A hacksaw

First the material might be treated to create a structure that is suitable for processing to shape; then it might be treated to produce a structure suitable for service. Elementary examples of this technique are concrete, where a mix of stones, sand, cement and water is moulded to shape before the chemical reactions that produce hardening take effect, and earthenware, in which firing produces a new structure and property profile. At a more sophisticated level, the technique is an important item in the materials engineer's toolkit. ▼**A hacksaw blade**▲ is one instance, and you will meet others in later chapters.

This relationship between structure and properties is all embracing and operates over a very wide spectrum of structural scale. Figure 2.3 indicates the size ranges of a variety of structural features in materials. You may not be familiar with all of them, but they will be explained in this or later chapters.

Figure 2.3 covers ten orders of magnitude and can be broken down into three ranges: macrostructure, microstructure, and atomic structure. *Makros* is Greek for large, and macrostructure is usually taken to be structure that can be seen with the naked eye, that is down to about 1 mm. *Mikros* is Greek for small, and microstructure is that structure revealed by a microscope. To make these ranges of scales more real, let's look at an example. ▼**Reinforced concrete**▲ illustrates the structural detail found in reinforced concrete at widely different magnifications. Concrete is a mixture of cement and an aggregate made

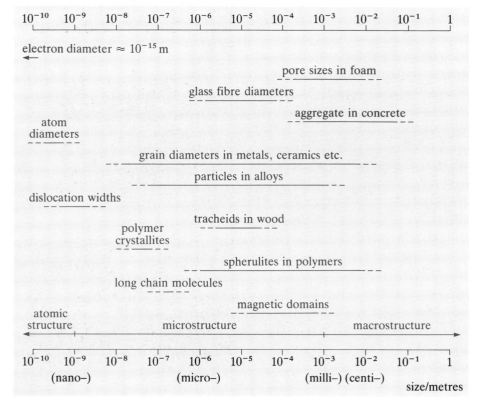

Figure 2.3 Size ranges of structural features

# ▼Reinforced concrete▲

Reinforced concrete at different levels of scale. The techniques for revealing structure are considered later.

Reinforced concrete (left) in action. It is a composite material on the grand scale. Concrete is weak in tension; steel is strong in tension. The reinforcing steel rods allow the use of concrete in components which carry tensile loads — beams, tall columns etc.

Concrete is a mixture of cement and an aggregate of sand and stone.

The reinforcing rods are a plain carbon steel (an alloy of iron with about 0.15% carbon).

Figure 2.4

Figure 2.5

(left) Broken concrete under a magnifying glass. The sand particles are not visible. Chunks of crushed rock can be seen, and the holes from which they have fallen. Clearly, they are not strongly bonded to the cement.

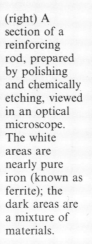

(right) A section of a reinforcing rod, prepared by polishing and chemically etching, viewed in an optical microscope. The white areas are nearly pure iron (known as ferrite); the dark areas are a mixture of materials.

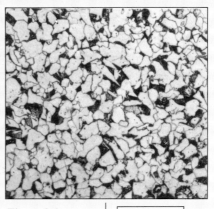

Figure 2.6 ⊢————————⊣ 0    50 µm

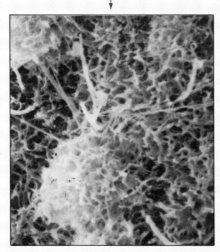

Figure 2.7 ⊢——⊣ 0  1 mm

(left) A stage during the hardening of cement as seen in a scanning electron microscope. The spiny growths of 'gel' form an interlocking network that binds the aggregate into a solid mass.

(right) Transmission electron micrograph of a dark area of Figure 2.6. Light areas are ferrite. Dark 'needles' are cementite, $Fe_3C$.

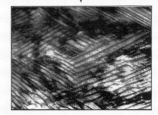

Figure 2.8 (a) ⊢——⊣ 0  400 nm

(right) An area of ferrite not from the reinforcing rod, at a higher magnification. The dark lines are structural defects called dislocations.

Figure 2.8 (b) ⊢——⊣ 0    1 µm

up of sand with crushed rock or gravel. When used as a structural member in buildings, bridges and so on, it is reinforced with steel rods — a modification to the macrostructure.

The pictures in 'Reinforced concrete' give important clues to the inner secrets of structures and their influence on properties. They show that, in one way or another, materials are mixtures. On a coarse scale materials may be mixtures of other materials, for example the aggregate and rods in concrete. On a very fine scale they may be mixtures of individual atoms, for example the phosphorus added to silicon to make it a semiconductor. They may be mixtures of order and disorder — of crystalline and amorphous regions, for instance — and of defects such as grain boundaries and dislocations.

The properties of a material depend on both the relative amounts of the ingredients and on how they are distributed. The challenge is to understand how particular mixes determine properties and to develop and control those mixes that give desired properties. In this chapter we shall look at a variety of mixtures, covering different levels of scale. Let's start by looking at the macroscopic scale, that seen by the naked eye.

# 2.2 Macrostructure

If you look at the materials in artefacts around you, and especially at the surface of a broken piece, most will reveal very little — the structural detail is on too fine a scale to be resolved by the unaided human eye. However, in some materials, such as concrete, a coarse structure can be detected. Can you think of other examples?

The list I came up with included the following.

• Wood and grasses. They are made up of aligned fibres bonded together. The fibres are easily seen in a broken piece.
• Plywood and chipboard. Pieces of wood are bonded together in an organized manner (plywood) or a disorganized manner (chipboard).
• Foam plastics. The cellular structure of the expanded polystyrene beaker of Figure 1.37; the 'foam' in pillows and upholstery; Aero chocolate.
• Glass fibre reinforced plastics (GFRP). Again they have a fibrous nature but the fibres are not necessarily aligned (as they are in wood). The fibres may be used as random short lengths or as woven cloth. GFRP is increasingly being used in car bodies, boat hulls, vaulting poles, and so on.
• Galvanized metal. Examples are metal dustbins, buckets and pipework. You can see large crystals in the coating.
• Plastics cloths. These are the fabrics impregnated with polymer used in some tablecloths, sou-westers and so on.

Figure 2.9 shows the macrostructure of softwood and GFRP. I have cheated a bit by using a small amount of magnification (about × 10). How does macrostructure determine properties?

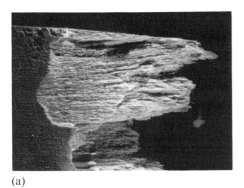 

(a)                              (b)

Figure 2.9 Macrostructures of (a) wood and (b) GFRP

## ▼A concrete mix▲

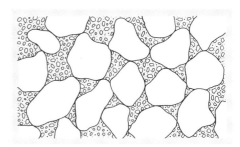

Figure 2.10 Packing of aggregate in concrete

You will notice that most of the materials in my list are **composites** — physical mixtures of materials which in combination have a better property profile than that of the individual constituents. Consider concrete again. Hardened cement is a brittle and relatively expensive material, but combined with gravel and sand it is stronger (in compression) and cheaper. See ▼A concrete mix▲. In many respects, concrete is a kind of artificial stone, with enormous processing advantages: before the cement hardens, the concrete mix can be moulded to shape. This is much more convenient than chipping away at a large lump of stone with a hammer and chisel. As concrete can be cast to virtually any size and shape, it has revolutionized civil engineering.

Let's consider another composite, GFRP. Neither glass nor thermosetting plastics are used to carry high loads in service: thermosets have relatively low stiffness and lack toughness; and glass, though stiff and strong in compression, also breaks readily. However, in composite form as GFRP, they are extremely useful. The glass fibres stiffen the polymer, and the polymer helps to exploit the strength of the fibres. It protects them and inhibits their catastrophic failure.

In materials which are mixtures, the term **matrix** is often used for the host material in which the other is embedded; thus in GFRP, the plastics is the matrix. The glass fibres can be incorporated in the matrix in a variety of ways, particularly as aligned continuous lengths, bundles woven into a mat, or random short lengths. The properties of the composite will obviously vary with the amount and configuration of the fibres it contains. Here, I will deal only with some simple examples.

## 2.2.1 The law of mixtures

At first sight, one would expect the properties of a composite to depend directly on the proportions of the constituent materials. So, for instance, a 50–50 mix, by volume, of fibres and polymer should have a density midway between those of the two components, and indeed this is the case. To see that this is so, consider a volume $v_C$ of composite made of fibres in a matrix. The mass $m_C$ of the composite is

$$m_C = m_M + m_F$$

where $m_M$ is the mass of matrix material and $m_F$ is the mass of fibre.

We can consider concrete in two ways: either the sand and gravel serves to strengthen and toughen the cement paste (if you have ever had cause to break up a bag of cement that has hydrated in a damp place, you will know that it is weak and brittle); or the paste is the adhesive that holds the sand and gravel together. The truth is probably a mixture of the two: cement paste alone is not a useful building material, nor are raw sand and gravel in their basic form. In combination they are.

Compared with other materials cement is cheap, but aggregate is cheaper. This is because cement is an energy intensive product — typically the production of five tonnes of cement consumes a tonne of coal or its fuel equivalent (largely in the high temperature kilns needed to react the limestone and clay from which cement is made). So there is financial incentive to minimize the amount of cement in the mix; but there must be sufficient cement to coat all the aggregate particles. In practice the optimum is achieved by using an aggregate of different sizes of particle in the ratio of about 5 volumes of aggregate to 1 volume of cement. The small particles fit in the spaces left between large particles and the cement paste has to flow only into the remaining spaces, Figure 2.10.

Now, mass is volume ($v$) times density ($\rho$) so

$$v_C \rho_C = v_M \rho_M + v_F \rho_F$$

hence

$$\rho_C = \frac{v_M}{v_C} \rho_M + \frac{v_F}{v_C} \rho_F$$

The terms $v_M/v_C$ and $v_F/v_C$ are called volume fractions. Using $V_M$ and $V_F$ to denote them, $\rho_C = V_M \rho_M + V_F \rho_F$

By definition, $V_M = (1 - V_F)$, so

$$\rho_C = (1 - V_F) \rho_M + V_F \rho_F$$

This is an example of the **law (or rule) of mixtures**. The law is frequently invoked to represent a range of composite properties. But not all the properties of a composite are in simple proportion to those of its constituents. ▼**Modelling the elastic modulus of a composite**▲ looks at one case where the law applies and one where it doesn't.

SAQ 2.1   (Objective 2.3)
You are considering substituting a rod made of a composite of polypropylene and aligned glass fibres for a rod made of polypropylene in a structure subjected to tensile forces along the direction of the fibres. The volume fraction of glass fibre $v_F$ in the composite is 0.093. The only criteria, artificially, are that the rods must have the same unloaded lengths, and that in service they must extend elastically by the same amount for the same load. Estimate the mass saving from using a composite by calculating

(a) the density of the composite $\rho$,
(b) the composite modulus,
(c) the relative mass of the two articles.

Use the following data.

Density of glass fibre

$$\rho_F = 2200 \text{ kg m}^{-3}$$

Density of polypropylene matrix

$$\rho_M = 900 \text{ kg m}^{-3}$$

Young's modulus of glass fibre

$$E_F = 72 \text{ GN m}^{-2}$$

'Modulus' of polypropylene matrix

$$E_M = 0.5 \text{ GN m}^{-2}$$

(Strictly speaking polymers do not obey Hooke's law. Measurement of their elastic modulus is considered in Chapter 6.)

Sometimes the properties of a composite are far superior to those of its individual components. The whole is better than the sum of the parts. An important example I want to look at is toughness. Before doing so I need to define toughness and explain why some materials are tough and others aren't.

# ▼Modelling the elastic modulus of a composite▲

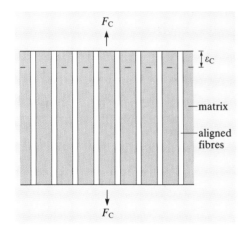

Figure 2.11 Fibres parallel to load

First, consider a composite with continuous aligned fibres under a tensile load $F$ parallel to the fibres, Figure 2.11. We will assume that the fibres are perfectly bonded to the matrix and, therefore, that under a load both the fibres and matrix undergo the same elastic strain $\varepsilon_C$. For this reason it is called the **homogeneous strain model**. Throughout we will use subscript M and F for matrix and fibres respectively, and C for composite. For the composite, stress $\sigma_C$ is

$$\sigma_C = \frac{F_C}{a_C}$$

where $a_C$ is the cross-sectional area of the composite. But

$$\sigma_C = E_C \varepsilon_C$$

where $E_C$ is the Young's modulus of the composite and $\varepsilon_C$ its strain. So

$$F_C = E_C \varepsilon_C a_C$$

The same arguments apply to the matrix and fibre, so

$$F_M = E_M \varepsilon_M a_M$$

and

$$F_F = E_F \varepsilon_F a_F$$

But the load on the composite must be the sum of the loads on the matrix and fibre, that is $F_C = F_M + F_F$, so

$$E_C \varepsilon_C a_C = E_M \varepsilon_M a_M + E_F \varepsilon_F a_F$$

As mentioned at the beginning, we are using a homogeneous strain model, that is $\varepsilon_C = \varepsilon_M = \varepsilon_F$, so

$$E_C a_C = E_M a_M + E_F a_F$$

whence

$$E_C = \frac{E_M a_M + E_F a_F}{a_C}$$

or, in terms of area fractions,

$$E_C = E_M A_M + E_F A_F$$

where $A_M = a_M/a_C$ and $A_F = a_F/a_C$.

Because the composite and constituents are parallel sided, areas $a_M$ and $a_F$ are proportional to volumes $v_M$ and $v_F$. Thus volume fractions $V_M$ and $V_F$ could be substituted for area fractions $A_M$ and $A_F$ in the last equation:

$$E_C = E_M V_M + E_F V_F$$

Since by definition $V_M = 1 - V_F$, this is often also written as

$$E_C = E_M (1 - V_F) + E_F V_F$$

So for this configuration of fibres the law of mixtures applies. But what about other configurations?

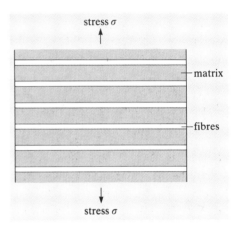

Figure 2.12 Fibres perpendicular to load

EXERCISE 2.1   Consider a composite with aligned fibres lying perpendicular to the load. In this case the fibres and matrix are subjected to the same stress as the overall stress on the composite; this is called the **homogeneous stress model**. The configuration is shown in Figure 2.12.

Derive an expression for the elastic modulus of the composite.

Hints
(a) The stress is homogeneous so $\sigma_C = \sigma_M = \sigma_F$.
(b) It follows that, under a given stress, the total extension of the composite $\delta_C$ is

$$\delta_C = \delta_M + \delta_F$$

# 2.3 Toughness

Sooner or later, when subjected to an increasing stress, a piece of material will break — a crack moves though the single piece and produces two or more pieces. This process is called **fracture** and the stress at which it happens defines the **strength** of the material — the **tensile strength** to pull it apart and the **compressive strength** to crush it.

Do not confuse strength with **toughness.** Toughness is to do with the nature of the fracture process and, in particular, with how much resistance a material offers to a crack attempting to grow through it. The materials in most products either already contain cracks or have structural flaws that, under stress, can readily turn into cracks. A material with a high toughness makes crack growth difficult, whereas in one with a low toughness cracks grow readily. You can see that toughness is an extremely important property. It accounts for why aeroplanes and ships can operate happily with cracks in them (provided, as you will see, the cracks are not too long), why the branch of a tree is difficult to break even when it is part-sawn, and why glass breaks catastrophically.

Formally, toughness is considered in terms of the different forms of energy involved in deformation and fracture, and how one form is converted to another. Since deformation is produced by stress, work is done when a material is deformed (work is force times distance) and toughness can be assessed in terms of the work that has to be done in propagating a crack. It is defined as the energy absorbed in creating unit area of crack and has the symbol $G_c$. ▼**Work of deformation and fracture**▲ derives its units and looks at another energy term we shall use, **elastic strain energy**. Table 2.1 gives typical values of $G_c$ for a variety of materials tested under similar conditions.

Table 2.1

| Material | Toughness* $G_c/(\text{kJ m}^{-2})$ (approximate values) | |
|---|---|---|
| pure copper | | 1000 |
| mild steel | | 100 |
| GFRP, stressed along aligned fibres | | 100 |
| GFRP, random fibres | | 40 |
| wood: stressed across grain | | 5–15 |
| stressed along grain | | 0.5–2 |
| polyethylene | | 5–10 |
| cast iron | up to | 3 |
| concrete: reinforced | up to | 4 |
| cement | | 0.01–0.03 |
| epoxy | | 0.3 |
| polyester | | 0.1 |
| alumina | | 0.02–0.06 |
| porcelain (electrical) | | 0.01 |
| glass | | 0.001–0.01 |

*The wide range of values for some materials shows that $G_c$ is very sensitive to variations in internal structure of the materials.

# ▼Work of deformation and fracture▲

The energy put into a material during deformation (that is, the work done) has to be either stored or used in some way. How much work is done and what happens to it?

To start with, how much energy is expended by the testing machine that produced the load–extension curve of a material shown in Figure 2.13? (Note that the material does not obey Hooke's law.)

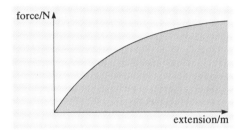

Figure 2.13 Load–extension curve

The work done is the area under the curve (mathematically, $W = \int F \, dx$). The area is measured in

[units of force] × [units of distance]

as may be confirmed by supposing the force to be constant over the extension, in which case $W = Fx$. So the units of the area are N × m, or N m, which is the same as the joule, J.

Normally, stress–strain curves rather than load–extension curves are used. What are the units for the area under a stress–strain curve?

They will be the product of the units of stress (N m$^{-2}$) and the units of strain (m m$^{-1}$), that is N m$^{-2}$. These units may be written as

$$N\ m^{-2} = (N\ m)\ m^{-3}$$
$$= J\ m^{-3}$$

So what is the significance of the area under a stress–strain curve?

The units tell us that it represents the work done during deformation per unit volume of the deformed material.

Figure 2.14 shows the stress–strain curve for a material that does not obey Hooke's law. The area under the curve, representing work done per unit volume, is shown shaded. The dashed line shows

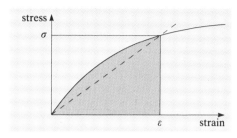

Figure 2.14 Stress–strain curve

the stress–strain curve for a material that does obey Hooke's law.

EXERCISE 2.2   A material has a Young's modulus $E$ and is deformed in accordance with Hooke's law to a stress $\sigma$ and a strain $\varepsilon$. Derive an expression for the work of deformation per unit volume, $W$, in terms of

(a) $\sigma$ and $\varepsilon$,
(b) $\sigma$ and $E$,
(c) $\varepsilon$ and $E$.

Where does this work of deformation go? As always, energy (or work) is conserved; it is merely transformed from one kind to another during deformation. First, consider elastic deformation. By definition, this strain is reversible, so all the work of elastic deformation can be recovered from the material when it is unloaded. This is called **elastic strain energy**. It is what you produce when you wind up your old mechanical alarm clock; you store energy in the spring in order to power the mechanism.

What about the work of plastic deformation? If you've ever burnt your fingers picking up the off-cuts from hacksawing you will have a pretty good idea of where some of the work of deformation goes. In fact most of the work done in plastic deformation is transformed to heat (90% in metals, 60% in polymers) and leaks away into the environment. No wonder the plastic strains are not reversible if the energy is dissipated in this way.

Not all of the work of plastic deformation is transformed to heat. The amount remaining is described in Figure 2.15.

The energy corresponding to the triangle ABC is elastic strain energy and is

recoverable when the material is unloaded. The energy equal to the remaining area under the stress–strain curve (OYAB) is either dissipated as heat or absorbed by the material, largely in the creation of defects known as dislocations during plastic flow. (Dislocations are considered in Section 2.6.2)

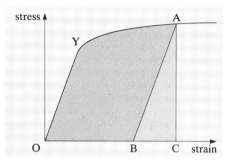

Figure 2.15 Stress–strain curve for plastically deformed material

Now what about the toughness $G_c$? Well, this is the work needed to break a piece of material which already contains a crack. $G_c$ can be measured from the area under the stress–strain curve obtained by testing a sample that contains a crack, that is, the work done is associated directly with the propagation of a crack. But what about the units? You will notice that plastic work of deformation has units of J m$^{-3}$ and $G_c$ has units of J m$^{-2}$.

If the sample already contains a crack that will propagate, the work done will be concentrated around the tip of the crack as the material is torn apart. So, the energy absorbed by the sample is given by

| plastic work of deformation (unit J m$^{-3}$) | × | volume of plastically deformed region around the crack (unit m$^3$) |
|---|---|---|

This energy (unit J) is expended over an area (unit m$^2$) ahead of the crack, so $G_c$ has units of J m$^{-2}$.

In general the values of $G_c$ obtained may depend on the size of the crack, the size of the specimen under test and the way the specimen is stressed.

The thermosetting plastic epoxy is commonly used as the matrix material in GFRP. Compare its values of $G_c$ with that of glass and GFRP; you will see that toughness does not obey the law of mixtures. Notice, also, the enormous range of values, with GFRP well up the list and at best as tough as mild steel. Note however that in steel the toughness is the same in all directions; it is said to be **isotropic**. GFRP is only very tough when stressed parallel to the fibres. It is **anisotropic**.

If the similar values of $G_c$ for mild steel and GFRP suprises you, it is probably because you associate lack of toughness with brittleness. In a nutshell, materials with some degree of toughness — most metals, GFRP, wood (stressed across the grain), thermoplastics and so on — have crack-stopping mechanisms 'built-into' their structures. Many of these materials are ductile and it is the plastic flow that they can undergo which provides the crack-stopping mechanisms; more about these in Section 2.6. Such mechanisms are not available in **brittle** materials, that is those which do not exhibit plastic flow. If possible they have to invoke other crack-stopping mechanisms. Most can't, but fibre-reinforced composites such a GFRP, wood and bone do, with considerable success. To understand how, we need to establish the conditions under which a crack grows in a material. Before doing this, I suggest that you have a go at SAQ 2.2 to check that you can distinguish between the four important terms I have just used: strong, tough, brittle, ductile. See also ▼**Ductility and brittleness**▲

SAQ 2.2   (Revision)

Are there errors in the statements below? If there are, correct them using one of the words from each of the following pairs:

- strong or weak (low strength)
- brittle or ductile
- tough or not tough

Note: I am asking you to make relative judgements, and I suggest that you don't just restrict yourself to tensile properties.

(a) A steel car body panel is strong, brittle and tough.
(b) A GFRP car body panel is strong, brittle and tough.
(c) A pottery mug is weak, brittle and not tough.
(d) A polyethylene bag is weak, ductile and tough.
(e) Raspberry jelly is weak, brittle and tough.

# ▼Ductility and brittleness▲

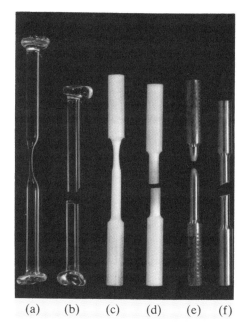

(a)  (b)  (c)  (d)  (e)  (f)

Figure 2.16 Tensile–tested samples

It is important not to assume that particular materials are always ductile and that others are always brittle. Figure 2.16 demonstrates the point for three very different types of material. It shows samples tested under various conditions to give brittle and ductile responses. Samples of the same material had the same original length.

The samples shown are:

(a) Glass tested at 800 K (ductile).
(b) Glass tested at 273 K (brittle).
(c) Polypropylene tested at a slow strain rate (ductile).
(d) Polypropylene tested at a fast strain rate (brittle).
(e) Mild steel tested at 273 K (ductile).
(f) Mild steel tested at 200 K (brittle).

## The ductile–brittle transition

From Figure 2.16 the distinction between a ductile and a brittle material is clear: a ductile material exhibits plastic flow, a brittle material doesn't. The appearance of the broken surfaces is also very different as Figure 2.17 illustrates. It shows the fracture surfaces of pieces of glass and mild steel broken at ambient temperature. The glass surface is relatively smooth and shiny. It is characteristic of the easy

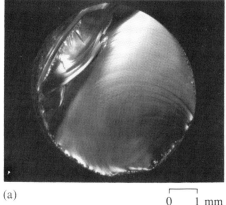

(a)                          0    1 mm

(b)                          0    5 mm

Figure 2.17 Fracture surfaces of (a) glass and (b) mild steel

fracture that occurs in, for example, the cleaving of slate. On the other hand, the steel surface has been disturbed extensively; a great deal of plastic flow has obviously occurred as the material has been torn apart. A lot more energy is expended breaking the steel compared with the glass.

When mild steel is brittle, as in Figure 2.16 specimen (f), each broken grain on the fracture surface actually looks fairly smooth and shiny.

For a given method of testing, the change from ductile to brittle behaviour occurs around a particular temperature, the **ductile–brittle transition temperature**. It can be determined by a variety of methods. The actual temperature depends not only on the condition of the material but also on the size and shape of the testpiece, the rate of deformation (strain) and the type of applied stress system.

## Assessing ductility

Ductility is the amount of plastic flow a material has undergone at fracture. It is usually measured as the percentage (plastic) elongation obtained in a tensile test. That is:

$$\text{\% elongation} = \frac{\text{final length} - \text{original length}}{\text{original length}} \times 100$$

Although very useful, this measure sometimes may not tell the full story. For instance although the mild steel specimen in Figure 2.16 extended much less than the polypropylene and the glass specimens, it actually underwent extensive plastic flow. Many metals behave like this with the flow being concentrated in the region in which the metal eventually breaks. It is called **necking**. Figure 2.18 shows a close-up of a neck in a mild steel test piece and the resulting fracture.

When necking occurs it is more appropriate to measure ductility as percentage reduction in cross-sectional area rather than elongation. That is

$$\text{ductility} = \left( \frac{\text{original area} - \text{final area}}{\text{original area}} \right) \times 100$$

Notice that the polypropylene testpiece, Figure 2.16(a), has also necked, but in this case, instead of continuing to be drawn down and fracture, the neck has stabilized at a particular width and is extending along the specimen. This process is called **cold drawing** and is common in ductile thermoplastics.

(a)

(b)

Figure 2.18 (a) Neck developing in mild steel. (b) Fracture

## 2.3.1 A model for toughness

One reason why strength and toughness are often confused is that they are intimately linked. Indeed, current knowledge of the behaviour of cracks and the importance of toughness stems from studies of the strength of solids started some 70 years ago by A. A. Griffith.

How tough should a material containing a crack be (theoretically)? Griffith proposed a model for fracture based on a balance of the different forms of energy involved. Basically, the idea is that the growth of a crack requires the provision of the work needed to create the two new surfaces of a growing crack, and that this has to be provided by the elastic strain energy released by the crack growth. It turns out that when these energies are in balance, an internal crack propagates under a stress of

$$\sigma = \sqrt{\frac{EG_c}{\pi l}} \qquad (2.1)$$

where $2l$ is the length of the internal crack, $E$ is the Young's modulus and $G_c$ is the toughness.

This equation is of fundamental importance for any structure under load. Now, $G_c$ and $E$ are determined by the structure of the material concerned; see Table 2.1 for $G_c$ values. So the variables in Equation (2.1) are $\sigma$ and $l$, and they can be viewed in two ways. First, if a material already contains a crack, the length of the crack determines the stress the material can withstand. Second, for any given stress, there is a critical length of crack $2l_c$ which will propagate catastrophically.

SAQ 2.3   (Objective 2.3)
Calculate the critical crack length in a mild steel component operating at a working stress of 200 MN m$^{-2}$. Take $G_c$ as $10^5$ J m$^{-2}$ and $E$ as 200 GPa.

Equation (2.1) clearly demonstrates the importance of the value of $G_c$. Some materials, such as alumina and glass, have a low $G_c$ and a crack propagates easily. Many metals have a high $G_c$ and are difficult to break. Polymers are in between and can be improved by reinforcing them with fibres. So, what are the crack-stopping mechanisms, and why are they lacking in some materials?

Before we can answer that question we should look at crack initiation and growth, and for that we require two new ideas.

First, we need the idea of the **ideal tensile strength** of a solid. This is the stress that can be withstood by a perfect solid, that is one without a crack or any other defect. A model of atomic bonding is needed to derive this, which we shall do in Chapter 3. In general, the ideal strength is at least a factor of ten greater than the tensile strength found in real materials, and often much more.

Second, we need the idea of **stress concentrations**. These are geometrical irregularities that cause the stress applied to a material to be magnified locally. The magnification may be so great that the local stress reaches the ideal tensile strength. The irregularities can be large or small, from holes, notches and sharp corners to inclusions and cracks (see ▼**Stress concentrations in practice**▲). The vitally important point about a stress concentration is its shape, rather than its size. A small round hole magnifies stress in its locality just as much as, say, a round porthole in a ship.

In some transparent materials, the effect of a stress concentration can be seen in polarized light. Figure 2.19 shows 'photoelastic' fringes around sharp notches in a sheet of transparent polycarbonate. The fringes are actually contours of equal maximum shear stress (shear stress and strain are considered later). Just as on a land map contours close together denote a steep gradient, the contours in Figure 2.19 show that there is a steep stress gradient around the notches.

For round or elliptical holes in an elastic material, the applied stress is enhanced locally by a factor of

$$1 \; + \; 2\sqrt{\left(\frac{l}{r}\right)}$$

where $2l$ is the length of the major axis of the hole (see Figure 2.20) and $r$ is the radius at the sharper end. For a circular hole, $l = r$ and the stress concentration factor becomes three. As $r$ gets smaller compared to $l$ the elliptical hole becomes crack-like and the stress concentration factor increases. For small values of $r$ the expression approximates to

$$2\sqrt{\left(\frac{l}{r}\right)} \tag{2.2}$$

To see how dramatic the effect is, imagine a sharp crack $1\,\mu m$ ($10^{-6}$ m) long and with a tip radius of $10^{-10}$ m (of the order of the distance between atoms in a solid). The stress concentration is

$$2\sqrt{\frac{10^{-6}}{10^{-10}}} \; = \; 200$$

Now let's consider the toughness of some typical materials. In a ductile metal, that is one that can undergo extensive plastic deformation, the stress concentration at the tip of a crack means that the yield stress is reached locally when the overall applied stress is much lower. So plastic flow occurs around the crack tip and has two effects. First, of course, it consumes a lot of energy which is why $G_c$ is high and why the surface of the steel in Figure 2.17(b) is so distorted. Second, it blunts the crack, that is it increases the radius of curvature of the crack tip, which decreases the stress concentration. So, plastic flow is an important toughening mechanism.

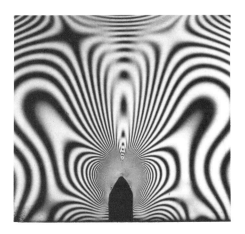

Figure 2.19 Photoelastic fringes in polycarbonate

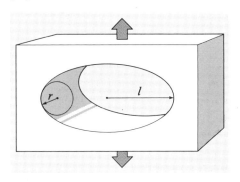

Figure 2.20

# ▼Stress concentrations in practice▲

This fact [that stress concentrations can lead to the failure of structures thought to be perfectly safe] has been known, of course, in a general sort of way, to the people who put the grooves in slabs of chocolate and to those who perforate postage stamps and other kinds of paper. A dressmaker cuts a 'nick' in the selvedge of a piece of cloth before she tears it. In the early 1900s serious engineers, however, had not shown much interest in these fracture phenomena, which were not considered to belong to 'proper engineering'.

(From J. E. Gordon, *Structures*, Penguin, 1978)

On the other hand, in their normal condition ceramics and glass do not possess toughening mechanisms. Since they do not exhibit significant plastic flow, the stress at a crack tip is much magnified and reaches the ideal strength for relatively low values of overall applied stress. With no mechanism to resist it, which is reflected in a low value of $G_c$, the crack propagates through the material very easily. The low value of $G_c$ for these materials makes them very fragile. The smallest crack or flaw can lead to fracture. This fact is exploited in the cutting of window glass and has to be guarded against in the production of glass optical fibres for communication systems.

## 2.3.2 Toughening mechanisms in composites

The commonly used matrices for GFRP are the thermosetting plastics polyester and epoxy, both of which are fairly brittle and have low toughness (see Table 2.1). How is it that embedding brittle glass fibres in them produces a tough composite?

What happens depends very much on the strength of the bonding between the fibres and the matrix. If the bond is strong, a crack propagating in the matrix at right angles to aligned fibres will pass through the fibres with very little hindrance. The toughness of the composite will probably obey the law of mixtures and lie somewhere between those of the two components (that is, low).

If, on the other hand, the bond is relatively weak, **debonding** can occur at the interface between the fibres and the matrix. Figure 2.21 shows a highly simplified situation, with a crack approaching a single fibre. In a real fibre composite the fibres would be packed closely together. As a crack approaches a fibre, the magnified stress ahead of the tip separates, or debonds, the fibre from the matrix and the crack is diverted harmlessly along the interface, parallel to the applied stress. Again, there are two effects. Energy is consumed in the creation of the new debonded surfaces — hence a fairly high value for $G_c$. And the crack is blunted (or even deflected in another direction) and therefore needs a higher stress to continue moving. So a strong, tough composite needs some bonding between fibres and matrix, but not a strong bond. The required bond is called a **weak interface**.

There is a further contribution to $G_c$ from the debonded fibres. For the composite to fail completely, the fibres and matrix must fracture. The fibre fractures do not all occur in the same place, so the fibre ends have to be pulled out of the matrix and work is done against friction. (Figure 2.22 shows the result.) A combination of these two mechanisms is responsible for the high toughness of GFRP. Bear in mind, however, that these processes cannot operate for cracks running parallel to the fibres; the toughness is very much less in this direction.

The strategy of providing weak interfaces is a powerful way of producing an anisotropically tough material. It is ubiquitous in nature

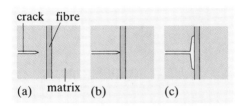

crack   fibre

(a)   matrix   (b)   (c)

Figure 2.21  Crack stopping at a weak interface

0      0.1 mm

Figure 2.22  Fibre 'pull-out', in this case carbon fibres from a glass matrix. Courtesy of Harwell Laboratories, AEA Technology

— in wood, tooth and bone for example. Also, you can see how the aggregate in concrete serves to toughen the cement matrix and hence to make a stronger material.

To see the composite structure in these other materials we need to look rather more closely at the grain than in Figure 2.9. Figure 2.23 is a micrograph of a section of a softwood. It has been viewed with an optical microscope at a magnification of 100 times. The structure consists of a close-packed array of tubes (rather like a box of drinking straws) which lie parallel to the axis of the trunk. The toughness of wood across the grain is due to the weak interface between these fibres.

In Figure 2.23 the structure was revealed by an optical microscope, and in Figure 2.22 by ▼Scanning electron microscopy▲. Structures revealed by microscopes are known as **microstructures**. Let us now investigate the microstructure of some other materials and its effect on some of their properties.

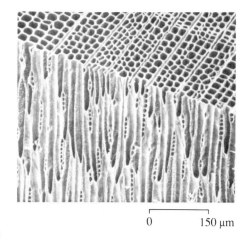

Figure 2.23 Close-up of piece of spruce

# ▼Scanning electron microscopy▲

This is the electron equivalent of the optical reflecting microscope. A beam of electrons (see Figure 2.24) is emitted from a filament held at about 20 kV. The beam is focused by a series of electromagnetic lenses. A deflector coil makes the beam scan the specimen. When the beam strikes the specimen, some electrons are scattered and collected by a detector, which converts them into a current. The current is amplified and used to control the brightness on a CRT display.

When the specimen is struck by the electron beam, the 'lee' sides of the features do not emit electrons (Figure 2.25).

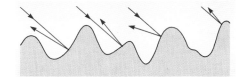

Figure 2.25

SEM is used to examine surfaces at magnifications up to about × 50 000. It is useful for examining fracture surfaces and etched specimens at higher magnifications than are possible with optical microscopes. Figure 2.26 shows two examples. Figure 2.26(a) shows part of the broken surface of a toothbrush handle. Figure 2.26(b) shows an etched specimen of an aluminium alloy. The grain boundaries and alloy particles can be seen.

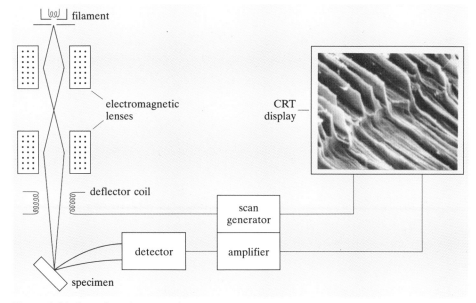

Figure 2.24 Scanning electron microscope

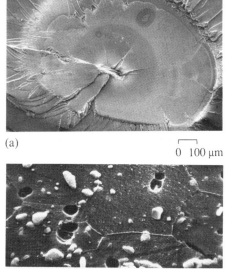

(a)
0  100 µm

(b)
0  . 50 µm

Figure 2.26 Micrographs from scanning electron microscope

SAQ 2.4   (Objectives 2.2 and 2.3)
Imagine a piece of GFRP with aligned fibres being pulled in the
direction of the fibres (as in Figure 2.11) and that a crack is initiated
on a plane at right angles to the fibre direction. Would you expect
the toughness of the composite to obey the law of mixtures if the
fibres are

(a) strongly bonded to the matrix,
(b) weakly bonded to the matrix?

Explain your conclusions.

## Summary

The strength of a material is measured by the stress at which the
material fractures. The ideal tensile strength is the tensile strength of a
solid with no defects (such as cracks).

Toughness relates to the ease or difficulty of crack propagation, and is
measured by $G_c$, the energy absorbed in creating unit area of crack
$(J\,m^{-2})$.

The stress $\sigma$ at which a crack propagates is given by

$$\sigma = \sqrt{\left(\frac{EG_c}{\pi l}\right)}$$

where $2l$ is the crack length and $E$ is the Young's modulus.

Irregularities such as holes, corners and cracks are called stress
concentrators because they locally magnify an applied stress. They can
cause cracks to propagate even when the overall applied stress is less
than the ideal tensile strength.

Tough materials have crack stopping mechanisms, for example a weak
interface between fibres and matrix, or the propensity for plastic flow.

# 2.4  Microstructure

By definition, microstructure is revealed by microscopes. From Henry
Sorby in the 1860s, who was the first to undertake systematic
examination of metals through a microscope, to the 1930s, microscopy
meant optical microscopy. It is still extremely important but, nowadays,
materials can also be investigated using electron microscopes, which can
provide structural information on a much finer scale (approaching the
level of atoms). Essentially, this is because electrons can have a much
smaller wavelength than light and, as a result, an electron microscope
can resolve detail on a much finer scale. See ▼The resolution limit of a
microscope▲

# ▼The resolution limit of a microscope▲

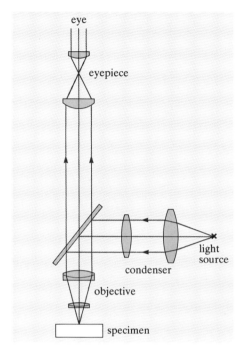

Figure 2.27 Optical microscope

The basic form and optical system of a reflection optical microscope used for the study of opaque materials is shown in Figure 2.27.

What level of detail can we see clearly using a microscope? Formally, this is measured by its resolution, which is defined as the closest spacing $d$ of two points that can still be seen as two distinct entities. It can be shown that

$$d = \frac{0.5 \, \lambda}{\sin \alpha}$$

where $\lambda$ is the wavelength of the light, and $\alpha$ is the half-angle of the light entering the objective lens (see Figure 2.27). Since $\sin \alpha$ cannot exceed 1.0, the smallest value of $d$ is half the wavelength of light, say $2 \times 10^{-7}$ m.

Now, the unaided human eye can just about resolve 0.2 mm ($2 \times 10^{-4}$ m). So the highest useful magnification of an optical microscope is about

$$\frac{2 \times 10^{-4}}{2 \times 10^{-7}} = 1000$$

In electron microscopy, because the angles through which the electron 'rays' are deflected by the objective are very small, $\sin \alpha = \alpha$ (in radians). So, with a wavelength of, typically, 0.004 nm and $\alpha = 0.1$,

$$d = \frac{0.5 \, \lambda}{\alpha}$$

$$= 0.02 \text{ nm } (0.02 \times 10^{-9} \text{ m})$$

This is smaller than single atoms but, unfortunately, because of the distortions produced by the lenses of the electron microscope, this resolution cannot be achieved. In practice, values of 0.2 nm are achievable — some 1000 times better than an optical microscope. So electron microscopes can offer magnifications of up to × 1000 000. On the other hand, many important structural features have dimensions in the range $10^{-3}$ m to $10^{-6}$ m and here the optical microscope is very useful, as is the scanning electron microscope (described in 'Scanning electron microscopy').

# ▼Polarized light▲

A light source emits a series of oscillating electric and magnetic fields. The variation of field strength with distance can be represented by a sine wave, like that shown in Figure 2.28(a). The electric field strength (the electric vector) is plotted as a function of distance. The diagram shows only one wave in the plane of the paper, but a light source normally sends out very many waves, and there is no reason why all the waves should lie in the same plane. If you were to look along the direction in which the wave was travelling (end on) you would find that the electric vector oscillated in all the planes shown by the arrows in Figure 2.28(b).

Many substances, however, will only transmit light if the oscillations are in a particular plane; an example is polaroid plastic. Therefore, if a series of waves such as those shown in Figure 2.28(b) were passed into such a material, the emergent

beam would have oscillations in one plane only, as shown in Figure 2.28(c). This transmitted light is called **plane polarized light**.

Some materials rotate the plane of polarization of polarized light. When two pieces of polaroid plastic are arranged so that their polarizing directions are perpendicular, no light will normally be able to pass through the combination. If, however, a slice of material which rotates the plane of polarization is placed between them, some light will get through. Spherulites in some polymers possess this rotational property and so polarized light is used to examine them. Roughly speaking, spherulites in polymers are the equivalent of grains in polycrystalline materials.

Some polymers exhibit the same rotational effect because of the alignment of their

molecular chains by stress. This **photoelasticity** was used to reveal the stress concentrations in Figure 2.19.

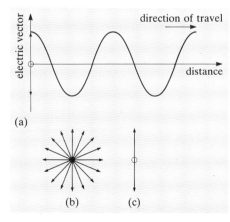

Figure 2.28

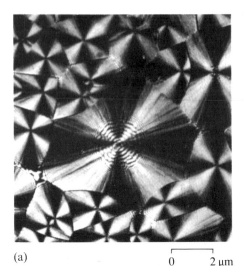

(a)    0    2 µm    (b)    0    0.1 mm

Figure 2.29 Structures shown by optical microscope

Electromagnetic radiation that is visible light, X-rays, γ-rays and so on, can be used to examine the structure of materials in two modes: transmission and reflection. The structural information obtained depends on the wavelength of the radiation and how it interacts with the material. If the material is transparent to light, transmission optical microscopy can be used. Figure 2.29 shows two examples. In Figure 2.29(a), polyethylene is shown under ▼Polarized light▲. The spherulites shown are typical of a polymer with a degree of crystallinity. (Polymer structures will be considered in more detail in Chapters 3 and 5.) Figure 2.29(b), also viewed under polarized light, shows a mixture of feldspar (striped) and hornblende in a sample of igneous rock.

To view the structure of opaque materials such as metals, using reflection optical microscopy, it is necessary to prepare the surface in some way. This is usually done by using a chemical reagent such as an acid to etch the surface in a specific and controlled manner. See ▼Chemical etching▲. These chemical techniques (and others) can yield much information about microstructures.

▼A gallery of microstructures▲ shows a variety of micrographs. (Several of the terms in the accompanying text will be new to you. They will be explained later in this chapter.) Notice the different types of 'mixture' and the effects of processing. The importance of the structure–property–process links are amply demonstrated at this level of structural scale.

To understand these links and to put them to practical use, we need to develop the appropriate theoretical models. In the remainder of this chapter I am going to introduce two: magnetic domains, which provide a basis for interpreting ferromagnetic behaviour and for producing different sorts of magnetic materials; and dislocations, which are the means by which plastic flow occurs in crystalline materials and are therefore a determinant of the yield stress, ductility, toughness, and so on, of such materials.

# ▼ Chemical etching ▲

Before being etched, a surface has to be prepared so that it presents an accurate picture of the microstructure. This usually involves creating a flat surface by cutting and then grinding and polishing with successively fine grades of abrasive, typically from coarse emery paper to diamond paste, to produce an optically flat surface from which the damage created by the cutting and grinding has been removed.

Etchants can be developed for most materials. They work in one of two ways:

(a) They preferentially attack certain planes of atoms in crystals. In a polycrystalline material, identical planes of atoms in different grains lie at different angles to the surface and, therefore, to the etchant.

(b) They preferentially attack structural discontinuities such as grain boundaries

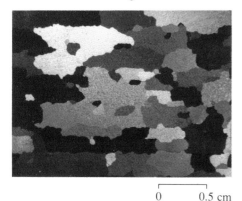

Figure 2.30 Grains in silicon steel used in transformer core.

and the interfaces between a matrix and particles of a second material.

An example of the effects of an etchant of the first type is shown in Figure 2.30, and Figure 2.31 shows a schematic illustration of why the grains appear differently in the microscope. In Figure 2.30 the grains in a sheet used in the core of a large power transformer are shown. The material is iron alloyed with silicon and has been etched with nitric acid. Figure 2.31 is a schematic representation of the unetched and etched surfaces with the dotted lines representing the regular packing of atoms in each grain. Some grains will reflect light back through the microscope objective,

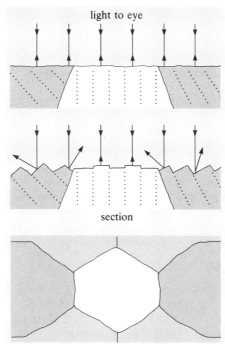

Figure 2.31

others partially or not at all. Thus they appear light, grey and dark.

The second type of etchant produced the micrograph in Figure 2.32. The material is a sample of polycrystalline alumina at a magnification of × 1500. The light lines are the boundaries between the grains and have been revealed by etching with

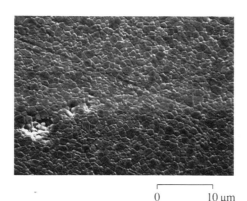

Figure 2.32 Micrograph of etched specimen of alumina showing grain boundaries

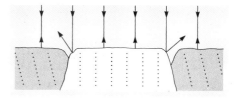

Figure 2.33

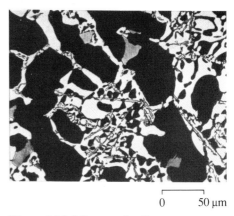

Figure 2.34 Micrograph of aluminium–copper–silicon alloy

hydrofluoric acid. Figure 2.33 is a similar schematic to Figure 2.31 and demonstrates that the grain boundaries appear black because the light reflected from the etched grooves does not return through the microscope objective.

When a material contains particles of other materials, the constituents can be distinguished by finding an etchant that attacks each of them at different rates. Figure 2.34 shows an example.

It shows the microstructure of an aluminium–copper–silicon alloy at a magnification of × 220. The light areas are aluminium rich, the grey areas are silicon rich and the black areas are the compound $CuAl_2$.

In practice, microstructures are often coloured, either naturally or by etching, and the colours are an aid to distinguishing observed features.

# ▼A gallery of microstructures▲

The techniques and terms are considered later in the chapter.

(left) The intimate mixture of Pb-rich and Sn-rich regions found in electricians' solder.

(right) A powerful permanent magnet alloy. It is made of iron together with aluminium, nickel and cobalt (an alloy called Alnico). The two materials present (light and dark) form elongated regions because the alloy was cooled from the liquid in a magnetic field. Seen in an electron microscope.

Both of these microstructures are of a sample of tin bronze (copper + 4% tin).

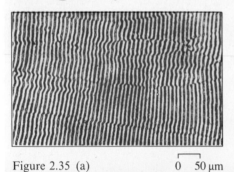

Figure 2.35 (a)    0   50 µm

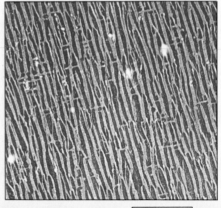

Figure 2.35 (b)    0    0.3 µm

(left) The surface after plastic deformation. The dark lines are slip lines — a manifestation of plastic flow in crystals. Notice the lines are at different angles in each grain.

(right) A polished and chemically etched section of the same materials after heating to about $0.4\,T_m$. The deformed material of Figure 2.36(a) recrystallizes to form a set of completely new grains.

Figure 2.36 (a)    0    0.25 mm

Figure 2.36 (b)    0    0.25 mm

Two sorts of cast iron.
(left) 'Grey' cast iron. The black areas are flakes of graphite. This iron is hard and wear-resistant but brittle — the flakes are a ready source of cracks.

(right) 'Spheroidal' cast iron. The addition of small amounts of magnesium or cerium (about 0.5%) leads to spherical particles of graphite rather than flakes. It is stronger than grey cast iron and has some ductility.

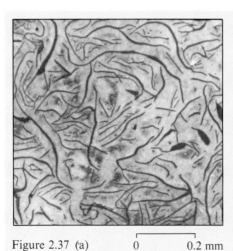

Figure 2.37 (a)    0    0.2 mm

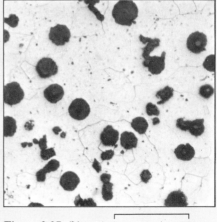

Figure 2.37 (b)    0    200 µm

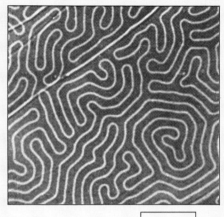

(left) Magnetic domains in a ferromagnetic oxide ceramic viewed in polarized light.

(right) Growth spirals on the surface of silicon carbide crystals grown from the vapour phase. The axes of the spirals are screw dislocations.

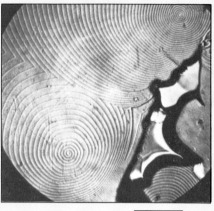

Figure 2.38 (a)          0          15 μm

Figure 2.38 (b)          0          0.25mm

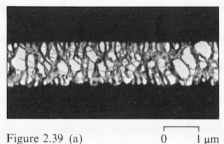

Two transmission electron micrographs showing toughening mechanisms in polymers.

(left) Increased toughness is achieved by energy being absorbed in the crazes produced by the slow tearing apart of thin strands (micro-fibrils) of the polymer.

(right) High impact polystyrene. Here energy is absorbed by very small particles of rubber.

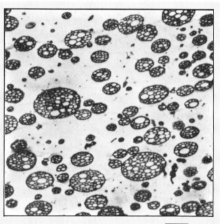

Figure 2.39 (a)          0     1 μm

Figure 2.39 (b)          0     1 μm

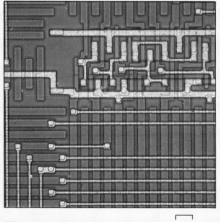

(left) Part of an integrated circuit seen under an optical microscope.

(right) A crack in a nickel superalloy used in jet engine turbine blades. The neat pattern of particles is designed to give high-temperature strength.

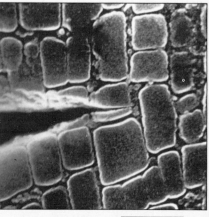

Figure 2.40 (a)          0   10 μm

Figure 2.40 (b)          0          0.5 μm

# 2.5 Soft and hard magnetic materials

I assume that, from earlier studies, you are familiar with the terms symbolized in the magnetization curve of a ferromagnetic material in Figure 2.41. If you want to refresh your memory, tackle SAQ 2.5.

SAQ 2.5 (Revision)
Define the terms denoted by the symbols $B$, $H$, $B_s$, $B_r$, $H_c$, $BH_{max}$, $\mu_r$ in Figure 2.41 and explain what is meant by hysteresis loss.

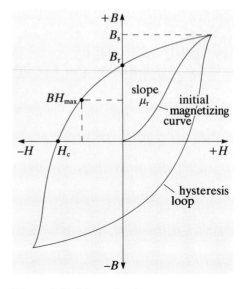

Figure 2.41 Magnetization curve

The shape of the hysteresis loop of a magnetic material is of great engineering importance. The largest use of magnetic materials is in transformers, electric motors and generators, and they are important in devices such as tape recorder heads and computer memories. Let's consider a transformer as an example.

A transformer consists of two coils wound on a magnetic circuit called a core. The changing magnetic flux set up in the core by the changing current in the primary coil induces a changing e.m.f. in the secondary coil. Materials for the core must be easy to magnetize and demagnetize, that is they should have high permeability and low coercivity. With standard alternating current of frequency 50 Hz (60 Hz in the USA), the core traverses the whole hysteresis loop 50 times a second. To minimize the energy lost during a cycle, the area within the hysteresis loop should be as small as possible. Materials having small hysteresis losses are called **soft** magnetic materials.

There is also a practical need for permanent magnets, that is for materials which remain magnetized. The requirements are for a large $B_r$ (that is, a large flux density in the absence of a magnetizing field) and a large $H_c$ (so that the magnet is not easily demagnetized). This combined requirement is usually assessed by $BH_{max}$, which should be large. Materials with large $BH_{max}$ are called **hard** magnetic materials.

Figure 2.42 provides a comparison of hysteresis loops for (a) soft and (b) hard materials. The material in (a) is an alloy of iron containing 4% (by weight) of silicon, and in (b) it is an Alnico alloy rather like that in Figure 2.35(b).

The iron–silicon alloy is used in the core of power transformers; Alnico is a fairly powerful permanent magnet material used in such devices as motors and loudspeakers. I chose them deliberately to demonstrate an important point. Both are based on iron but have different alloying elements. The striking difference between the two is in the coercivity $H_c$ (a factor of $10^3$). So, how does the difference in magnetic behaviour arise? As you will have anticipated, the answer lies in their very different microstructures — compare that of the Alnico alloy, Figure 2.35(b), with the piece of iron–silicon shown in Figure 2.30. In order to relate the properties to the structures, we need to know about magnetic domains.

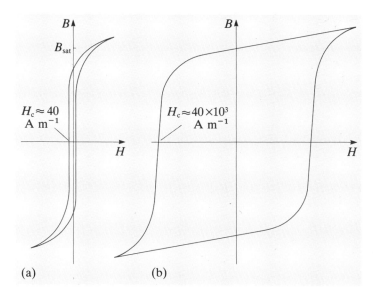

Figure 2.42 Hysteresis loops for (a) soft and (b) hard magnetic materials

## 2.5.1 Magnetic domains and the magnetization curve

Explanation of why only a small group of elements and compounds exhibits ferromagnetism (Fe, Co, Ni, Gd, $Fe_3O_4$ and various oxides based on them, and a few others) rests on two theoretical ideas. First, that ferromagnetism is caused by the spin of the electron. Each electron behaves like a minute magnet which points its north pole up or down the magnetic field $H$ according to the direction of its spin. Oversimplifying a bit, we can picture the electron as a small electrically charged sphere spinning about its axis; the rotation of the charge produces its own magnetic field about this axis. In ferromagnetic atoms, more electrons spin one way than the other and the atoms themselves behave like magnets.

According to this model, some other atoms should also be ferromagnetic, notably chromium and manganese. That they are not is explained by invoking the second idea, namely that in atoms like Fe, Co and Ni these are forces between electrons which favour the alignment of electron spins in neighbouring atoms, and give ferromagnetism. In, say, manganese the spins are not aligned in parallel but in groups of parallel and anti-parallel combinations which cancel out. More about this in Chapter 3.

These theoretical models account for the magnitude of the magnetic saturation $B_s$, but not for why ferromagnetic materials can exist in an unmagnetized condition, nor for the wide variation in other magnetic characteristics which can be obtained from one magnetic material — in our case iron containing silicon on the one hand and Al, Ni and Co on the other. Enter the idea of **magnetic domains**, first proposed by Weiss 1907.

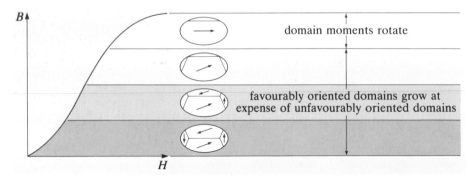

Figure 2.43 Domain growth and rotation during magnetization

He postulated that each crystal in an unmagnetized ferromagnetic material consisted of small domains, or regions, in each of which all the atomic magnets were aligned in a particular direction. So over several crystals oriented at random there is no net magnetism. He argued that the magnetization curve was to do with domains being lined up, rather than the behaviour of individual atoms. Subsequent development of techniques for observing domains (▼Seeing magnetic domains▲) provided the experimental basis for explaining the wide variation observed in coercivity, hysteresis loss, and other properties.

When a ferromagnetic material is fully magnetized (at $B_s$), all the domains have the same direction of magnetization. This could come about by the growth of domains with their direction of magnetization parallel to the applied magnetic field, or by rotation of the direction of magnetization in unfavourably oriented domains. Both occur, as Figure 2.43 illustrates schematically for an initial magnetization curve.

# ▼Seeing magnetic domains▲

Almost 25 years after Weiss proposed the existence of domains, Bitter showed they could be seen with a microscope. He polished the surface of a ferromagnetic specimen and sprinkled a colloidal suspension of iron particles onto it. The ferromagnetic particles formed patterns. Figure 2.44 is a simple example of what can be found. It is a photograph of the domain structure in a single crystal of iron. The arrows indicate the direction of magnetization in each domain. The domains are typically in the range $10^{-3}$ m to $10^{-5}$ m across.

Figure 2.45 shows how the image in Figure 2.44 is formed. At a domain wall there is a change in orientation of atomic magnets. Within a domain, all the atomic magnets are aligned, as indicated by the arrows. In the wall shown between domains 1 and 2 the orientation of atomic magnets changes by 180°. The

magnets are partly pointing out of the surface, and it is to these 'free poles' that the iron particles are attracted.

Nowadays more refined techniques are available which allow domains to be

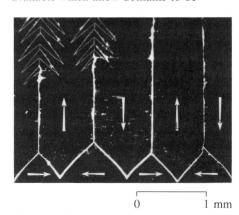

0          1 mm

Figure 2.44 Magnetic domains

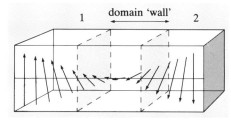

Figure 2.45 Change of domain orientation at domain wall

observed in motion. One relies on the Kerr magneto–optic effect. Here polarized light is reflected from a magnetized surface. Two adjoining domains, being magnetized in opposite directions, produce oppositely directed rotations of the plane of polarization. When the reflected light is viewed through a microscope, it can be arranged that one domain will appear light and its neighbour dark. Figure 2.38(a) is an example.

As the magnetizing field $H$ is increased, the favourably oriented domains grow and the net magnetic flux density increases rapidly. When this is complete, further magnetization can occur only by the rotation of the direction of magnetization within a domain, that is each atomic magnet is reoriented. Since this is a more difficult process, the slope of the magnetization curve decreases.

Hysteresis loss means that energy is expended in going around a hysteresis loop. Where does the energy go?

It must be expended in rearranging the domains, that is in moving the domain walls and reorienting the atomic magnets within them. So, ferromagnetic materials must contain structural features that impede wall movement. Basically, any feature which produces disorder in the regularity of a crystal will reduce the mobility of domain walls. Important examples are particles of a second material, grain boundaries, pores, and dislocations (which we consider in the next section).

SAQ 2.6  (Objective 2.3)
In terms of domain wall movement and microstructural features, specify the requirements of

(a) an ideal hard magnetic material,
(b) an ideal soft magnetic material.

Relate your answers to the micrographs of Figures 2.35(b), and Figure 2.30.

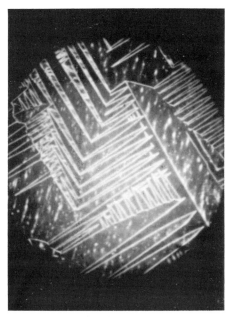

0        0.1 mm

Figure 2.46 Domains in transformer core alloy

The iron–silicon alloys used extensively for the core of power transformers are not pure single crystals and they produce hysteresis loss, albeit small as Figure 2.42(a) shows. In fact, the domain structure of a piece of transformer-core alloy shown in Figure 2.46 demonstrates that the actual material is not as simple as the rather ideal ones we have been considering. In addition to a minimal hysteresis loss, the properties required of the material for function and processing are of major importance. The choice of material is an interesting example of product–property–process (PPP) interactions, as ▼**Choosing the iron–silicon alloy for a transformer core**▲ describes.

Magnetically soft metal alloys like iron–silicon are also mechanically soft. They have a low yield stress and are ductile. Hard magnetic metals are also mechanically hard and of low toughness. The alloy of Figure 2.35(b) has to be cast or sintered and then shaped by grinding. Clearly then, their microstructures determine mechanical and magnetic properties in analogous ways. The reasons are revealed in the next section, in which we explore the origin of plastic flow in crystalline materials and some of its consequences.

# ▼Choosing iron–silicon alloy for a transformer core▲

The core of a transformer has to be a soft magnetic material to minimize the hysteresis loss. Development of a suitable material involves consideration of the factors covered in SAQ 2.5. In addition, interactions between product design and the properties of the core material affect the choice of a suitable alloy.

Because iron is an electrical conductor there are **eddy currents** in the core. These arise because the changing magnetic flux induces currents in the core, and energy is dissipated as heat.

The core material should minimize these losses. A partial solution is to build the core from thin sheets that are electrically insulated from one another. Each sheet carries a fraction of the total magnetic flux in the core, so the eddy currents within each sheet are less than would circulate in a solid core. Because the heating effect is proportional to the square of the current, the losses are reduced. The thinner the sheets the better.

Eddy currents can also be reduced by increasing the electrical resistivity of the core material. This is the main reason for using an alloy. But an alloy with which element?

Figure 2.47 shows the effect of various alloying elements on the resistivity of iron. The front runners for a transformer core are silicon and aluminium. In practice, aluminium is rarely used. It is very reactive and during its preparation it readily reacts with air to form aluminium oxide and aluminium nitride. These both form non-magnetic inclusions and so increase hysteresis loss.

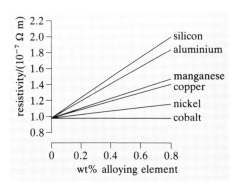

Figure 2.47

Figure 2.48 shows how selected properties are altered by changing the silicon content of an iron–silicon alloy. Note that iron–silicon alloys undergo a ductile–brittle transition, just as mild steel does (this was described in 'Ductility and brittleness'). Figure 2.48(c) shows that the transition temperature is very sensitive to composition.

SAQ 2.7 (Objective 2.4)
(a) Specify the properties for performance and processing you would use as criteria for choosing the silicon composition of a transformer core alloy.

(b) Deduce from Figure 2.48 the optimum composition for an iron–silicon alloy for a transformer core.

(c) Draw a product–process–property–principle tetrahedron for the transformer core, entering it with the principle of minimum eddy current loss.

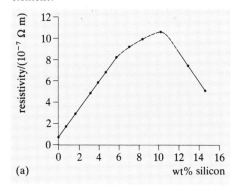

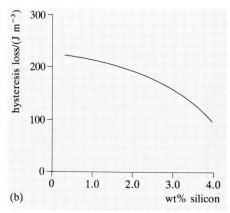

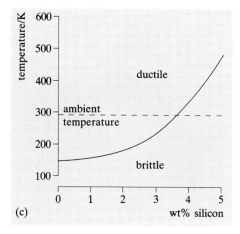

Figure 2.48 Effect of silicon concentration on properties of iron–silicon alloy

## Summary

Ferromagnetic materials consist of small regions, called magnetic domains, in each of which the atomic magnets are aligned. Overall, in an unmagnetized material, the magnetism of the domains balances out.

Magnetization involves the growth and rotation of domains.

A soft magnetic material requires easy domain wall movement; a hard magnetic material relies on immobilizing domain walls.

Selection of an alloy for a transformer core illustrates product–process–property–principle interactions.

# 2.6 Plastic deformation in crystals

By definition, all flow processes involve a permanent change of shape, hence the 'flowing to shape' processes such as injection moulding and rolling you met in Chapter 1. These shape changes are achieved by parts of a material sliding over one another. This is how glaciers and treacle flow and is what has happened in the examples of Figure 2.49. Sliding is produced by **shear forces** acting on the plane on which the sliding occurs, as explained in ▼**Shear deformation**▲. The dark areas in the workpiece of Figure 2.49(a) have been etched relative to the white areas. These areas have been sheared into flow lines in the chip. The shape change becomes clear when you compare the shapes of the shaded pink areas in Figure 2.49(b) before and after shearing. At this stage we will concentrate on the predominant mechanism of plastic flow in crystalline materials, that produced by the structural defect known as a **dislocation**. You will meet other mechanisms for flow in solids in later chapters.

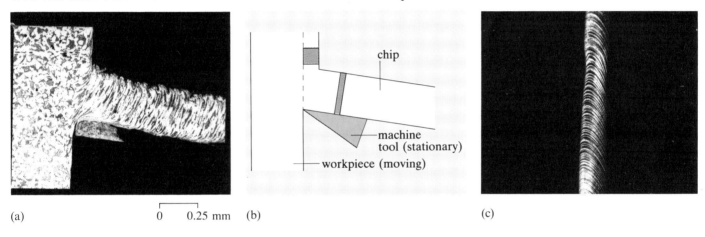

(a)  0  0.25 mm  (b)  chip — machine tool (stationary) — workpiece (moving)  (c)

Figure 2.49 (a) Chip being machined from etched mild steel. (b) Geometry of the machining operation. (c) Single crystal of copper–aluminium alloy after being pulled along its axis

## 2.6.1. Slip in crystals

Figure 2.36(a) shows the surfaces features on a polycrystalline sample of a tin bronze (that is, copper with 4% tin) that has been plastically deformed. Each of the dark lines in the various sets of parallel ones actually corresponds to a small step on the surface. We will return to this rather complex picture later, but to start with it is easier to study single crystals — hence the deformed single crystal in Figure 2.49(c). Different parts of the crystal have sheared over one another. It looks rather like a pack of playing cards that has been pushed sideways. This type of deformation is called **slip** and Figure 2.50 is a schematic representation of an increment of slip in a simple crystal structure, in which it is assumed that atoms are hard spheres packed together in a cubic array. Imagine that this is a minute volume of atoms plucked out of Figure 2.49(c) with the slip plane being the plane on which the single crystal has sheared.

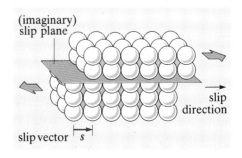

(imaginary) slip plane

slip direction

slip vector *s*

Figure 2.50 Slip in simple crystal structure

Now, slip has a number of specific features, which can be defined with reference to Figure 2.50.

**Slip plane**. The (imaginary) plane between the planes of atoms on which slip occurs.

**Slip step**. The step produced on a crystal surface. You can see them in Figure 2.49(c). The edges of slip steps produced the dark lines in Figure 2.36(a).

**Slip direction**. The direction in which slip occurs.

**Slip vector**. The direction and magnitude of an increment of slip.

**Slip system**. The combination of a slip plane and a slip direction. More about this shortly.

EXERCISE 2.3

(a) What is the necessary geometrical relationship between a slip plane and a slip direction?
(b) What do you notice about the crystal structure in Figure 2.50 before and after an increment of slip?

So slip in crystals is, in fact, very different from the sliding of cards in a pack because the direction and amount of slip is specific. In some types of crystal it is possible for planes of atoms to be sheared over one another by distances less than that between atoms. This leads to a local change in the crystal structure. The process is called a **shear transformation** and you will meet examples later.

A slip system defines the geometry of slip. For instance, the crystal portrayed in Figure 2.50 has undergone slip on a particular plane and in a particular direction in that plane. This plane and direction constitute a slip system. Now a crystal does not slip an any arbitrary plane or direction. It is very selective. The slip direction is always that in which atoms are closest together. This means that the slip vector is the smallest possible. So, in Figure 2.51(a), which shows the front face of a simple cubic structure, slip vector *a* is chosen rather than, say, *b*. The favoured slip plane is usually that on which the atoms are most closely packed. This is because the most closely packed planes are wider apart than others. In simple terms, on an atomic scale they are smoother and can slide over one another more easily — see Figure 2.51(b).

Since many crystals are highly symmetrical, they possess a number of equivalent slip planes and directions, and therefore slip systems. Under the appropriate shear stress, the crystal in Figure 2.50 could slip on the same plane but in a direction at 90° to the one shown. The crystal also has identical planes of atoms lying vertically and mutually perpendicular to one another (think of the other faces of a cube of atoms). So, how many different slip systems has this structure?

Six. There are three equivalent sets of slip planes, each of which contains two slip directions.

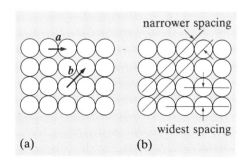

(a)    (b)

Figure 2.51

Slip in polycrystalline materials is more complicated than in single crystals because the individual grains cannot behave as if they were single crystals. We know that polycrystalline metals can be deformed by large amounts — bars can be drawn, into wire, sheets pressed into oven trays, for example. This does not cause the metal to fall apart at the grain boundaries, so each grain must deform 'in sympathy' with its neighbours, that is, in such a way that it remains in contact with them at its boundary. Each grain must slip on a number of slip systems, and Figure 2.36(a) provides good evidence of this. (Note that the intersecting slip steps imply intersecting slip planes.)

On the assumption that in a perfect crystal slip occurs in a single movement by one whole plane of atoms sliding bodily over another, it is possible to calculate the theoretical shear stress (and hence yield stress) at which this should happen. The answer turns out to be of the order of $10^{10}$ N m$^{-2}$; this compares with measured yield stress of unalloyed metals of $10^6$ N m$^{-2}$ to $10^7$ N m$^{-2}$. Clearly, a different theoretical model is needed.

# ▼Shear deformation▲

The **shear force** producing sliding motion acts in the same direction as the resulting motion (think of shearing a pack of playing cards). It gives rise to a **shear stress** in the material and produces a **shear strain**. Imagine a block of material with its base rigidly fixed to which we apply a shear force $F_s$ parallel to the top surface of area A, as in Figure 2.52.

The shear stress $\tau$ is given by force divided by area:

$$\tau = \frac{F_s}{A}$$

If the shear stress displaces the top of the block from the position ABCD to A′B′C′D′ as shown, the shear strain $\gamma$ is defined as

$$\gamma = \frac{x}{y}$$

Just as with tensile forces, for small strains Hooke's law is obeyed, in which case

$$\frac{\tau}{\gamma} = G,$$

and G is called the **shear modulus**.

Shear stresses are usually present in a body under load; the only exception is when the loading is **hydrostatic** (equal stresses exerted in all directions, for

example by the pressure of water on submerged objects). Figure 2.53 illustrates a shear force created by a tensile force. The shear force $F_s$ is the resolved component in the plane of shear of the tensile force $F_t$.

If the rod in Figure 2.53 has a cross-sectional area of A, what is the shear stress produced by the shear force $F_s$?

The area of the shear plane is $A/\sin \theta$, so in terms of $F_s$,

$$\tau = \frac{\text{force}}{\text{area}} = \frac{F_s \sin\theta}{A}.$$

Since $F_s = F_t \cos\theta$, in terms of $F_t$,

$$\tau = \frac{F_t \cos\theta \sin\theta}{A}$$

Of course $F_t/A = \sigma$, the tensile stress, so

$$\tau = \sigma \cos\theta \sin\theta$$

From this expression you can see that the shear stress produced by a tensile stress is a maximum when $\theta = 45°$. In this condition

$$\tau = \frac{\sigma}{2}$$

which is a simple expression well worth remembering. It is in this configuration that the single crystal of Figure 2.49(c) has undergone sliding.

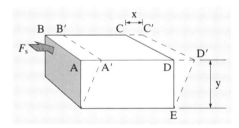

Figure 2.52 Shear deformation produced by shear stress

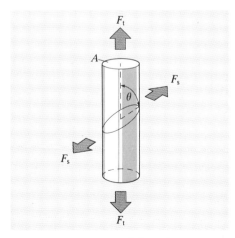

Figure 2.53 Creation of shear force by tensile force

## 2.6.2 The dislocation model of slip

In real crystals, planes of atoms do not slip bodily over one another. Instead slip is an incremental process. A few neighbouring atoms in a plane suddenly undergo slip by jumping simultaneously into their next sites in the slip direction — pushing the atoms before them out of their normal positions. At this stage, slip has occurred over a small area of a plane relative to the neighbouring plane. This slipped area then grows by the atoms on its outside — those that were pushed out of position — undergoing slip, pushing atoms ahead of them out of position and so on. Eventually, all of the atoms in the plane will have slipped and only then will it look as if the whole plane had slipped bodily.

The simple analogy of trying to move a large heavy carpet may help here. Rather than trying to slide the carpet by pulling on one edge (that is, sliding a whole plane of atoms simultaneously), it is easier to put a ruck in the carpet and push the ruck across the carpet, Figure 2.54. Large movement can be achieved by repeating the process.

Now let's picture, schematically, what happens. Imagine a block of crystal under a shear stress $\tau$ and containing a plane ABCD on which slip occurs by an increment described by the slip vector $s$. Figure 2.55(a) and (b) represent the position before and after slip occurs. I shall be using two sorts of model of the material undergoing slip: the 'block' representation of Figure 2.55(a) and (b), which is trying to give an overall picture; and a model which represents the centres of atoms by small circles and in which, to illustrate the regularity of the structure, straight lines connect atoms that are near-neighbours. Figure 2.55(c) is the 'atomic model' of the slipped crystal of Figure 2.55(b). I am using this way of representing atoms, rather than that of Figure 2.50, because it is easier to visualize changes.

Suppose that slip starts at a stress concentration at the point D and begins to spread across the plane ABCD; that is, that part of the top half of the block slips by an amount $s$ relative to the bottom half. Figure 2.56(a) shows the situation when slip has occurred over the area DEF and hasn't yet occurred over the area FEABC.

The line in Figure 2.56(a) that marks the boundary between the slipped and unslipped areas is called a **dislocation line** (usually simply

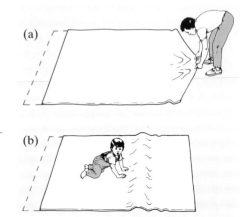

Figure 2.54 (a) The hard way. (b) The easy way

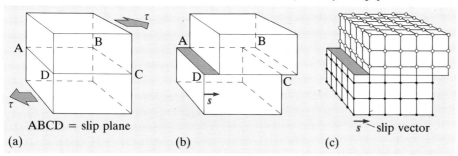

ABCD = slip plane

(a)        (b)        (c)

$s$ slip vector

Figure 2.55 (a) Block model, (b) Atomic model

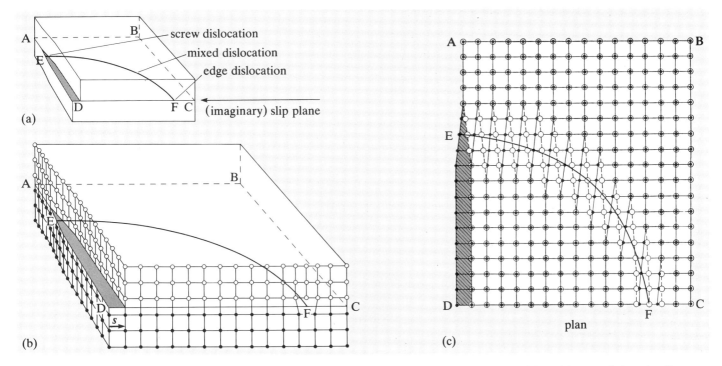

Figure 2.56 (a) Slip part way across the plane ABCD in block model. (b) and (c) show atomic model. In (c) atoms below the slip plane are represented by a dot. Those above are represented by a circle. A circle with a dot indicates atoms in register

**dislocation**). In Figure 2.56(a) it is the line EF. As the slipped area grows, the dislocation line moves across the plane.

Now since nothing has yet happened on the far side of the crystal, the line EF must be a region of some disorder. And so it is, which is why it is called a dislocation. The plan view of the atomic model, Figure 2.56(c), shows clearly the slipped area, the unslipped area and the transition region in between. The disorder produced in the crystal depends on the angle the dislocation line makes with the slip vector *s*. Two extremes are possible.

- The dislocation line lies at 90° to the vector. Such a segment is called an **edge dislocation** and is the one that emerges at point F in Figure 2.56(a). The atomic disorder around it is illustrated in Figure 2.56(b).
- The dislocation line lies parallel to the vector. Such a segment is called a **screw dislocation** and emerges at point E in Figure 2.56(a).

Of course, most of the dislocation line EF is in some orientation other than edge and screw. Such orientations are called **mixed dislocation** because they can be considered as a mixture of small segments of edge and screw dislocation. As the dislocation EF continues to move across the slip plane, it will change its configuration. Figure 2.57 shows a later stage in its movement. It has moved nearly across the whole of the slip plane.

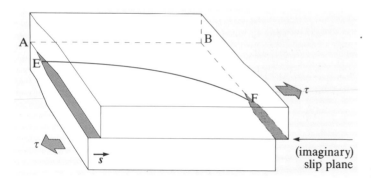

Figure 2.57

Is the dislocation predominantly edge, screw or mixed?

It is mainly screw dislocation, that is with the dislocation line parallel to the slip vector. When the dislocation passes through points A and B, the block will have slipped to that shown in Figure 2.55(b).

Dislocations can be seen by ▼**Transmission electron microscopy**▲, and Figure 2.58 shows an example. Figure 2.58(a) shows dislocations moving on slip planes in stainless steel. The dislocations are the wiggly black lines. Figure 2.58(b) shows why they appear as short lines. Do not be misled by the illusion that the dislocations appear to end *inside* the material. The appearance of Figure 2.58(a) arises because the electron beam of the microscope is fired through a very thin slice of the metal about 100 nm thick. Figure 2.58(b) illustrates, very simply, the geometry of the situation.

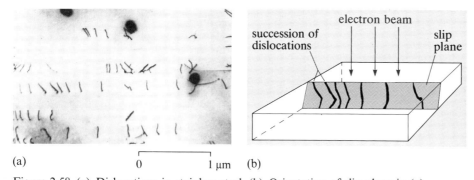

(a)     0   1 μm (b)

Figure 2.58 (a) Dislocations in stainless steel. (b) Orientation of slip plane in (a)

It is important to appreciate that, since a dislocation bounds a slipped area, it must be continuous within a crystal. Dislocations can exist as loops or join up with one another. The electron micrograph of Figure 2.59 shows these features in a plastically deformed specimen of copper containing small particles of silica. (The dislocations are lying on a plane at a different angle from the one in Figure 2.58a.) Of course, dislocations can terminate at a surface, as do those in Figures 2.56 and 2.58.

Whether a dislocation is mainly edge or screw in character is important because, as you will see in Chapter 6, the two types behave very differently. At this stage, we will just consider a few of their features.

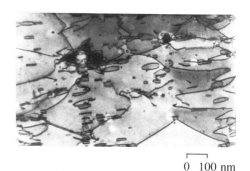

0 100 nm

Figure 2.59

# ▼Transmission electron microscopy▲

Unlike scanning electron microscopy, TEM uses electrons that have passed through the specimen to yield structural information. This requires a high voltage ( > 100 kV) and very thin specimens (about 100 nm to 200 nm). Special techniques have been developed for thinning most materials.

Figure 2.61 shows that, above the specimen, transmission and scanning electron microscopes are similar. The objective lens forms a first image, a small part of which is then magnified to produce an intermediate and then a final image on a fluorescent screen.

When the planes of a crystal lie at a particular angle to the electrons passing through, the electrons are diffracted. They do not pass through the crystal. The planes can be considered to act rather like mirrors, as illustrated in Figure 2.60(a). (Diffraction is considered further in Chapter 3.)

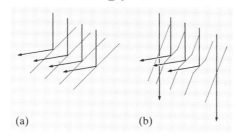

(a)                          (b)

Figure 2.60

Now, to obtain details of internal structure, one of the tricks is to tilt the specimen very slightly so that the electron beam is not diffracted and passes straight through, except for regions of the crystal where there are slight misorientations of the plane. These will diffract the beam, as in Figure 2.60(b), and will therefore appear dark on an otherwise bright image. This is how the dislocations in, for instance, Figure 2.8(b) and the particles in Figures 2.39(b) and 2.40(b) were revealed.

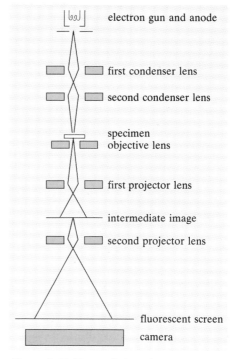

Figure 2.61 Transmission electron microscope

## 2.6.3 Screw and edge dislocations

The screw dislocation segment of the dislocation EF in Figure 2.56(a) emerges from the crystal at point E. The distribution of atoms on the crystal surface around the dislocation is shown in Figures 2.56(b) and (c). Notice that the surface has been converted into an atomic sized spiral ramp by the screw dislocation which runs into the crystal at the tip of the slip step; the dislocation creates the step as it moves towards A. Because they create a step on a surface, screw dislocations provide an important mechanism for the growth of crystals, as
▼Screw dislocations and crystal growth▲ explains.

The edge dislocation segment of Figure 2.56(a) emerges from the crystal at point F. From Figure 2.56(b) you can see that above the slip plane the atoms have been squeezed together. In fact, there is an extra row of atoms above F. This results in disorder in the crystal centred around F and means that, on an atomic scale, a dislocation line has a width. The disturbance decreases with distance from the centre of the dislocation; and the width can be defined as the number of atom spacings over which the atoms are significantly disturbed from their normal positions; it varies from about ten in copper to about two in silicon. The width is extremely important, as you will see shortly. As Figure 2.56(c) illustrates, there are many planes parallel to and behind the one through which the edge dislocation threads. So, the extra row of atoms in Figure 2.56(b) becomes an extra plane of atoms in three dimensions. It is called the **half plane** of the dislocation.

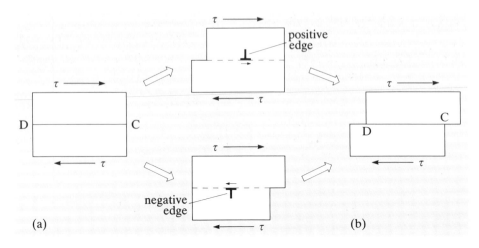

Figure 2.62 Same slip deformation produced by edge dislocations of of opposite sign

Can you see now why the dislocation is called an 'edge'?

Because it lies along the edge of the extra half plane. Conventionally, the edge dislocation emerging at E is symbolized by ⊥, which indicates which side of the slip plane the half-plane lies. It is called **positive edge dislocation**. The need for the distinction can be seen from Figure 2.62; here we are just considering the front face of the 'block' model.

Under the shear stress τ shown in (a), the slipped block of (b) can be achieved by a positive edge dislocation moving from D to C or a negative edge dislocation moving from C to D. The dislocations are said to be of opposite sign, and screw dislocations can be of opposite sign too. Generalizing a bit from this, the slipped block of Figure 2.55(b) could be produced by a dislocation of opposite sign to that of EF in Figure 2.56(a) which starts at point B and moves across the plane and through A, D and C.

Another feature that can be described more readily by reference to the edge rather than the screw, is why it is relatively easy for the dislocation to move. Rather than all the atoms sliding simultaneously, as is assumed by the earlier theoretical model, for a dislocation to move by an increment requires the extensive movement of relatively few atoms. This is indicated, very schematically, in Figure 2.63 for an edge dislocation moving one atomic spacing to the right.

The wider the dislocation the more the distortion produced by the dislocation is spread out and the lower is the applied stress needed to move it through a crystal. The ease of motion of a dislocation is called its **mobility**; it is of fundamental importance in determining the mechanical properties of material. The next section introduces you to some of the consequences of dislocation mobility and the role it plays in conflicting properties.

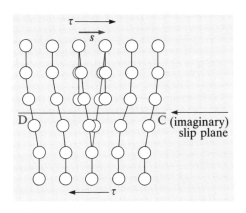

Figure 2.63 Atom adjustment in region of a moving edge dislocation. Red atoms show positions prior to slipping

# ▼Screw dislocations and crystal growth▲

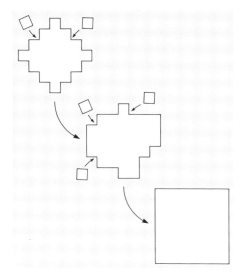

Figure 2.64 Amount of growth of nucleus by atoms lodging on available steps is limited

Screw dislocations were postulated, by Frank in 1948, not to account for plasticity in crystals but to explain how crystals grew in the first place. As you will see in Chapter 4, crystals usually solidify from the liquid (or vapour) by atoms (or molecules) attaching themselves to small solid nuclei. Now, if you think about what is going on at the atomic level, growth is easier if the atoms are deposited at steps on the surface of a nucleus. An atom is more likely to be captured at a concave corner than by a flat surface. The problem is that such steps are soon used up, as Figure 2.64 demonstrates (very schematically).

However, if a screw dislocation emerges at the surface of a nucleus it creates a perpetual step. The screw dislocation converts the flat surface into a spiral ramp; so atoms can be added to the step indefinitely and the crystal grows 'round' the dislocation. The process is illustrated

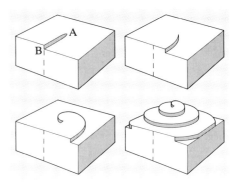

Figure 2.65 Growth spiral produced by slip dislocation

in Figure 2.65. The step will complete one rotation round the dislocation line with the addition of fewer atoms near the point A than near the point B. So, if atoms join each site along the step at the same rate the step become a spiral. Figure 2.38(b) shows an example of silicon carbide crystals that have grown in this way.

SAQ 2.8 (Objective 2.3)

(a) If the ruck in the carpet in Figure 2.54(b) is a dislocation line, what sort of dislocation is it?
(b) Figure 2.66 represents a single crystal undergoing slip on two planes. For each edge dislocation shown, indicate by an arrow the direction in which the dislocation must move in order to produce an elongation of the crystal.

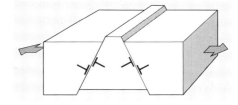

Figure 2.66 Slip on two planes

## 2.6.4 Dislocation mobility and properties

Crystalline materials with mobile dislocations are ductile; they can be deformed plastically. This is the main reason why metals are so useful. It has been said (by a metallurgist, of course) that without dislocation mobility, we would still be living in a stone age! Polymer technologists might not agree. The main consequences of ductility in metals are:

• Metals can be deformed to shape in the solid state — hence rolling, forging, wire drawing and so on. Ductility is a crucial *property for processing*.

• Metals are tough, so ductility is also a *property for performance*. In addition, the microstructure can be manipulated by plastic deformation to produce better performance properties, such as greater tensile strength, hardness, or reduced magnetic hysteresis loss.

SAQ 2.9 (Objective 2.3)
Explain, in terms of stress concentrations, slip and dislocations, why a ductile metal is tough. You will find it helpful to look back at Figure 2.17.

Crystalline materials in which dislocations are immobile are brittle. Generally, in such materials a lower stress is needed to propagate cracks than to move dislocations. The dislocation immobility is a consequence of two inter-related factors: the type of bonding between the atoms and the complexity of the crystal structures compared with those of metals. The bonding usually leads to narrow (immobile) dislocations and the types of crystal structure often mean that the disorder in the locality of a dislocation is so complex that the simple slip motion we have been considering is not possible. For these reasons, diamond, silicon, ceramics such as alumina and silicon carbide, and crystalline minerals such as mica and asbestos are brittle. Some dislocation motion can occur under stress in many of these materials at elevated temperatures, typically at about $0.75 T_m$ where $T_m$ is the melting temperature in kelvins. Unfortunately, this plastic flow at high temperatures is very restricted and therefore of little practical use as far as shaping processes are concerned, hence the need to use alternatives such as slip casting, pressing and sintering.

Pure metals like gold, copper and iron, are so soft — dislocations move so easily — that they are not of much use an engineering materials. Gold, for instance, is so malleable that it can be beaten into foils so thin that they can be seen through; high purity aluminium is good for cooking foil but not as a load-bearing material. Much of the science and technology of metals is concerned with making them stronger *whilst retaining useful ductility*. The task isn't easy because these two requirements are in conflict. In terms of dislocation mobility, can you see why?

Ductility relies on the extensive movement of dislocations; increasing the yield stress means retricting dislocation movement.

Essentially, any structural disorder or 'misfit' in a crystal acts as an obstacle to dislocation motion and its size, shape and distribution all contribute to its effectiveness. Important examples are:
• Grain boundaries. They stop dislocations passing from one grain into the next; so polycrystalline metals are stronger than single crystals.
• Other dislocations. As plastic deformation proceeds, the number of dislocations increases rapidly. Figure 2.67 shows the result of many dislocations moving on intersecting slip planes. Clearly, it is difficult for dislocations to move within these tangles. This is why metals work-harden.
• Particles of a second material. Figures 2.37 and 2.40(b) are examples of a whole variety of possibilities.

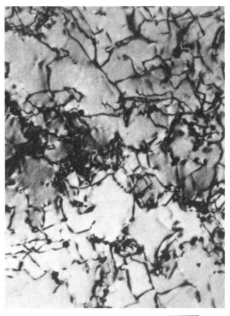

0    100 nm

Figure 2.67 Tangled dislocations in stainless steel

- Mixtures of all sorts. The mechanisms by which these obstacles affect dislocation movement, their effectiveness and uses will be dealt with in later chapters. But remember that of those properties that rely on dislocation mobility, some can be improved only at the expense of the others.

SAQ 2.10   (Objective 2.4)
Comment on the validity and consequences of the following statement:

> Increasing the yield stress of a metal decreases its toughness and affects the way it can be processed to shape.

Clearly, considerations such as these have important implications for the property profile of a material, both for performance and for processing. In practice, microstructures are manipulated to provide an optimum set of properties for both processing and performance. The alternative is, as I said earlier, first to produce a microstructure that is suitable for processing (for example, one that provides ductility) then to change the structure to one suited for function (for example, one with high strength).

In this chapter I have tried to demonstrate that the properties of materials are determined at different levels of scale, and I have concentrated on the macrostructural and microstructural levels as portrayed in Figure 2.3. But I am sure that you will have realized that trying to categorize the structure of materials in this way, and the theoretical modelling of properties to go with it, doesn't work. This is really because to understand how structure determines properties, we need to consider structure over a range of levels of scale. For instance, we can think about toughness in terms of visible sharp corners and notches, but the crucial processes occurring at the tip of a crack are at the microstructural level and, furthermore, these processes — the mobility of dislocations, for example — depend on the types of atom/molecule present and the forces that hold them together. So, our theoretical models span the scales from macrostructure to atomic structure.

Also, you will have noticed that often I have hinted at the importance of structure at the atomic level: the electronic structure of atoms determines their magnetic characteristics; the mobility of dislocations depends on the crystal structure and the type of bonding between atoms, and so on. Invariably, we need to explore atomic structure in the search for adequate theoretical models. We start to do this in Chapter 3.

SAQ 2.11   (Objective 2.1)
In this chapter we have considered a variety of mixtures. List the general types of ingredient that can be added to make a mix under the headings of:

(a) macrostructure,
(b) microstructure.

For each type, indicate their role in influencing properties.

## Summary

Slip in crystals is produced by shear stress. It occurs on a slip plane and the slip vector describes the direction and magnitude of an increment of slip.

A dislocation line bounds an area which has undergone slip. Dislocation lines must form closed loops or end on the surface.

A dislocation line lying at 90° to the slip vector is called edge dislocation. A dislocation line lying parallel to the slip vector is called screw dislocation.

An edge dislocation is associated with an extra half plane of atoms on one side of the slip plane.

Ductility arises from mobile dislocations; brittleness from immobile dislocations.

Structural disorder, such as grain boundaries, embedded particles, and so on tend to reduce dislocation mobility and therefore to increase yield stress.

## Objectives for Chapter 2

You should now be able to do the following.

2.1 Show, by means of examples, that materials are mixtures of ingredients at some level of scale. (SAQ 2.11)

2.2 Explain, with examples, how specified properties of a material depend on the absence or presence of particular ingredients and their size and distribution. (SAQ 2.4)

2.3 Explain the principles and uses of the following theoretical models:
(a) homogeneous strain for stiffness of a composite
(b) crack propagation,
(c) dislocations and plasticity,
(d) magnetic domains in hard and soft magnetic materials.
(SAQ 2.1, 2.3, 2.4, 2.6, 2.8, 2.9)

2.4 Describe and give examples of the ways in which materials development can lead to improved properties (such as processability) and thus to the optimization of a property profile; also describe and give examples of the way compromises may have to be struck between properties when there are conflicting requirements. (SAQ 2.7, 2.10)

2.5 Define and use the following terms and concepts:
macrostructure
microstructure
law of mixtures
homogeneous stress and strain models for composite properties
toughness
work of fracture
ductile–brittle transition temperature
critical crack length
stress concentration
weak interface
elastic strain energy
soft and hard magnetic materials
magnetic domains
domain walls
slip plane, slip vector, slip system
edge and screw dislocations
dislocation width and mobility
scanning and transmission electron microscopy
resolution of a microscope
hardness measurement
polarized light

# Answers to exercises

EXERCISE 2.1 Extension $\delta$ is strain $\varepsilon$ times original length $l$. So for the composite,

$$\delta_C = \varepsilon_C l_C$$

Strain is stress divided by Young's modulus, so

$$\delta_C = \frac{\sigma_C}{E_C} l_C$$

The same arguments apply to matrix and fibres, so

$$\delta_M = \frac{\sigma_M}{E_M} l_M$$

and

$$\delta_F = \frac{\sigma_F}{E_F} l_F$$

The total extension $\delta_C$ is $\delta_M + \delta_F$ (see the hints in the question), so

$$\frac{\sigma_C}{E_C} l_C = \frac{\sigma_M}{E_M} l_M + \frac{\sigma_F}{E_F} l_F$$

We are considering the homogeneous stress model, in which $\sigma_C = \sigma_M = \sigma_F$, so

$$\frac{l_C}{E_C} = \frac{l_M}{E_M} + \frac{l_F}{E_F}$$

whence

$$E_C = \frac{E_M E_F l_C}{E_F l_M + E_M l_F}$$

$$= \frac{E_M E_F}{(E_F l_M/l_C) + (E_M l_F/l_C)}$$

or in terms of length fractions $L_M$ and $L_F$, where $L_M = l_M/l_C$ and $L_F = l_F/l_C$,

$$E_C = \frac{E_M E_F}{E_F L_M + E_M L_F}$$

Because the block is parallel sided, length fractions are equal to volume fractions, so

$$E_C = \frac{E_M E_F}{E_F V_M + E_M V_F}$$

In this example the law of mixtures is not obeyed.

Exercise 2.2   Hooke's law is plotted as the dashed line on Figure 2.14.

(a) In Figure 2.14, the area under the dashed (Hookean) line is equal to the work of deformation per unit volume $W$. It is

$$W = \frac{1}{2} \sigma \varepsilon$$

(b) Hooke's law requires that

$$\sigma = E\varepsilon$$

Rearranging this,

$$\varepsilon = \frac{\sigma}{E},$$

and substituting for $\varepsilon$ in the equation for $W$,

$$W = \frac{1}{2} \sigma \frac{\sigma}{E}$$

$$= \frac{1}{2} \frac{\sigma^2}{E}$$

(c) Substituting $\sigma = \varepsilon E$ into $W = \frac{1}{2}\sigma\varepsilon$,

$$W = \frac{1}{2} (E\varepsilon)\varepsilon$$

$$= \frac{1}{2} E\varepsilon^2$$

EXERCISE 2.3

(a) The slip direction must lie in the slip plane.
(b) The crystal structure is the same before and after slip. So the magnitude of the slip vector must be equal to the distance between atoms in the slip direction.

# Answers to self-assessment questions

## SAQ 2.1

(a) From the law of mixtures, the density of the composite $\rho_c$ is given by

$$
\begin{aligned}
\rho_C &= V_M \rho_F + V_F \rho_F \\
&= (0.907 \times 900 \\
&\quad + 0.093 \times 2200) \text{ kg m}^{-3} \\
&= 1021 \text{ kg m}^{-3}
\end{aligned}
$$

(b) The law of mixtures applies in this case, so

$$
\begin{aligned}
E_C &= V_M E_M + V_F E_F \\
&= (0.907 \times 0.5 \\
&\quad + 0.093 \times 72) \text{ GN m}^{-2} \\
&= 7.15 \text{ GN m}^{-2}
\end{aligned}
$$

(c) The composite rod is about 14 times stiffer than the polypropylene rod. So, for the same elastic extension under the same load, its cross-sectional area should be one-fourteenth that of the polypropylene rod. Since the rods have the same length, the volume of the composite must be one-fourteenth that of the pure polypropylene.

Let subscript P denote polypropylene. Then, representing mass by $m$ and volume by $v$,

$$
\frac{m_C}{m_P} = \frac{v_C \rho_c}{v_P \rho_P}
$$

But $v_C/v_P$ is $1/14$, so

$$
\begin{aligned}
\frac{m_C}{m_P} &= \frac{\rho_C}{14 \rho_P} \\
&= 0.08
\end{aligned}
$$

That is, for the same stiffness, the glass-filled structure is about 12.5 times lighter than the pure polypropylene.

## SAQ 2.2

(a) Incorrect. Steel is a ductile material, which is why most cars dent in collision rather than breaking into pieces.
(b) Correct. It will absorb energy in a collision, but break rather than dent.
(c) Correct.
(d) Correct. However you may not agree that a polyethylene bag is weak. Indeed, some plastics carrier bags seem to be quite strong.

(e) Incorrect. Jelly is not tough. Find a flaw in it, or introduce a small crack, and it will break very easily.

## SAQ 2.3 From Equation (2.1)

$$
\begin{aligned}
l_c &= \frac{E G_c}{\pi \sigma^2} \\
&= \frac{200 \times 10^9 \times 10^5}{\pi \times (200 \times 10^6)^2} \text{ m} \\
&\approx 16 \text{ cm}
\end{aligned}
$$

This is half the critical crack length. You can see that mild steel is very tough and very 'forgiving'. No wonder it's so popular.

## SAQ 2.4

From the consideration of toughness in this chapter, you should expect that in (a) the law of mixtures will be obeyed. The toughness of the composite should be the weighted average of the values of $G_c$ for the matrix and the glass fibres. That is

$$
G_{cC} = (1 - V_F) G_{cM} + V_F G_{cF}
$$

where $G_{cC}$ is $G_c$ for the composite, $G_{cM}$ is $G_c$ for the matrix, $G_{cF}$ is $G_c$ for the fibre and $V_F$ is the volume fraction of fibre.

In (b) the law of mixtures will not be obeyed. The composite will be tougher than the law predicts. This is because the energy absorbed in creating the interfaces between the fibres and the matrix is an additional contribution to $G_c$.

## SAQ 2.5

The magnetizing field strength $H$ induces a magnetic flux density $B$ in a material. The slope of the initial magnetizing curve is a measure of the ease of magnetizing a material. It is called the **relative permeability** $\mu_r$, where $\mu_r = B/H$. The steeper the slope, the higher the permeability, and the easier the magnetization. In practice the initial magnetization curve is not linear, and it is usual to measure the maximum slope.
With increasing $H$, the flux density reaches a maximum level $B_s$ which is called **magnetic saturation**.

As the field strength is reduced to zero, some flux density $B_r$ remains; it is called the **remanence**. To produce zero flux, a specific field strength has to be applied in the opposite direction; this is the coercivity $H_c$.

The curve never retraces the original magnetizing curve, but traces out a loop. It is called a **hysteresis loop** and the area it encloses is a measure of the energy dissipated in a complete magnetization cycle, that is the **hysteresis loss**.

$BH_{max}$ is simply the maximum value of the product of $B$ and $H$. In practice, the maximum value in the second quadrant is taken, rather than the absolute maximum for the whole loop.

## SAQ 2.6

(a) Hard magnetic material. Here the aim is to restrict domain wall movement so that, once magnetized, the magnet stays magnetized. This can be achieved by making the microstructure a complex mixture of fine particles, as in Figure 2.35(b).

(b) Soft magnetic material. Here domains walls must be able to move freely in order to allow easy magnetization and demagnetization. Such materials should be free of particles of other materials and free of structural irregularities such as grain boundaries. The ideal would be a completely pure, structurally perfect single crystal! But at least the grains in Figure 2.30 are quite large.

## SAQ 2.7

(a) The performance properties required are low hysteresis loss and high electrical resistivity. The processing requirement is ductility, so that the alloy can be rolled into sheets.

(b) The hysteresis loss decreases progressively with the addition of silicon and, up to about 11%, the resistivity increases with the addition of silicon. The problem is the ductility. Figure 2.48(c) clearly indicates that iron–silicon alloys become brittle with silicon contents above about 4%. On the present data, this fixes the optimum composition. In practice, other factors to do with processing requirements and the presence of impurities in the iron mean that the silicon is typically 3.25%.

(c) See Figure 2.68.

SAQ 2.8
(a) The direction in which the carpet is moving indicates that the 'slip vector' runs from left to right and is at 90° to the line of the ruck. The ruck is an edge dislocation. Incidentally, it is a positive edge dislocation.

(b) In Figure 2.69 the arrows indicate the direction in which the edge dislocations must move in order to produce slip under a tensile stress. The easiest way to follow this is to imagine which way the 'extra' half-plane of each dislocation has to move to produce the slip step.

SAQ 2.9 Metals are tough when they are ductile. Slip is the main process by which plastic flow occurs in metals. The toughness arises for two reasons. Primarily it is because of the high value of $G_c$ due to the large amount of energy absorbed by plastic flow as the material is torn apart. The steel fracture surface in Figure 2.17(b) shows that there has been extensive pastic flow, unlike the glass in Figure 2.17(a). Secondly, the toughness of metals arises because slip by dislocation motion increases the radius of a crack tip and therefore reduces its stress concentrating effect — the crack is blunted. The way slip achieves this is illustrated, very simply, in Figure 2.70.

SAQ 2.10 Manipulating the microstructure of a metal to increase the yield stress means decreasing the dislocation mobility which, in turn, decreases ductility. Ductility provides the mechanism for toughness in metals. The lower the ductility the less potential a metal has for relieving stress concentrations and, therefore, the lower the toughness.

In metals of high strength, and therefore low ductility, there would clearly be problems of deforming them to shape in the solid state, especially by cold working. Much higher forces would be needed to deform them in rolling, for example, and the shape change that could be achieved would be more limited.

This encourages the idea of trying to shape a material when it is in a soft, ductile condition and subsequently changing its microstructure to provide the properties required in service.

SAQ 2.11
(a) Macrostructure.
 (i) Lumps, for example gravel in concrete. Increase compressive strength, lower cost.
(ii) Fibres, for example in GFRP and wood. Increase stiffness and toughness of matrix; produce processing problems.

(b) Microstructure.
 (i) Grain boundaries. Increase yield strength and hysteresis loss, decrease ductility.
 (ii) Dislocations. Agent for ductility and toughness in metals, increase hysteresis loss.
(iii) Particles. Increase yield stress in metals, hysteresis loss, remanance and coercivity.

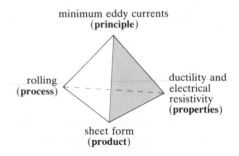

Figure 2.68

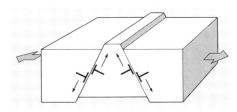

Figure 2.69 Movement of edge dislocations

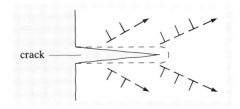

Figure 2.70 Blunting of a crack by dislocation movement

# Chapter 3  An atomic view of solids

## 3.1  Introduction

### 3.1.1  Stick and rattle (the theory of kinetic atomism)

1500 BC Chinese casters of bronze knew that putting lead in their casting alloys made the metal follow the finest detail. Some 2500 years later compatriot potters were using ingredients which, with uninstrumented skills, could be fired to translucent porcelain. Earlier (around AD 50) swordsmiths supplying the Roman troops in Britain and elsewhere were producing carburized and heat-treated 'steel' cutting edges (Figure 3.1). None of these ancient craftsmen (materials engineers?) had any theoretical concepts to guide their actions or to explain the consequences and certainly none would have imagined the possibility of a single conceptual framework in which rational approaches to all three processes could be discussed. But in recent times, that is within the past 200 years, our science has provided just such an all-embracing view of what makes matter behave as it does; let's call it the 'theory of kinetic atomism'. Given this understanding, materials technology has changed from happy chance to directed research. We now expect materials to be tailored to meet production or service needs.

The new view of matter has two key ideas. The first is that everything is made of 'atoms' of a fairly small number of 'chemical elements' joined together in a great variety of ways. Specific patterns make the characteristic unit of each substance — molecules as you know them. The existence of solids and liquids in bulk further implies cohesion between matter at the atomic/molecular scale. This is the sticking side of the theory. The second key idea is that one form of the energy we recognize as heat is the rapid random motions of the atoms. In solids these will be vibrations which you can visualize as tending to shake the structure apart. The changes of state, solid to liquid to gas as temperature is raised, reflect the increasing effectiveness of thermal energy in overcoming the forces holding atoms together. This is the rattling principle. Vigorous rattling not only breaks bonds between atoms but is also liable to spoil patterns and get them into a muddle. Both effects have far reaching consequences.

Much of the behaviour of matter can be seen as a conflict between sticking atoms together and rattling them apart! As a broad engineering principle we can now understand that adjusting the strength of stick or the amount of rattle is often what we are doing as we try to modify materials in order to coax service or process performance from them. (See ▼**Glass for bottles**▲, ▼**Hot material for jet engines**▲ and ▼**The right semiconductor**▲.)

Figure 3.1  Roman sword and sheath (by kind permission of the Trustees of the British Museum)

# ▼Glass for bottles▲

Glass bottles have a hard time in the market with all the competition from alternative containers. Are they made as cheaply as possible? They are blow-moulded at a temperature high enough for rattle to soften the glass to a workable state. It is the nature of the sticking in glasses which makes them soften over a range of temperature rather than melting suddenly like ice. Can the degree of stick be reduced to give a lower softening temperature? That could save process energy costs . . . you pay for less rattle! Figure 3.2 shows the variation of viscosity with temperature for various glasses. At the softening temperature the glass flows appreciably under its own weight. It has been known in glassworking for centuries that alkali-rich glasses, containing sodium or potassium ions for example, soften at low temperature. You

will see why in due course. But they are also easily scratched and prone to acid attack. The scientific development of glass making in the last few decades has produced various improvements — scratch resistant coatings and internal surface treatments which extract alkali ions from the surface after the bottle has been formed.

Rattle energy costs are not always paramount. The alkali metal addition is not cheap and improvements in process technology now permit forming to take place at a higher temperature — paying for rattle with savings in raw material costs. Chemical resistance is improved both by the lowering of the alkali metal content and by other addition such as alumina (a source of aluminium ions $Al^{3+}$).

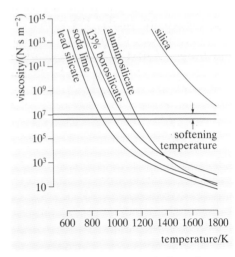

Figure 3.2 Temperature variation of viscosity for glasses

# ▼Hot material for jet engines▲

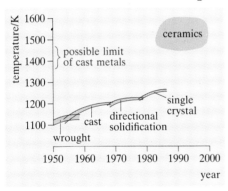

Figure 3.3 Limits on turbine blade temperatures

In a jet engine, incoming air is compressed by a fan and then heated to about 1600 K by burning fuel in it, so that it can do work on the turbine. The turbine itself

extracts only enough energy to drive the compressor, the rest of the energy being turned into kinetic energy by the jet nozzle. The performance of gas turbine engines gets better as higher temperatures are used at the first stage of the turbine. So, to get the power to fly bigger aircraft, the turbine blade material has to be able to withstand higher temperatures. Figure 3.3 shows how the temperature limits on turbine blades have risen over the years, making use of new materials technology. If you know something about aircraft you may be able to trace the results in terms of bigger and faster planes.

Beyond about half way (on the absolute temperature scale) to its melting temperature, a metal is generally prone to creep, that is, it will stretch under tension and eventually break. Creep is caused by

the random atomic rattle helping the defects such as vacancies and dislocations within crystals and at crystal boundaries to move in response to a load. How can the sticking of metals be enhanced to hold atoms in place even when rattling is intense? The military importance of air power has ensured that this question has been deeply researched.

Complex nickel-based alloys which do not creep out of specification at 80% of their melting temperatures have been devised. On the whole this has been done by judicious metallurgy rather than by haphazard trial and error — that is, by understanding at an atomic level what is happening within an alloy during creep. Ceramics now offer the possibility of even higher temperatures.

# ▼The right semiconductor▲

The earliest transistors were made of germanium but since the late 1950s silicon has been preferred. What can stick and rattle say about this? Semiconductor devices had to be made in single crystal material of great purity. To begin with, the low melting temperature (weak stick) of germanium ($T_m = 1210\,K$) made the

processes of purification and crystal growing much easier. But the weaker stick also implied that extra electrons would rattle free from the atoms and join in the conduction of electricity if there were a modest rise of temperature such as might be caused by a current flowing through the device. Thus germanium device

characteristics were unhelpfully current and temperature sensitive. In silicon ($T_m = 1680\,K$) the atoms are stuck tighter together and hang on to their electrons in the face of more rattle. So processing is more difficult but the devices are much more stable. As soon as the manufacturing technology allowed, silicon took over.

The idea is not limited to the sticking of atoms. Electrons are rattled off atoms to make the beam in a cathode ray tube. What sorts of materials give electrons easily? We could also think of atomic magnets stuck in some kind of orderly array within magnetic materials in spite of thermal jostling. At first the magnetization falls off gradually with rising temperature, then more and more rapidly until at the 'Curie' temperature the magnetic order is rattled to pieces. Figure 3.4 shows the temperature dependence of the saturation flux density for iron, nickel and cobalt. All other magnetic materials behave in much the same way, but what sorts of materials make strong magnets?

The goal of the next two chapters is to work up the basic idea of sticking and rattling atoms to a level where you can begin to use it to understand and describe a wide range of phenomena and perhaps to project that understanding into engineering actions. In this chapter you will learn about the structure and generally the 'cold' properties of materials, considered on the atomic scale ($\approx 10^{-10}$ m) — how atoms stick. The next chapter will deal with temperature-related properties — the effects of rattle.

(See ▼Historical aside▲.) The following sections start with some revision of what you may already have learned but remember to read carefully the instant you encounter something unfamiliar. The first is a review of what atoms 'are' and of the raw materials, the chemical elements, from which all our working materials are assembled.

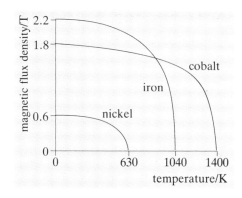

Figure 3.4 Saturation magnetization curves

# ▼Historical aside▲

'The idea that the material world at our fingertips is made of atoms which we have the power to rearrange in new ways is the master-concept of our age.' So wrote a biographer of John Dalton who is honoured as the first scientist to express atomic principles clearly, early in the nineteenth century. Neither this nor the kinetic theory of heat had easy acceptance into scientific thought. Embryonic forms of both ideas had at various times come to mind — to Robert Boyle for example, when he explained gas behaviour, and to Rumford to account for the inexhaustibility of friction as a supply of heat. But at that time, these ideas were less developed, and less appealing than the prevailing views of matter as a continuum and heat as a pervasive material fluid.

Chemistry provided the first impetus for accepting the idea of atoms and molecules. First, Lavoisier described the role of oxygen in combustion, to show that mass is conserved in a chemical reaction. At

that time, most quantitative experiments were not accurate enough to prove this decisively. Later Wollaston pursued the thought of shaped atoms stuck together to produce crystal shapes. But these 19th century atomistic explanations had to make do with a general assertion that the atoms just (inexplicably) stuck together. Nevertheless Davy and Faraday showed, through electrolysis, that there was some connection between electrical phenomena and chemistry.

Early in the twentieth century physicists established the 'nucleus and electron cloud' model of atomic structure. Since then, the quantum mechanics of the electron cloud has developed, first into descriptions of atoms of elements and then into perceptions of what makes atoms stick together. This product of many high powered minds is a 'splendid and elegant elucidation', as one of them proudly wrote, of the properties of matter in terms of electron dynamics.

## 3.2 Atoms and elements

The elements possess their particular properties by virtue of how their atoms are constructed out of the basic building blocks, protons, neutrons and (especially) electrons. Of course any tangible piece of engineering material contains a vast number of atoms crammed close together and interacting busily. You will be better able to manipulate materials, to control and modify their properties to suit product needs, or even just to choose the right stuff if you know something of the mechanisms, variety and consequences of those atomic interactions. The concepts and nomenclature of single atom models are an essential starting point for understanding how atoms interact in greater numbers. As far as possible I am going to use pictorial models of atoms, talking of electrons as orbiting particles, making sparing reference to the quantum mechanics which are used in solid state physics with significant quantitative success. The particle-orbit images put the qualitative descriptions on more familiar ground and are widely used conversationally. You will find that from time to time I switch between images from classical particle mechanics and images from quantum mechanics. This reflects the inadequacy of our understanding and the general desire to work with the least complicated model we can.

The map of how all the different atoms are constructed is the periodic table (see Appendix 1) of the elements. Although engineers rarely use pure elements to make things, a brief study of this scientific classification of the ultimately simple chemicals is justified because it gives a firm base from which to understand more complex and useful materials.

## 3.2.1 What are atoms like?

SAQ 3.1 (Revision)
Scribble notes to remind yourself about the structure of atoms. What can you say about:
(a) the size of an atom and its nucleus,
(b) where protons, neutrons and electrons are to be found,
(c) the distribution of mass in atoms,
(d) the distribution of charge in atoms,
(e) what distinguishes one element from another?

The most essential feature of our model of atoms is that each electron possesses a very specific total energy and is associated with a particular orbit. To jump between energy levels an electron must collect or reject exactly the amount of energy which is the difference between its initial state and its final state. This hypothesis was originally formulated to explain the characteristic light emissions (atomic spectra) from hot atoms. Nowadays atomic electron systems are understood in terms of 'quantum numbers' (reflecting the discrete energy levels), subject to strict rules deduced from studies of atomic spectra by theoretical physicists working in the early twentieth century.

SAQ 3.2  (Revision)
Remind yourself of the notation of atomic electron shells and the orbitals within them using the following prompts.

(a) The principal quantum number $n$ refers to . . .
(b) The azimuthal quantum number $l$ and the letters s, p, d, f, refer to . . .
(c) The electronic configurations in carbon, oxygen and sodium (with atomic numbers 6, 8 and 11 respectively are . . .
(d) What is meant by spin-up and spin-down?

The electron energy levels are described by a hierarchy of states organized in shells around the nucleus. Each state, for example an s orbital or one of the three p orbitals, can accommodate just two electrons which must have oppositely aligned spins. When a state contains a pair of electrons with opposite spins it is full — other electrons are excluded. This is referred to as the Pauli exclusion principle. Within shells, electron orbitals differ in energy and spectroscopic evidence (that from the study of spectra) points to an increase in energy up the sequence s p d f. The energy levels of all atoms seem to follow a sequence which fills the first shell, which has one s orbital only, before the second shell, which has s and p orbitals, is begun. Likewise the second is filled before the third. The third shell, which has s, p and d orbitals, is unable to comply with this elementary pattern and apparently contains higher energy states than the lowest states of the fourth shell. Before we can proceed to understanding atoms in solids we must attempt to account for this break in the elementary sequence, as it results directly in a variety of elements of great engineering importance — transition elements.

## 3.2.2  Just where are the electrons?

If we think of electrons in atoms as particles whizzing around a nucleus (the Bohr model of the atom) then an electron gets round its orbit so fast that before anything interesting, such as a collision, could happen it has done several laps. See ▼Orbit period and collision duration▲

This is why the images of atoms deduced from quantum mechanics, which show the volume of space where an electron is most likely to be, are often more useful than attempts to pin it down to a precise orbit. So we have pictures of where, on average, the electron spends most of its time. Figure 3.5 traces the idea showing how there are different spatial distributions for each kind of orbital (s, p, d and so on). These time–space 'maps' of electron density are useful tools for envisaging how atoms interact when they stick together. Here are a few noteworthy points.

# ▼Orbit period and collision duration▲

Consider a Bohr atom with an electron, mass $m$, orbiting a nucleus at radius $r$ with tangential velocity $v$.

(a) Bohr's angular momentum quantum rule for circular orbits lets us do a quick estimate of the time it takes an electron to get round once. He said that for the simplest orbit the angular momentum $mvr$ should be equal to $h/2\pi$, where $h$ is Planck's constant.

$$mvr = h/2\pi$$

Now, the orbit period, $t_1$ (time to do one circuit) is

$$t_1 = 2\pi r/v.$$

Using the first equation to replace $v$ in the second

$$t_1 = (2\pi)^2 mr^2/h$$

Taking $r$ to be 0.1 nm (typical atomic radius) and using $m \approx 10^{-30}$ kg and $h \approx 6.6 \times 10^{-34}$ J s:

$$t_1 = \frac{4 \times 10 \times 10^{-30} \times 10^{-20}}{6.6 \times 10^{-34}} \text{ s}$$

$$\approx 6 \times 10^{-16} \text{ seconds}$$

(b) The mean kinetic energy ($\frac{1}{2}mv^2$) of atoms depends on their temperature $T$ and is equal to $3kT/2$, where $k$ is Boltzmann's constant. (This is explained in more detail in the next chapter.) It is reasonable to say that a collision is actually happening while two atoms are within an atom radius of each other. The time for this is at least distance divided by speed and that's not allowing time for slowing down and accelerating away. At room temperature (300 K) the numbers give an average speed $v(=\sqrt{(3kT/m)})$ of about 2500 m s$^{-1}$ for a hydrogen atom ($m \approx 1.7 \times 10^{-27}$ kg). At this speed the time $t_2$ taken to move 0.1 nm is

$$t_2 = \frac{10^{-10}}{2500} \text{ s}$$

$$= 4 \times 10^{-14} \text{ s}$$

and in that time the electron has done almost 70 circuits.

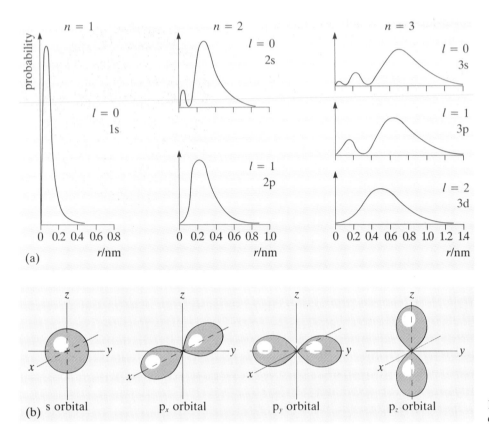

Figure 3.5 Electron probability distributions

(a) The probability distribution of an s state is spherically symmetrical around the nucleus. Figure 3.5(a) shows, for various s and p states, the radial variation of the probability of finding an electron. Notice that for each state there is a most likely distance from the nucleus, corresponding to the highest probability peak for that state.

Some of the states have a narrow but significant secondary peak near the nucleus, indicating that they penetrate fairly deeply into the atom. The simple (classical) particle orbit view reflects this pattern by supposing that, from time to time, the particle penetrates deep into the body of the atom — in much the same way that Halley's comet nips quickly round the part of its orbit close to the sun, where we can see it for about a year, and then saunters slowly round the rest of its orbit for the next 75 years. But an electron orbit would be repeated about two million times every nanosecond. In an s orbital, its path would be something like the shape shown in Figure 3.6.

(b) The p states are not spherically symmetrical. The space occupied by a p state electron has a distinctive lobe pattern, as shown in Figure 3.5(b). Later we shall see that similar lobe-shaped electron orbitals account for the strong directionality of covalent bonds.

In the particle orbit view, an electron in a p state spends a little time penetrating close to the nucleus, but not as much as an s state electron.

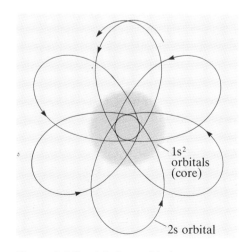

Figure 3.6 Particle in s orbital

(c) The filled inner electron states shrivel to a 'core' which occupies only a tiny volume as the nuclear attractive charge increases. The bulk of the atomic volume holds just the outer shell of a few electrons. The outer electrons are attracted by the nuclear charge but repulsion by the core electrons weakens the degree to which outer electrons are held in the atom. Figure 3.7 gives a 'visualization' of a sodium atom. The density of shading reflects the probability of finding electrons in any region.

(d) The d and f orbits barely penetrate the core and experience the positive nuclear charge mostly through a screen of negative charge due to other electrons in lower energy states.

This last point is particularly important in accounting for the breakdown of the elementary filling pattern for shells and orbitals. The angular shapes of d and f states are not so directional as p states and we won't need to worry about their detail.

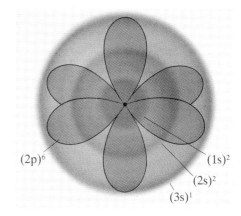

Figure 3.7 A sodium atom

You should now be able to work across the periodic table (Appendix 1) with some appreciation for the sequence of energy level filling. Exercise 3.1 will lead you through the first eighteen elements to argon.

EXERCISE 3.1
(a) What shell and which orbital is being filled between H and He?
(b) What shell and which orbitals are being filled between Li and Ne?
(c) What shell and which orbitals are being filled between Na and Ar?

Progressing further through the periodic table, the elementary pattern would put electrons into the third shell ($n = 3$) d orbitals next. But remember d orbits don't penetrate too deeply and so may be less strongly bound than the deeper penetrating s orbits. Thus the fourth shell s orbitals (4s) fill next and then in the transition elements (Sc to Zn) the higher energy 3d states begin to fill up.

## 3.2.3 Spin and pairing

In physics, you will have come across the notion of angular momentum associated with the curved (angular) part of any motion. Now the modern description of electrons in atoms (quantum mechanics) requires electrons to have two sources of angular momentum. The first is associated with the 'orbit' — an electron circulating an atom can certainly have angular momentum, as the Earth does orbiting the sun. The second source can be imagined as due to the electron spinning on its own axis just as the Earth's daily spinning gives it a second source of angular momentum. The amount of spin angular momentum can be one of only two values $+1$ unit (or 'up') and $-1$ unit (or 'down').

The rules of the quantum mechanical model state that within a set of orbitals (that is, a given $l$) we should first fill energy levels with electrons of the same spin ('up' or 'down') as far as possible before beginning to add electrons of opposite spin. The electron configuration for iron in Figure 3.8 shows the spin state by means of an arrow. Note the excess of 'up' spin states in the 3d orbitals. We shall see the importance of this later.

So much for isolated atoms. We will now review the general notion of bonding before developing a fuller atomistic view of the solid state.

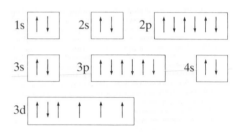

Figure 3.8 Electron configuration in iron

## 3.3 The sticking principle

### 3.3.1 Elementary ideas of bonding

Stories about bonding usually start by drawing attention to the column of 'noble' gases in the periodic table — He, Ne, Ar, Kr, Xe and Rn, which chemically are virtually inert. The electron structures of these are $(n\text{s})^2(n\text{p})^6$, save for He which has the $n = 1$ shell filled with $(1\text{s})^2$. As you will remember, elementary ideas of bonding assert that other elements' atoms are striving to achieve this chemically inert configuration by gaining, losing or sharing electrons. The question is: how? SAQ 3.3 revises the simplest answers to this question.

These harshly separated statements of atomic bonding types are just a beginning. You will learn of covalent bonds which spread to more than two atoms and of metallic bonds which are restricted to a few atoms. There are ionic bonds which are somewhat directional and covalent bonds with distributed charge giving electrostatic effects; in a metal, the free electrons have been stripped from the atoms leaving positive ions. Figure 3.9 shows a less sharply differentiated view of bonding with some important examples to convince you that better models are needed to account for many materials. However, given that atoms are stuck together, we can begin to appreciate some of the mechanical properties of solids and for the remainder of this section we'll consider bonded atoms without further distinction. We will further investigate the nature and consequences of specific types of bond in later sections.

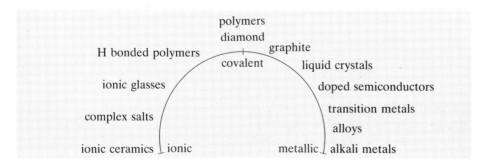

Figure 3.9 A continuum of bonding types

SAQ 3.3 (Revision)

Fill in the gaps in the following passage using the words listed in the key. Each word is used only once.

There are three mechanisms by which atoms bond strongly together, namely _____ , _____ and _____ bonding. Metals obviously use the third of these, in which some _____ are detached from individual atoms and spread throughout the spaces between them. This separation of _____ produces a _____ , _____ force. The presence of free electrons means that metals are electrical _____ . The interatomic forces in metals are non-directional and the atoms are spheres, so _____ crystal structures are often formed.

Ionic bonds are typical of _____ such as magnesium oxide. In this type of bond electrons are _____ from one atom to another, making _____ charged _____ which attract each other _____ to form the bond. These materials are electrical insulators until they melt when the ions become _____ . The forces between ions are also nondirectional but because ions of like sign _____ ionic crystals are of more _____ structure than metals.

Polymers are made of distinct _____ which _____ each other only weakly but within which much _____ bonds join atoms. In these covalent bonds, electrons are _____ between two atoms and the attraction is _____ (along the line of atom centres). Because all the electrons are locked into bonds, polymers are electrical _____ . Polymer molecules are generally _____ and so get tangled and only partially achieve the _____ needed to be crystalline.

Key word list — attract, ceramics, charge, close-packed, cohesive, conductors, covalent, directional, electrons, electrostatically, insulators, ionic, ions, non-directional, long chains, metallic, mobile, molecules, open, oppositely, organization, repel, shared, stronger, transferred.

## 3.3.2 Forces, energy and atom spacing

The difference between gases and the condensed states of matter (solids and liquids) is seen in atomic terms as a difference of atom spacing; in gases the particles arc well separated, in solids and liquids they are (somehow) held close together by attractive forces. That solids and liquids are relatively incompressible tells us also that if we try to push atoms still closer, some kind of repelling force comes into play. So without appeal to any mechanism of how atoms stick, we can propose the kind of variation of force $F$ between atoms with atomic separation $r$ shown in Figure 3.10(a).

The associated potential energy diagram $U(r)$ is sketched in Figure 3.10(b). The equilibrium spacing (zero force, minimum potential energy) is $r_0$. Remember that work is force times distance, and that work will be done if the atoms are pulled apart. We can relate the force and potential energy more formally by $F = \mathrm{d}U/\mathrm{d}r$. Here positive $F$ implies an attractive force — one tending to decrease separation. Negative $F$ implies a repulsive force — one tendency to increase separation. Conventionally the energy curve is set with zero potential energy for infinite spacing, sensibly acknowledging that atoms which can't 'feel' each other have no 'potential' to influence one another). The potential energy curve goes negative (below the axis) as atoms come together. The potential energy curve represents the minimum energy of a block of atoms; any 'rattle' energy has to be added above the curve. So on this diagram we would see the gaseous state as where enough kinetic energy has been added, that is the stuff is hot enough, to reach an overall positive energy which has no upper limit to the separation between atoms. Atoms in solids and liquids have less kinetic energy and are trapped in the 'potential well'.

The bottom of the well, where $F = \mathrm{d}U/\mathrm{d}r = 0$, corresponds with a stable equilibrium spacing, since it implies that both tension and compression would be resisted. Any change of spacing provokes a force opposing the change. At equilibrium, the centre of one atom is at $r = 0$, its nearest neighbours are at $r = r_0$ and this position is usually taken to represent the solid. Really the bottom of the well refers to a cold solid as the addition of thermal (kinetic) energy lifts the total atom energy above the minimum and the atom separation is on average greater than $r_0$. The deeper the well, the more resistant the material to the randomizing effects of thermal rattle and the higher the melting temperature. In this chapter we neglect the rattle, restricting the discussion to solids a long way below their melting temperature. This very qualitative picture can immediately be put to work to interpret some mechanical properties of materials. But with what success?

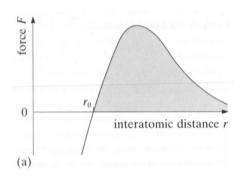

(a)

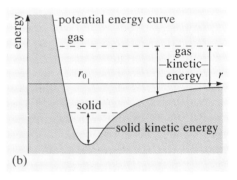

(b)

Figure 3.10 Force and energy curves for atoms

### 3.3.3 Young's modulus and Hooke's law

It was early appreciated that crystal shapes imply orderly stacking of atoms so the environment of each atom is very similar. In that case, taking Figure 3.10(a) to represent the force between just two atoms, the interactions between millions of atoms might be calculated as the sum of many interactions of pairs of atoms. Experiments on metal wires do show that the initial stretching of a specimen (containing over a billion atoms) is in proportion to the applied force (Hooke's law), just as Figure 3.10(a) now conjectures for a pair of atoms, that is the force–separation curve near $r_0$ is almost a straight line. Young's modulus is then taken to be a description of the way in which the interatomic forces of attraction all add together to resist an imposed strain, see ▼An atomic view of Young's modulus▲. In terms of the

# ▼An atomic view of Young's modulus▲

The loaded rod pictured in Figure 3.11(a) has undergone a tensile strain $\varepsilon$ in order to generate an internal stress $\sigma = F/A$ which balances the applied stress. It is then in equilibrium. The Young's modulus $E$ is the stress per unit strain

$$E = \sigma/\varepsilon$$

The magnified view, Figure 3.11(b), shows the cross-section where atoms in a crystalline arrangement are imagined to be tied to the next plane by springs (bonds) of relaxed length $r_0$ linking the nearest atoms in a simple cubic array. Each atom accounts for an area $r_0^2$ so that the total cross-sectional area $A$ contains $A/r_0^2$ atoms. Similarly each atom is presumed to take

an equal share of the total load, $F\,(=\sigma A)$ so that each bears a load

$$dF = \sigma A/(A/r_0^2)$$
$$= \sigma r_0^2$$

This is the increment of interatomic attractive force $dF$ which is generated when adjacent atoms are pulled $dr$ further apart. The strain in the rod is spread uniformly along its length so we'll assume the microscopic strain $dr/r_0$ for the two layers of atoms to be the same as the observed macroscopic strain $\varepsilon$. So the equation for Young's modulus is:

$$E = \frac{\sigma}{\varepsilon} = \frac{dF/r_0^2}{dr/r_0} = \frac{1}{r_0}\frac{dF}{dr}$$

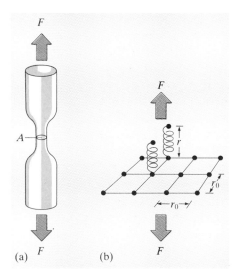

Figure 3.11 (a) Rod in tension. (b) Atomic view

potential energy–separation curve shown in Figure 3.10(b) the work done on the solid during elastic strain is stored in stretched bonds (for $r > r_0$) — see 'Work of deformation and fracture' in Chapter 2. Young's modulus is related to the slope of the force–separation curve near the equilibrium position ($dF/dr$ near $r = r_0$). Also, the force curve is related to the slope of the potential energy curve ($F = dU/dr$). It then follows that large elastic moduli imply deep narrow potential wells with large curvature at the bottom ($d^2U/dr^2$ is large near $r = r_0$). I have already suggested that this would give high melting points too. SAQ 3.4 invites you to check this but first, let's look at just what it takes to pull a solid apart.

## 3.3.4 Tensile strength

Fracture occurs at some limiting strain when these interatomic forces are no longer sufficient to sustain the load. A theoretical estimate of the strength of solids can be based upon the 'pair potential' picture together with the idea that material surfaces have an energy per unit area. This energy is associated with the surface atomic layer of a solid or liquid which could be bonded to a further layer, but isn't. Surface atoms do not have as low a potential energy as those in the bulk, which occupy potential wells with the maximum number of neighbours. On fracturing a material, work must be done to create new surfaces since bonds must be broken to produce them. $G_c$ (representing fracture energy) introduced in Chapter 2 accounted for the work done in pulling a material apart by propagating a crack through it. This fracture energy includes the surface energy but in ductile materials it is dominated by the energy associated with plastic deformation. ▼An atomic view of fracture▲ will lead you to the result that the ideal tensile strength $\sigma_f$ can be expressed

# ▼An atomic view of fracture▲

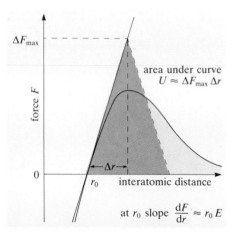

Figure 3.12 Approximations to the force–separation curve

We are going to estimate the tensile strength of a material by considering the force–separation curve for a pair of atoms. To simplify the calculation we'll approximate the real curve by two straight lines, as shown in Figure 3.12. The energy required to separate two atoms from their equilibrium position to infinitely far apart is the depth of the potential well in Figure 3.10(b). But this is also represented by the area under the force curve in Figure 3.10(a) between $r_0$ and infinity, which is approximately equal to the area of the triangle in Figure 3.12.

$$U = \Delta F_{max} \Delta r$$

The slope of the line in Figure 3.12 at $r = r_0$ is proportional to Young's modulus

$$E = \frac{stress}{strain} = \frac{\Delta F_{max}/r_0^2}{\Delta r/r_0}$$

which may be rearranged to give $\Delta r = \Delta F_{max}/(Er_0)$.

Use this equation to substitute for $\Delta r$ in the expression for $U$:

$$U = \frac{(\Delta F_{max})^2}{Er_0}$$

A single atom–atom bond is associated with a cross-sectional area $r_0^2$ in a solid (see Figure 3.11(b)) so the maximum local stress is $\Delta F_{max}/r_0^2$. I can now replace $\Delta F_{max}$ in the last equation by $\sigma_f r_0^2$ (where $\sigma_f$ is the tensile strength), by presuming that the local stress is the same as the macroscopic stress which we would measure in a real experiment. We can therefore relate the energy $U$ to fracture a bond between two atoms to the macroscopically observed stress $\sigma_f$ and the modulus $E$ as follows:

$$U = \frac{\sigma_f^2}{E} r_0^3$$

Next we consider surface energy. Once separated, each atom will sit in a surface occupying an area $r_0^2$. The surface energy $\gamma$, measured in $J\,m^{-2}$, associated with the two atoms in their new surfaces $(2\gamma r_0^2)$ is just that energy required to separate the atoms, so using the last equation:

$$2\gamma r_0^2 = \frac{\sigma_f^2 r_0^3}{E}$$

Rearranging,

$$\sigma_f = \sqrt{\frac{2\gamma E}{r_0}}$$

Do you suppose we have under or over estimated the fracture stress?

in terms of Young's modulus $E$, surface energy $\gamma$ and atom spacing $r_0$ by

$$\sigma_f = \sqrt{2E\gamma/r_0}$$

Surface energy can be measured in 'zero creep' experiments, where the work done in stretching idealized crystal shapes (increasing surface area) is counteracted by the natural tendency of solids to minimize surface area by diffusion. So an independent assessment of this 'ultimate strength' equation is possible — SAQ 3.4.

SAQ 3.4  (Objective 3.1)
Complete the final column in Table 3.1 and then compare:
(a) trends in Young's modulus and melting temperature,
(b) observed and theoretical tensile strengths.

Let us look at reality to see if this extension of the model is useful. Iron in the form of the strongest high tensile steel (piano wire) can reach $3\,GN\,m^{-2}$. Alumina, as engineering ceramic, has reached $1.5\,GN\,m^{-2}$. Both are less than 10% of the theoretical limit. Something rather important is missing! What we have missed out is all the structural imperfections in the solids. The defects mean that we cannot simply scale macroscopic stresses and strains down to atomic dimensions.

Table 3.1

| Substance | Young's modulus $E/\mathrm{GN\,m^{-2}}$ | Melting temperature $T/\mathrm{K}$ | Surface energy $\gamma/\mathrm{J\,m^{-2}}$ | Atom spacing $r_0/\mathrm{nm}$ | Tensile strength $\sigma/\mathrm{GN\,m^{-2}}$ | |
|---|---|---|---|---|---|---|
| | | | | | observed (in buk) | calculated |
| alumina | 345 | 2320 | 1.0 | 0.22 | 0.7 | |
| copper | 190 | 1360 | 1.7 | 0.21 | 0.4* | |
| diamond | 1200 | > 4000 | 5.4 | 0.16† | 50 | |
| glass (pristine fibre) | 70 | 1400 | 0.5 | 0.35 | 3.6 | |
| graphite | 200 | > 3770 | 0.07‡ | 0.36‡ | 0.1 | |
| iron (wrought) | 200 | 1810 | 2.0 | 0.29 | 0.20* | |
| salt (NaCl) | 44 | 1073 | 0.12 | 0.27 | 0.1 | |
| silicon (single crystal) | 190 | 1680 | 1.2 | 0.22 | 3.3 | |
| tungsten | 360 | 3650 | 3.0 | 0.16 | 1.5* | |

* Substantial yield before failure. †Deduced from hardness measurement. ‡Between atom planes

SAQ 3.5 (Revision)
Give two examples of a structural defect and discuss how it limits the atomistic ideas just presented.

## 3.3.5 The cold shoulder

In an atom, everything except the incomplete outer shell of electrons forms the core. In all atoms (except hydrogen) the core is the atomic nucleus surrounded by a very tight-knit ball of electrons, and it usually has a net positive charge. So one source of repulsion between adjacent atoms is the electrostatic force between the cores. But that is not the whole story. Within the core, all the quantum electron states are full, so any attempt to push electrons from one core into the space occupied by those of another will meet a problem — there are no states available for the new arrivals to take up. This quantum mechanical exclusion of intruders generates an even stronger repulsion than the classical electrostatic force. The combined effect is the abrupt rise of the potential energy curve as atoms are pushed so close together that the cores start to overlap. The steep rise in potential energy is often modelled by scaling the repulsive energy $U$ as $1/r^{12}$ (where $r$ is the distance between centres of two adjacent atoms) and so the force $(dU/dr)$ varies as $1/r^{13}$.

EXERCISE 3.2 If the repulsive energy curve follows $A/r^{12}$, what increase in repulsive force would result from a 5% compressive strain, that is from having $r$ decreased by 5%?

This exercise should make you realize that we can consider atoms to be hard spheres.

## 3.3.6 The energy trade-off

From an energy point of view, cohesion is the consequence of a reduction of potential energy as atoms come together. The evidence of incompressibility is that atoms eventually repel if we try to push them too close and that represents an increase of potential energy. Sticking (cohesion) must therefore, first and foremost, be recognized as a trade-off between whatever energy reduction mechanism can be devised for atoms 'quite close' and the necessary energy increase when they are 'too close' (the cold shoulder). There has to be an overall reduction in energy if the atoms are to stick. The origin of the repulsion is to be found in the interaction of the atoms' core electrons while the attraction comes from several of the interactions of the outer parts of the electron cloud where there are incompletely filled shells.

Arrangements of atoms which offer particularly large reductions of energy produce molecules or crystal patterns able to hold their integrity to high temperature in spite of there being a lot of thermal energy (rattle) tending to shake them to pieces. These are the 'primary' (strong) bonds, such as the three fairly distinct types; ionic, covalent and metallic with which we began this section. Other arrangements offer more modest overall energy reductions and result in weaker structures which relatively little rattle can disrupt. Substances we experience as liquids at room temperature more or less cohere through these 'secondary' interactions. You will recognize that within a material both weak and strong mechanisms may be at work simultaneously. Thus the molecules of a thermoplastic stick to one another by weak forces, while the atoms making up those molecules are interacting through the strong covalent mechanism. That is why many thermoplastics can flow at quite low temperatures without decomposing the chemical substance (the molecules). Thermosets by contrast are hardened by making primary bonds between the molecules after they have been put where we want them, that is when the 'artefact' has been shaped.

## Summary

Electron interaction binds atoms together in different ways, often divided into ionic, covalent and metallic bonding. In many materials the bonding falls between these distinct categories.

The forces between, and therefore the potential energy of, adjacent atoms accounts for thermal and mechanical properties.

Young's modulus and tensile strength can both be described in terms of interatomic forces and energies. Because defects in the structure of a material play an important role in fracture, the tensile strength predicted by the atomic separation model is an order of magnitude higher than the measured tensile strength.

There is a very strong (quantum mechanical) repulsion between atoms which get close enough for their cores to interact. The energy trade-off between repulsion of the atomic cores and the attraction due to the outer electrons determines the amount of 'sticking'.

# 3.4 Ionic bonds and ionic crystals

Ionically bonded crystals are within a wider group of engineering materials classed as ceramics, being generally hard, brittle and insulating. Sodium chloride (common salt) is often the first ionic compound to which people are introduced, although it is not well known for its engineering properties and after a brief scientific comparison we will deal with it no further. We will however examine various oxides which form structures with significant engineering implications arising from their refractory or electrical nature. Many compounds are not purely ionic. For example silicates contain both ionic and covalent bonds. But here we will discuss materials which have a predominant ionic component. Let's begin by looking at the nature of the bond.

## 3.4.1 The ionic bond mechanism

In ionic solids the atoms are bound together as charged ions. The electrostatic forces involved are not directional so a feature of all ionic crystals is that the nearest neighbours of any ion are several ions of opposite sign. The resulting electrostatic attraction balances the repulsive forces. What are the repulsive effects? There are two:

(a) The inevitable core repulsion. This dictates how close the nearest neighbours can get.
(b) Ions of like sign repel. Electrostatic forces vary as $1/r^2$ — the attraction of (oppositely charged) nearest neighbours is tempered by a repulsion of (similarly charged) next nearest neighbours.

Balancing up all these forces for ions of different sizes and charges determines the geometry of the crystal.

> EXERCISE 3.3  The ions in common salt are $Na^+$ and $Cl^-$, the spacing between nearest neighbours is 0.28 nm and the melting temperature is about 1000 K. Magnesium oxide has a similar crystal structure, spacing 0.21 nm, the ions are $Mg^{2+}$ and $O^{2-}$ and it melts at about 3200 K. Can you account for MgO being so much more strongly bonded? Which will have a higher Young's modulus?

## 3.4.2 The size of ions

The energy needed to pull one electron off any isolated atom is of the order of several electron volts, for example helium 24.5 eV, caesium 3.9 eV, and is generally least for metal atoms. (If you are not familiar with electron volts, read ▼Energy units▲.) The removal of further electrons is progressively harder. In forming a positive ion an atom sheds an outer electron. Now when an electron is lost by a metal atom, the positive nuclear charge pulls the remaining electrons more tightly around itself, particularly if the electron is lost from a penetrating s

## ▼Energy units▲

The SI unit of energy is the joule. It turns out that on an atomic scale the joule is a bit too big, and the SI prefix required to achieve a more useful quantity is 'atto'.

$$1 \text{ attojoule } = 10^{-18} \text{ joule}$$

In practice, this is still rather large, and for historical reasons a different energy unit, the electron volt, is widely used.

One of the simplest experiments in atomic physics is the measurement of ionization energies. Ionization energy is the energy needed to knock electrons off an atom. It is measured by accelerating some electrons, firing them into a low pressure gas and then detecting the positive ions created when electrons are knocked off the gas atoms. The definition of the electron volt follows directly from this. It is the energy acquired by an electron, or any other particle with a unit of charge, when accelerated by an electrical potential of one volt. The electron volt (eV) is related to the joule by

$$1 \text{ eV } = 1.60 \times 10^{-19} \text{ J}$$

state, giving the positive ion a smaller radius (a tighter core) than the neutral atom.

In contrast only a few elements can accept an electron to form a negative ion without an exorbitant cost in energy. They are mostly light elements with a small number of unoccupied deeply penetrating (s or p) orbits into which the electron can go. The halogen elements of group 7 (F, Cl, Br, I and At) have an empty p state where the extra electron can form a spin pair, offering a slight energy compensation. These elements can take in one electron per atom — $F^-$ and $Cl^-$ cost the least energy. In group VI oxygen and sulphur can accept two electrons into their outer p shell forming $O^{2-}$ and $S^{2-}$. For most practical purposes you can forget about all other elements as negative ion formers. When we have discussed the mechanism of covalency some wider options will appear, for example silicate ions $(SiO_4)^{4-}$. When an atom accepts an extra electron in a penetrating orbit to become a negative ion, it swells considerably. This is because the time average negative charge inside the core is increased a little, so all the core electrons experience less attraction to the nucleus.

In building ionic crystals, big negative ions have to pack together with small positive ones — a feature of some importance in controlling the structure. Figure 3.13 shows the relative sizes of some elemental ions.

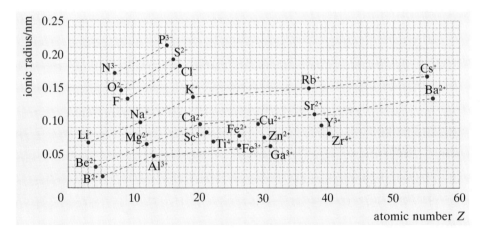

Figure 3.13 Radii of elemental ions

## 3.4.3 Ionic crystal patterns

I have now set up the facts that we need in order to explore ionic crystal patterns, at least if we stay with ions which can be thought of as hard spheres. A good way of developing your ideas of these structures is to think of the small positive ions as held in 'cages' made of the big negative ions, and then build up a whole 'prison' of cages. A slight oddity of this thought is that the captives are actually responsible for holding the walls of their cages in place by electrostatic attraction. That force is isotropic, that is, it is of equal strength in all directions. So however many negative ions form the cage, the simplest arrangements

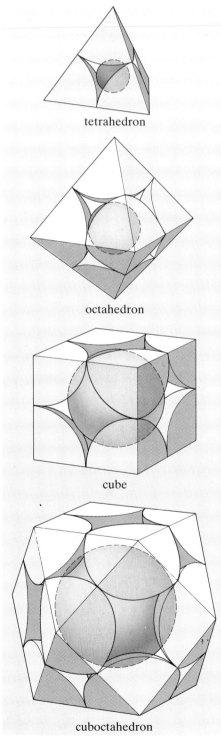

tetrahedron

octahedron

cube

cuboctahedron

Figure 3.14 Ionic crystal building blocks

# ▼Cage size and coordination number▲

Four spheres of radius $R$ can be stacked so that their centres lie at the corners of a tetrahedron — see Figure 3.14. The empty space (cage or void) at the centre could just accommodate a small sphere of radius $0.22\,R$. Larger cages could be made if the confining spheres were to stand apart a little allowing the space to be occupied by a larger sphere. However a cage of $0.41\,R$ is created when six spheres pack so that their centres define an octahedron. Likewise eight touching spheres can be arranged around a 'cubic void' of $0.73\,R$ and twelve would form a cuboctahedron around a void of $1.0\,R$. The number of surrounding spheres is referred to as the coordination number. Table 3.2 relates this number to the size and geometry of the cage.

EXERCISE 3.4   Read the radii of $Cs^+$ and $Cl^-$ from Figure 3.13. Could a cubic cage of $Cl^-$ imprison the $Cs^+$? If the $Cl^-$ ions all touch a $Cs^+$ ion, do they also touch one another? If not, how far apart are they? (See Figure 3.14)

Table 3.2

| Coordination number | Cage size | Geometry |
| --- | --- | --- |
| 4 | $0.22\,R$ | tetrahedron |
| 6 | $0.41\,R$ | octahedron |
| 8 | $0.73\,R$ | cube |
| 12 | $1.0\,R$ | cuboctahedron |

will have the negative ions equidistant from their prisoner. Electrostatic repulsion between the negative ions is trying to push them apart. So if we stay with simple spherical ions of one type, such as $O^{2-}$, the negative ions forming a cage will get as far from each other as their numbers and the attractive power of the positive ion will allow. That means in the first instance the cages will be regular three dimensional shapes. Figure 3.14 shows the four such shapes which could be built from 4, 6, 8 or 12 negative ions. In each case the outlined negative ions (pink) surround a trapped positive ion at the centroid. These, or slight distortions of them, will suffice for our purposes. See ▼Cage size and coordination number▲.

The cages have to fit together to form a whole prison and the reason for setting up the shapes in Figure 3.14 is precisely that combinations of these can assemble to fill 3D space in regular patterns, sharing faces as the dividing walls between adjacent cages. Obviously cubes can stack on their own. But the other shapes cannot. However, Figure 3.15 shows

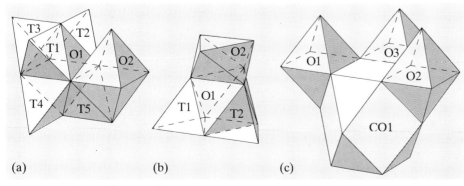

Figure 3.15 Composite crystal structures

three ways of stacking mixtures of octahedra and one of the other shapes. Figure 3.15(a) shows octahedral cells joined along the edges with tetrahedra in between. Figure 3.15(b) shows octahedra joined face to face, with tetrahedra. In Figure 3.15(c) the octahedra only meet at the corners and the remaining space is filled by cuboctahedra. These designs of prison are the prototypes for quite a wide range of technically interesting ionic compounds.

Cubic prisons   The easiest structure to see is probably that made of cubes, with common walls, floors and ceilings for every cage. The negative ion centres define a simple cubic structure. When every cage contains a body this is sometimes called the **body centred cubic** structure (BCC) although strictly BCC refers to the packing of indistinguishable spheres. Caesium chloride (CsCl) crystallizes like this.

A technically interesting example of a cube cage ionic crystal occurs where the prison is only half full: oxides of formula $Me^{4+}(O^{2-})_2$ with Me a heavy metal which forms a large ion in spite of having shed so many electrons. Uranium dioxide ($UO_2$) used to make refractory fuel pellets for high temperature nuclear power reactors is like this. Figure 3.16 shows the $UO_2$ structure. The $O^{2-}$ ions (not shown but centred at the cube corners) form cubic cages, half of which are filled by $U^{4+}$ ions.

Octahedra with tetrahedra   The crystal structure of magnesia exemplifies one of the combinations of octahedra and tetrahedra shown in Figure 3.15(a). The structure is shown again in Figure 3.17 where it has been sketched to highlight the cages in one part and the cubic crystal planes in another. The octahedral cages formed by (almost touching) negative ions are highlighted on the right. Each contains a positive ion. On the left the cubic structure is apparent. Study the figure carefully and notice that each positive ion is surrounded by six equivalent negative ions and vice versa.

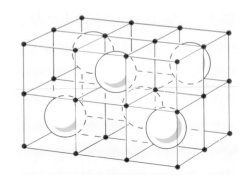

Figure 3.16  Uranium oxide structure

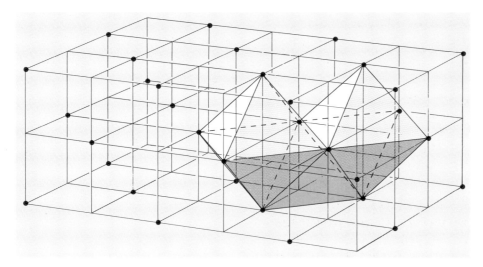

Figure 3.17  Magnesium oxide structure

Look up the sizes of $Mg^{2+}$ and $O^{2-}$ on Figure 3.13 and confirm the placing of the positive ions in octahedral cages.

The radii are 0.065 mm ($Mg^{2+}$) and 0.145 ($O^{2-}$). The ratio is 0.45 From Table 3.2 we can see that octahedral cages are appropriate for ratios between 0.41 and 0.73.

Industrially magnesia is used at high temperature in firebricks. In this application it must not 'creep', that is deform under imposed load over a long time. Any displacement of the crystal lattice which would involve ions of like sign becoming nearest neighbours will demand high energy. So slip is restricted to those few directions which contain only like ions as shown in Figure 3.18.

Octahedra and tetrahedra can also be combined in a different way to fill space with hexagonal symmetry. Alumina exists as this latter structure.

Perovskite . . . an easy one to predict!  The next structure to look at is called 'perovskite' after a mineral which adopts the same form. A whole family of electronically interesting ceramics with this structure is used in a host of applications (capacitors, loudspeakers, hydrophone sensors, gramophone pick-ups, electro-optic switches). The archetype is barium titanate, $Ba^{2+}Ti^{4+}O_3^{2-}$. SAQ 3.6 will let you deduce the cage structure for yourself.

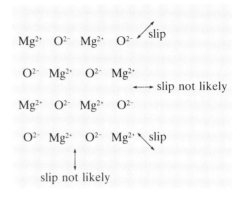

Figure 3.18 Slip planes in magnesium oxide

SAQ 3.6 (Objective 3.2)
Look up the sizes of $Ti^{4+}$, $Ba^{2+}$ and $O^{2-}$ on Figure 3.13 and choose the most suitable cage for each positive ion on the criteria of Table 3.2 bearing in mind that the two types of cage must stack together.

Sketches of crystal structure don't have to show the cages explicitly. Figure 3.19 in fact emphasizes the symmetry of perovskites by highlighting a cube formed by titanium ions (compare this with the answer to SAQ 3.6). As drawn, the cube represents a basic repeat unit (but not the cage structure) with which we could fill space.

Let's count atom fractions in the cubic structure cell shown in Figure 3.19 to confirm that it meets the stoichiometry (the proportions in which the chemical elements react) of $BaTiO_3$ — each cube corner is shared between eight cubes and so an eighth of a titanium ion at each corner 'belongs' to the cell and, since we have eight such corners, there is one $Ti^{4+}$ per cell. Similarly each cube edge is shared between four cubes and we have twelve edges. So there are three oxygen ions per cell. Finally, there is one whole barium ion in the middle of the cell so the cell contains one formula unit $BaTiO_3$.

In fact below 120 °C, barium titanate adopts a distorted form of the structure shown. The positive ionic charges don't remain symmetrically surrounded by negative charge and the crystal becomes a dipole — the charges balance over all but with excess positive charge at one end and excess negative charge at the other. The application of strain can affect

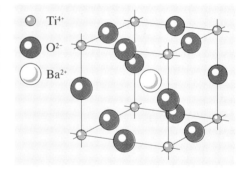

Figure 3.19 Perovskite structure

the strength of the electric dipole; mechanical pressure can produce electrical voltage and vice versa, the material is said to be 'piezoelectric', and this is the basis of its application in electronics. We will return to this rattle-related phenomenon later.

## Summary

Ionic crystals are generally hard, brittle, insulating and refractory.

Ionic bonds are formed between oppositely charged ions. In general, the greater the charge on the ions, the stronger the bonds holding the ionic crystal together.

Positive ions tend to be smaller than negative ions, because the electrons are pulled more closely into the core. Many regular structures can be understood in terms of large negative ions coordinated around smaller positive ions and voids.

The relative size of the different ions in a crystal determines the coordination number — four, six, eight or twelve, corresponding respectively to tetrahedral, octahedral, cubic and cuboctahedral cells.

There are several ways in which the cell shapes can combine to fill space. The mechanical properties of magnesia and the electrical properties of barium titanate owe much to the packing of ions in their crystal structures.

# 3.5 Covalency

Covalent bonds, unlike ionic bonds, are highly directional. So whereas positive ions can be surrounded by as many as twelve negative ions, the number of covalently co-ordinated atoms is much more restricted. The directional nature of covalent bonding leads to the tetrahedral structure of silica ($SiO_2$) and silicate ions ($SiO_4$)$^{4-}$

## 3.5.1 Two atom covalency: hybrid electron states

Viewed as a particle, an electron in an orbit (say an s orbit) around a single atom spends most of its time at the distant parts of its trajectory travelling slowly under minimal influence of its nucleus. While out there its orbit may overlap a similar orbit of another atom. If that orbit is not full the electron can switch onto it and penetrate deep into a second atom, Figure 3.20. An electron that 'visits' two nuclei should have lower energy than when it visits only one. For the electron to find both atoms repeatedly suggests that the new orbit can only persist if the two atoms stay more or less in the same position, being thus bonded together, provided this arrangement of two atoms offers an energy saving. The potential energy of the electron falls as the spacing of the two atoms decreases but core overlap (the cold shoulder) limits their approach.

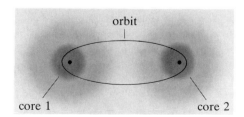

Figure 3.20 Orbital penetrating two atom cores

Two electrons doing the same thing give an even lower energy and the newly found orbit can accommodate two electrons as a spin pair. Remembering that electrons in isolated atoms singly occupy as many orbits as they can, it is quite likely that the overlapping orbits of the two bonding atoms initially will contain one electron each. This is the familiar story of a covalent bond being formed when atoms 'share' electrons: two unpaired electrons moving on their own s or p orbits in each of two atoms become a spin pair moving on a lower energy two-atom orbit.

However descriptions based on single atom orbits are too naive; the same time–space distribution of electrons cannot apply to electrons orbiting two nuclei, particularly when several atoms are close together. The outer regions of bonding atoms may in fact adopt completely new patterns. These are known as 'hybrid' orbitals, because in a mathematical sense they combine single-atom states. Three distinct hybrid orbit patterns can be devised by mixing the s orbit with either one, two or all three p orbits penetrating close to the nucleus, called $sp^1$, $sp^2$ and $sp^3$ respectively. The new spatial maps of electron distribution are in Figure 3.21. The numbers refer to the average distribution of electrons — 1 for a single electron and $\frac{1}{2}$ for an electron equally shared with a similar region in another atom. Notice the different shapes which are available. The $sp^1$, $sp^2$ and $sp^3$ hybrids have linear, planar and tetrahedral geometry respectively. None of these would be energetically favourable for any isolated atom. But by restricting the angular spread of the electrons' paths, they offer the possibility of persistent overlapping with orbits of another atom, and the consequent reduction of energy which we foresaw. Just as you may try to fit a piece of a jigsaw several ways before accepting or rejecting it, so we might consider an atom trying to fit into a bonded structure as in one of a number of electronic configurations, seeking the lowest energy configuration. Covalent bonds formed when these new orbital shapes overlap give strong directional bonds (leading to stiff structures).

Let's build the model further via some examples. Carbon is a convenient species to discuss, both because of the technical importance of its chemistry and because it displays all three hybrid forms.

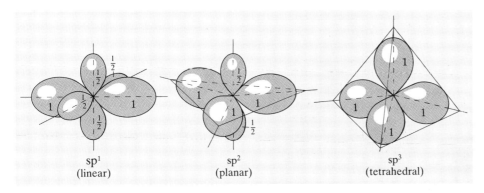

$sp^1$
(linear)

$sp^2$
(planar)

$sp^3$
(tetrahedral)

Figure 3.21 Hybrid orbitals

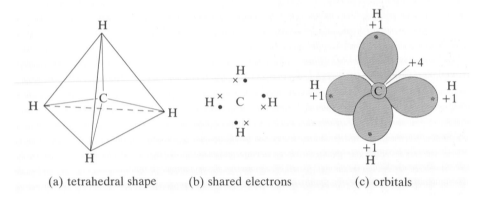

(a) tetrahedral shape     (b) shared electrons     (c) orbitals

Figure 3.22 Methane molecule

## 3.5.2 Methane, polyethylene and diamond: $sp^3$ hybrids

Methane ($CH_4$) molecules are regular tetrahedra, as shown in Figure 3.22(a). There are four equal covalent bonds each made by two electrons being shared by a C and an H atom. In Figure 3.22(b), dots represent shared electrons from H atoms and crosses represent shared electrons from the outer shell of the C atom.

The carbon takes on the $sp^3$ organization with hybrid orbitals reaching out to the four corners of a tetrahedron, each containing an unpaired electron. When the new orbitals overlap the space occupied by the 1s hydrogen states, all eight electrons move onto two-atom orbits, as in Figure 3.22(c). The complete energy budget shows a net saving, so a stable molecule forms.

The same $sp^3$ hybridization allows carbon atoms to join together to form long chains (as in polyethylene) or even the three-dimensional continuous structure of diamond. These bonds, where the electrons are confined to the straight line between the atoms, are called **sigma bonds** (sigma, $\sigma$, is the Greek letter s). They are strong bonds. Note that in the polyethylene chain molecule shown in Figure 3.23(a) the C—C bond is fairly free to rotate, whereas the diamond network structure shown in Figure 3.23(b) is more rigid.

## 3.5.3 Ethylene and acetylene: $sp^2$ and $sp^1$ hybrids

Ethylene ($C_2H_4$), often called ethene, is bound together rather differently. Four electrons provide a double covalent bond between the carbon atoms. Figure 3.24 compares (a) the $sp^2$ orbital picture, (b) the electron dot and cross representation (dots for H electrons, crosses for C electrons) and (c) the chemical formula for ethylene.

The disposition of the atoms in ethene tells us that the carbons are in the $sp^2$ hybrid condition (planar). Each atom proffers three electrons 120° apart on a plane, the fourth electron being available in an

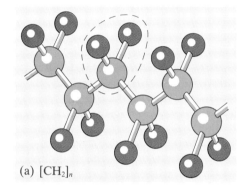

(a) $[CH_2]_n$

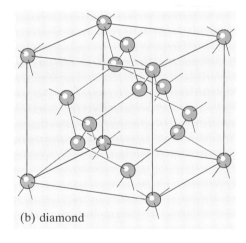

(b) diamond

Figure 3.23 Polyethylene and diamond structures

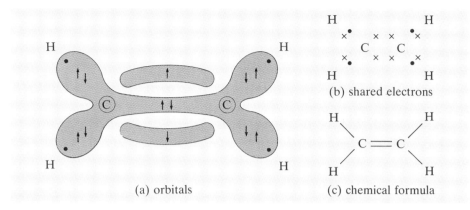

Figure 3.24 Ethylene molecule

unaffected p orbital normal to the plane (and directed to both sides of it). The molecule forms with a C—C σ bond in the plane whereupon the p orbits come into sideways overlap and another spin pair forms uniting the carbon atoms with blobs of electron density above and below the plane of the molecule. This latter arrangement is called a 'π-bond'. The resulting C═C bond involves four electrons, two in a σ bond and two in a π bond.

Molecules with π bonds tend to be reactive — you can visualize that electrons away from the straight line between atoms centres are more easily accessible to overlap from orbitals of nearby molecules. That's why ethylene is the popular reactive feedstock for many industrial chemical syntheses and reacts with itself to make $sp^3$ bonded polyethylene.

In the triple bonded ethyne molecule, H—C≡C—H, an $sp^1$ hybrid is involved, forming a single σ bond, while overlapping p states form two π bonds to link the carbon atoms as shown in Figure 3.25. Having to keep six electrons between a pair of carbon atoms limits the energy saved in forming the bond so there is plenty of scope for the formation of more stable bonds. Its more familiar name is acetylene. If you're a welder you'll know it as the gas used to get an incredibly hot flame — electrons losing lots of energy as they form new chemical bonds.

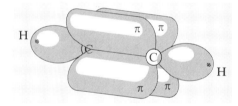

Figure 3.25 Acetylene molecule

## 3.5.4 Benzene and graphite: π bond rings

The 120° angle of $sp^2$ orbitals offers the special possibility of six carbon atoms joining together to form a hexagonal ring. Figure 3.26 illustrates a structure which uses σ and π bonds to join adjacent carbon atoms. The π bonds produce two continuous zones of electron density above and below the ring round which electrons can circulate. This is a strong structure because breaking any one π bond destroys the circulation and involves a big energy cost. With six $sp^2$ orbitals all linked to hydrogen atoms, a benzene molecule is formed. While it is easy to substitute other groups of atoms onto the side of the ring in place of one or more of

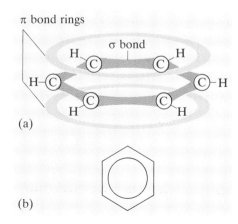

Figure 3.26 Benzene ring (a) structure (b) symbol

these hydrogens, it is much more difficult to break into the ring itself. The important thing to note is that this is our first example of atoms joining where electrons are shared between several atoms, not just two. The electrons are 'delocalized'.

The next step is to see that two hexagons can join edge to edge to form naphthalene ($C_{10}H_8$ used in mothballs), a two-hexagon molecule. The $\pi$ bond zones are continuous over both rings increasing the delocalization. The extreme case of a 2D network of joined hexagons is graphite (Figure 3.27), a sort of polybenzene in which near infinite sheets of rings bond weakly face to face to form a crystal of solid carbon with very different qualities from diamond. Graphite is soft and flaky while diamond is ultrahard. Graphite conducts electricity while diamond is an insulator. The bonding between sheets in graphite is due to secondary (van der Waals) forces which are discussed later. See ▼**Carbon resistors**▲

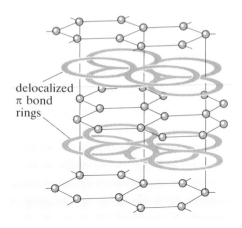

delocalized $\pi$ bond rings

Figure 3.27 Graphite crystal structure

EXERCISE 3.6   Look at the structure of graphite in Figure 3.27. Would you expect the conductivity to be isotropic (the same in all directions) for a crystal of graphite?

SAQ 3.7   (Objective 3.3)
Why is diamond hard and insulating in contrast with graphite which is soft and conducting?

## 3.5.5  Silica ($SiO_2$) and silicates (covalent ions)

We have already met an ionic oxide with a similar formula, $UO_2$, for which we developed a cage structure of edge-sharing cubes, containing $U^{4+}$ ions (Figure 3.16). The basis of silica structures is corner-sharing tetrahedra with a central silicon atom having a half share in four oxygen atoms, so the formula is really $Si(\frac{1}{2}O)_4$. There are two extreme explanations in SAQ 3.8.

SAQ 3.8   (Objective 3.2)
(a) Radii of $Si^{4+}$ and $O^{2-}$ are respectively 0.04 nm and 0.145 nm. With Table 3.2, derive an ionic description based on tetrahedral cages. Do the negative ions touch?
(b) Silicon is the element below carbon in the periodic table. What condition of covalency would give tetrahedral silica units?

Which has the lower energy, an ionic or a covalent structure? Well, in practice the covalent idea dominates so the bonding electrons orbit the silicon and oxygen cores but with the latter getting a larger share. The oxygen atoms adopt $sp^1$ hybrid forms to give linear $\sigma$ bonds (Si—O—Si) through the corners of the tetrahedral unit. Because the

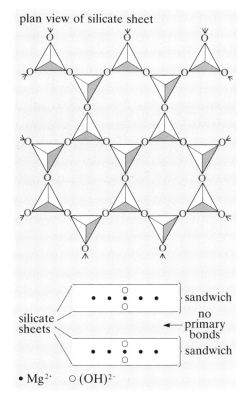

plan view of silicate sheet

• $Mg^{2+}$   ○ $(OH)^{2-}$

Figure 3.28  Structure of talc

# ▼Carbon resistors▲

For many years the principle of electrical conduction in graphite has been exploited for making electrical resistors (the product) with a wide range of ohm values (the property) while having the same physical size — how (what process)? Powdered graphite of chosen fineness is mixed with naphthalene and an insulating filler such as china clay, in appropriate amounts, and the mix is pressed into little cylinders. When the mixture is fired, most of the naphthalene boils off, leaving the structure porous, but some dissolves in the graphite and serves to assist these grains to sinter together where they chance to be in contact. A continuous electrically conducting path is thus assured but the resistance is determined by the narrow bridges between the graphite particles. There are so many process variables (proportions of mix, grade of graphite, firing temperature and time) that virtually any resistance can be obtained. Also the process is very difficult to control accurately so even within a single batch, quite a range of resistance values results. So, after end caps and wires and an outer insulator have been put onto the cylinder, every one is measured and marked with its value using the familiar code of coloured rings. If close tolerance resistors (5%) are selected from a wider tolerance set and sold at a higher price, then 10% tolerance resistors never have the right value among them, but there is no wastage.

For example, a batch may contain resistors between $70\,\Omega$ and $110\,\Omega$, to be sold as $82\,\Omega$ and $100\,\Omega$ resistors. All those falling in the $95$–$105\,\Omega$ range will be sold as $100\,\Omega \pm 5\%$. Those in the $90$–$95\,\Omega$ and $105$–$110\,\Omega$ ranges will be sold as $100\,\Omega \pm 10\%$. Similarly, $82\,\Omega \pm 10\%$ resistors will range from $73.8\,\Omega$ to $77.9\,\Omega$ and from $86.1\,\Omega$ to $90.2\,\Omega$.

> EXERCISE 3.5 A similar process might be used to make pencil leads. What principle and what property is involved?

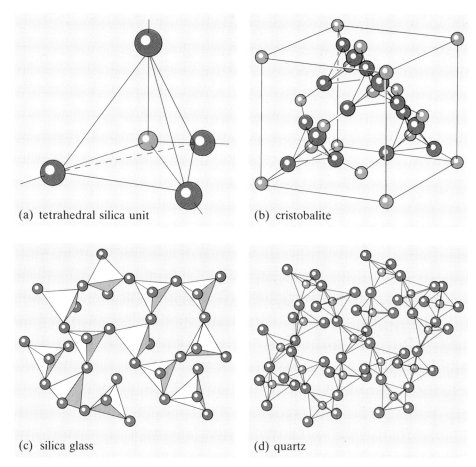

(a) tetrahedral silica unit     (b) cristobalite

(c) silica glass     (d) quartz

Figure 3.29 Silica structures

tetrahedra share corners only, there is a great deal of freedom in the structures which can be assembled. Figure 3.29 shows just a few of the many structures available. Cristobalite has tetrahedra with silicon atoms placed as in the diamond structure shown in Figure 3.23. Quartz has tetrahedra in helical formation. Silica glass is an amorphous arrangement of tetrahedral units. Molten silica owes its viscosity to the presence of strings of tetrahedra which persist in the liquid state. Notice that the oxygen atoms in an isolated $SiO_4$ unit have only seven electrons associated with their outer shell. This is what makes the units stick together so well. The oxygen atoms gain an eighth electron by bonding to a second silicon atom. However, an alternative source for the eighth electron could be ionized metallic atoms, which then form ionic complexes with the largely covalent silicate ion $(SiO_4)^{4-}$. Alloys of silica with metal oxides are abundant. They occur in glasses (see ▼Glasses based on silica▲) and cements. They also occur naturally, with a higher proportion of ions, in the widely exploited silicate minerals such as kaolin, asbestos and talc (Figure 3.28). When very few of the tetrahedra are linked covalently, silicates are water-soluble, like sodium silicate $3Na_2O.\ 2SiO_2$.

# ▼Glasses based on silica▲

On heating, fused silica softens at about 2000 K and then forms a very viscous liquid. Adding sodium oxide to molten silica lowers the viscosity and the softening temperature considerably. It does this by limiting the extent to which the $SiO_4$ tetrahedra link to each other. Sodium ions encourage an ionic end to a tetrahedral chain, preventing an oxygen bridge. This is illustrated in Figure 3.30. Divalent and trivalent metal ions may link tetrahedral chains through ionic bonds which are much less stiff (more easily twisted) than the covalent bonds they replace. Here is a principle at work controlling the property viscosity. What processes might need this control? (Refer back to 'Glass for bottles'.) Other oxides can be added to control properties such as colour and refractive index (see Table 3.3).

Note that in Figure 3.30, the fourth oxygen of each tetrahedron has been neglected for clarity. In soda–silica glass, there is one nonbridging oxygen for each $Na^+$ ion. In lime–silica glass there are two nonbridging oxygens for each $Ca^{2+}$ ion.

There is, however, a limit to the oxide additions. The reduction in viscosity brought about by reduction of the chain length in the melt aids the formation of the glass into useful shapes but it can have an adverse effect. The increased mobility of the ionized silica tetrahedra may enable the material to form regular crystalline arrays and become opaque.

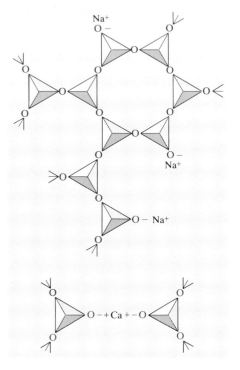

Figure 3.30 Structure of glass

Table 3.3

| Property required | Oxide to add |
| --- | --- |
| Colour | |
| green | $Fe^{2+}$ |
| blue/green | $Cu^{2+}$ |
| red | $Cu^+$ |
| blue | $Co^{2+}$ |
| yellow | $Ni^{2+}$ |
| Range of temperature over which glass can be formed to shape | |
| narrow | $K^+$ |
| wide | $Na^+$ |
| high refractive index | $Pb^{2+}$ |
| low thermal expansion | $B^{3+}$ |

## Summary

In covalent bonding, outer electrons are shared among two (or more) atoms. The resulting molecular orbitals can be described as mixtures or hybrids of atomic (s, p, d, . . .) orbitals. These hybrid orbitals are energetically favourable when several atoms are close together. The resulting bonds can be strongly directional.

Carbon exhibits three kinds of molecular orbital in its compounds. For example in polyethylene and diamond the $sp^3$ hybrid states are occupied, so that the carbon bonds tetrahedrally. By contrast, in benzene and graphite, a planar structure follows from an $sp^2$ hybrid state for the carbon atoms.

σ-bonds are strongly directional because of the directionality of the parent atomic/hybrid states. Electrons remain close to the line joining atom centres.

π-bonds are less directional and often result when atomic p states overlap off the line between atoms. This can give rise to delocalized electrons, for example in graphite.

Silicon also forms covalent bonds with electrons in hybrid orbits. Silicon in silica, like carbon in methane, bonds tetrahedrally. Silica mixed with ionic oxides is the basis of the various different types of glass. Silicate ions, which are themselves covalently bonded, occur naturally in minerals such as asbestos, kaolin and talc.

Do not lose sight of the scale of our current discussion — we are still considering small structures of atoms. Assemblages of atoms can then combine to form macroscopic structures. The next section begins to consider how.

# 3.6 Distributed charge

In discussing covalency I have described how electrons orbiting two nuclei reduce their energy, but I have said little about the distribution of these electrons relative to the two nuclei. When the bond is between identical atoms the shared electrons should spend equal time at each atom, but with different atoms maybe it's not such even handed sharing. The extreme of unequal sharing is when electrons leave one atom completely and take up states in another — ionic bonding. Where elements like oxygen and fluorine occur in covalent molecules, sharing is unequal, but short of actual transfer of electrons. This greed for electrons is called **electronegativity**. High electronegativity is a characteristic of elements in the top right corner of the periodic table, excluding noble gases. The bond acquires some ionic character. This produces 'polarized' molecules with positive and negative electrically charged regions. Such 'electric dipoles' can attract one another by getting positive and negative parts of adjacent molecules together and so can provide secondary bonding between groups of atoms.

## 3.6.1 Four simple molecules

To begin to grasp the origins of polarity I want you to consider the simplest molecules made when hydrogen combines with each of the elements carbon, nitrogen, oxygen and fluorine, atomic numbers 6 to 9.

Which electron states are being filled in this sequence? The 2p states are filling, approaching the full $n = 2$ shell at neon. Now work through the following.

(a) If the 2s and 2p states all mix to create four sp$^3$ hybrid states, what shape is the heavier atom anticipating for the molecule?

(b) Sketch the C, N, O and F atoms as cores plus lobes, marking outer electrons as ↑ ↓ in the lobes.

(c) Add the hydrogen atoms (as proton plus electron) and identify regions of high negative charge concentration and hence potential dipoles.

(d) Electrons which bond H atoms will orbit two nuclei. What is the effect of the increasing nuclear charge C $\rightarrow$ F?

The answers are:

(a) Tetrahedral geometry is anticipated.

(b) See Figure 3.31. The molecules they form are methane $CH_4$, ammonia $NH_3$, water $H_2O$ and hydrogen fluoride HF. There are strong similarities between them. In each molecule the heavier atom has the same core of $1s^2$ electrons and gets all its $n = 2$ states filled, making a total of ten electrons in every molecule. In each case hydrogen is making covalent bonds with sp hybrid states of the other species. The differences come from the increasing population of electrons in the sequence of elements which means that fewer hydrogens are needed (as the formulae show) to import extra electrons to fill their $n = 2$ states. In each case except $CH_4$ some of the hybrid orbit lobes contain spin pairs of electrons without an accompanying proton (hydrogen nucleus). Immediately you can see that the symmetrical methane molecule cannot be polar but that the other three must be positive on the side where the protons are embedded in the electron cloud, negative where the so-called lone (electron) pairs are. Only methane can be exactly tetrahedral with pure $sp^3$ hybrid states.

(c) See Figure 3.31.

(d) Less obvious is the effect of increasing nuclear charge on the orbitals which do contain a proton. An electron penetrating the core of oxygen, which has a nuclear charge of $+8$, is going to be more attracted than when penetrating nitrogen, which has a nuclear charge of $+7$. Increasingly therefore, the electron distribution in these bonds is going to close in upon the larger atom leaving the hydrogen nucleus starved of electrons and more exposed. This will further enhance the polarization of the molecule. In fact in HF the fluorine holds four of its electrons in unaffected p orbits so the molecule has only one lone pair, not three (see bottom of Figure 3.31).

The more electronegative elements induce nonuniform charge distribution wherever they are present by drawing electrons towards themselves. By contrast, metals in organic compounds can inject electrons and become sites of positive charge. An example of technical properties resulting from molecular charge distribution is the liquid crystal display, described later in this chapter.

More easily explained as a consistent phenomenon, and more directly important as a technically applied strategy, is the stronger intermolecular binding offered by hydrogen bonding.

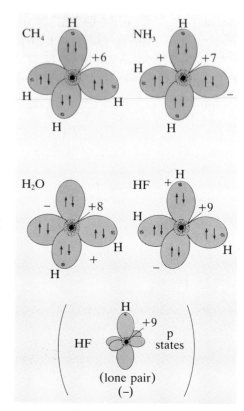

Figure 3.31 Molecule shapes and charge distributions

## 3.6.2 Hydrogen bonds

The covalent bonds made by hydrogen in the molecules just described are not 'hydrogen bonds' in the usual sense of the term. The hydrogen bond is an extra feature by which hydrogen forms a bridge between two molecules.

To make a hydrogen bond two conditions must be met:
• There must be hydrogen atoms which, being covalently bonded to an electronegative element, are significantly starved of electrons.
• There must be 'lone pair' electrons in a hybridized orbital on an electronegative atom in another molecule. Then when the hydrogen atom meets the lone pair an electrostatic bridge can form. Having no core electrons around it, the hydrogen nucleus is able to get in close enough to experience a significant attraction. The resulting bond has about a tenth of the strength of a covalent bond.

Figure 3.32 shows hydrogen bonds forming between water molecules to form ice. We have already noted in Figure 3.31 that the water molecule meets both these conditions and is polarized. Significant amounts of hydrogen bonding persist in the liquid state in which there are clusters of several molecules. Water has extraordinary properties as a consequence of its hydrogen bonding capabilities. Its melting and boiling temperatures are much higher than substances of comparable molecular mass, telling straight away of some extra sticking. It is almost a universal solvent, in particular revealing the acid or alkaline character of some solutes. And to the comfort of fish, if not of shipping, its solid floats on its liquid, a rare property implying a very open crystal structure.

In fact ice has several crystal structures, all of which are open lattices which make it less dense than liquid water. When the lattice has been rattled to pieces, the molecules pack tighter and in fact the maximum density is at 4 °C.

Water molecules don't just hydrogen bond to other water molecules. They can take part in hydrogen bonds with other molecules which offer hydrogen atoms starved of electrons to electron-rich electronegative atoms (so dragging them into solution). See ▼Detergents▲

Wood and allied topics   All the cell walls of plants are made of the natural polymer cellulose. Thus this is a major constituent of wood, the other ingredient being a substance known as lignin. Cellulose is a beautiful compound; exact, regular strings of glucose rings lie closely bound to each other by hydrogen bonds, Figure 3.34. The microstructure which packs these 'ropes' into load bearing units is also

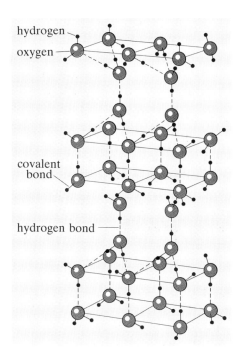

Figure 3.32 Hydrogen bonding in ice

## ▼Detergents▲

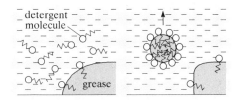

Figure 3.33 Detergent molecules in action

Detergents remove oil and grease (hydrocarbons) through the action of polar molecules. Soap is a detergent based on animal and vegetable products although many detergents are soapless being based on petrochemicals. In either case, the action is the same. A detergent molecule bonds at one end to water and at the other to the oil or grease. To do this the water bonding end must be something like oxygen, hoarding electrons (to promote hydrogen bonds), while the other end must be distinctly hydrocarbon (to promote bonding into the oil or grease). Figure 3.33 illustrates how, together with agitation, soap molecules break up grease into small water-borne droplets which can then be rinsed away.

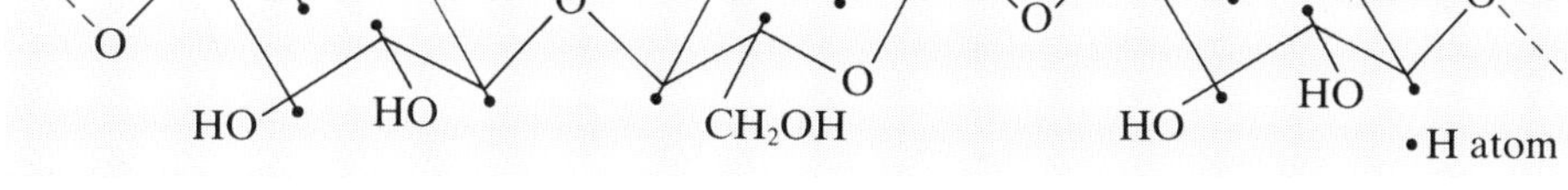

Figure 3.34 Cellulose

127

fascinating to see (Figure 2.23 in Chapter 2). So closely packed is this structure that very few digestive enzymes can get to work on it. However grazing animals have suitable digestive systems for which they have long been exploited. ▼Conservation of wood▲describes an exercise in applied hydrogen bonding, in marine archaeology.

# ▼Conservation of wood▲

When wooden objects were recovered from the Tudor warship *Mary Rose* the wood was treated to preserve it, using the following process. The first stage is to wash away salts from the wood by thorough treatment with pure water. Note that the water is hydrogen bonded to the cellulose molecules. Shrinkage cracks are a hazard if the wood simply dries out so water is removed by a slow substitution process. The wood is immersed in a bath of a polyethylene glycol (PEG, Figure 3.35) which is water soluble. The PEG solution is strengthened slowly over a period of a few weeks and gradually PEG replaces the water throughout the piece.

Why? Well hydrogen bonding makes PEG miscible with water. Diffusion, 'rattle' provoking muddle, is therefore able to carry PEG into the wood and adding fresh PEG maintains the concentration gradient so that eventually much of the water is extracted.

The next stage is to freeze the soaked wood and then freeze dry to remove the remaining water. The PEG hydrogen bonds to the cellulose and to itself and, being less volatile, remains behind. The wood is finally given a further surface coating of PEG. The result is a hard stable block of the same shape as the original waterlogged material.

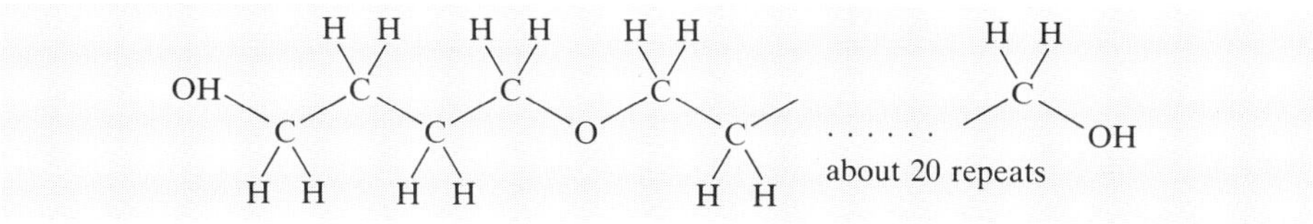

Figure 3.35 Polyethylene glycol

## 3.6.3 The ubiquitous attraction of van der Waals

The distribution of charge between atoms will vary with time. As a result there is an attractive force, though only a very weak one, which is always present. It is known as the van der Waals force (after the man who postulated its existence to account for gases like argon eventually liquefying at low temperatures — see ▼Van der Waals forces▲. It is sometimes the only cohesive force present and then it becomes technically important. For example it is the only attractive force between hydrocarbon molecules, so substances such as petrol (mainly octane, $C_8H_{18}$ which has molecular mass of 114) with quite heavy molecules are mobile, easily evaporated liquids at room temperature. Likewise the hydrocarbon polymers polyethylene, polypropylene and polybutadiene are more readily softened than polymers whose molecules cohere by stronger forces. More about this in Chapter 4.

# ▼Van der Waals forces▲

The ideal gas equation relates pressure $p$, volume $v$ and temperature $T$ of $n$ moles of a gas through

$$pv = nRT$$

where $R = 8.314\,\mathrm{J\,mol^{-1}\,K^{-1}}$ is the universal gas constant. Real gases only behave according to this equation at very low densities. To account for deviations from the ideal behaviour, the Dutch physicist, Diderik van der Waals proposed in 1873 a modified equation of state

$$\left(p + \frac{a}{v^2}\right)(v - b) = nRT$$

in which the term $a/v^2$ takes into account the attractive forces between the gaseous atoms or molecules, whilst $b$ represents the effect of their finite volumes.

Other than specify that the forces between the atoms or molecules should be attractive, van der Waals made no comment as to their nature. It wasn't until the 1930s, with the aid of quantum mechanics, that a fuller understanding of the different contributions to the van der Waals forces became possible.

Superficially one might say that two electrically neutral atoms (or nonpolar molecules) at a distance would exert no electrostatic force on one another. But the moving electrons vary the distribution of charge instant by instant. So from a distance an atom doesn't look like a neutral blob but a flickering dipole. Now distant dipoles do interact (think of magnets) and the electron motions can get into a synchronous pattern which always produces attraction. At a certain range it manages to balance the repulsion effects.

## Summary

Covalent bonds acquire an ionic character where electrons are shared unequally between atoms. The molecules formed are polar.

Electronegative elements tend to hold onto more of the electronic charge and become the negative part of the molecule. Electropositive elements like metals may become sites of positive charge.

Hydrogen bonds form between molecules in which the small hydrogen atom is in a polar covalent bond with a strongly electronegative element. Hydrogen bonding accounts for many of the properties of, for example, water and wood.

Van der Waals forces are a much weaker (but always present) attraction between atoms, caused by transient charge distributions. This accounts for the weak cohesion in volatile liquid hydrocarbons and some polymeric solids.

Now we proceed to consider large molecules which are internally bound by primary covalent forces but which stick to each other either through secondary (distributed charge) effects such as we have just seen, or else through other primary bonds.

# 3.7 Polymers: structure and properties

## 3.7.1 Crystal and amorphous mixtures

'Polymer', as you know means many pieces, referring to the fact that their molecules are formed by joining together hundreds or thousands of smaller molecules (monomers). In the simplest kinds, the pieces are all the same. But diverse properties can be obtained by complex combinations, for example proteins are all different organizations of up to 21 different monomers.

In this section I will discuss a few organic polymers, but you should realize that thermoplastics, thermosetting resins, elastomers, silicones, inorganic glass, proteins and nucleic acids are all polymeric substances.

The evidence of ▼X-ray diffraction▲ suggests that many polymeric materials can be part crystalline and part amorphous. We have already looked briefly at silica glass, which can be thought of as a fully

# ▼X-ray diffraction▲

So how do we know so much about crystals which involve the packing of atoms, especially since the atoms are only about $10^{-10}$ m in diameter? Well, 'Transmission electron microscopy' in Chapter 2 just told you that electron beams are diffracted by crystals and showed exploitation of this fact in viewing dislocations with transmission microscopy. Here we look at how the diffraction of waves by crystals occurs specifically with regard to X-rays.

Diffraction is a wave phenomenon. Sound waves diffract round corners of buildings. Water waves diffract round objects in their path which are comparable in size with the spacing between wave crests (wavelengths). Light waves diffract through slits a few wavelengths or so in width. What wavelength will be required to get diffraction from objects (planes of atoms within crystals) measuring about $10^{-10}$ m? $10^{-10}$ of course! So-called '10 keV' X-rays (that is, X-rays formed by bombarding a metal with electrons having 10 keV of kinetic energy) have just the right wavelength for this.

Let's look at diffraction in a two dimensional crystal (an array of dots) and see what happens to X-rays as they are 'scattered' from atoms lying in the planes shown in Figure 3.36.

A great simplification of this problem proposed by Bragg (senior) was to regard the rows of atoms with spacing $a$ as mirrors which reflect the X-rays. In a real crystal these rows would be sheets, or planes of atoms. Imagine that the rows of atoms were in fact sheets of glass stacked up on top of each other. Each piece of glass would partially reflect and partially transmit the light falling on it.

Bragg then analysed this reflection by considering the scattering of X-rays from two adjacent scattering centres (atoms), rather like the interference problem of 'Young's slits', which you may have come across. If the two rays, PP' and QQ' are to reinforce each other then the path difference between then must be equal to the wavelength $\lambda$ (or some multiple of $\lambda$). From Figure 3.36 it is quite easy to calculate the path difference. The waves are clearly in phase across the line AB. The path difference between the two rays is therefore the distance BD + DC. From the geometry of the figure

$$BD = DC = a \sin \theta$$

Therefore

$$BD + DC = 2a \sin \theta$$

For reinforcement,

$$2a \sin \theta = n\lambda$$

where $n$ is a whole number.

This is Bragg's law and is a fundamental law of crystallography.

It must be emphasized that Bragg 'reflection' is not the same as the normal reflection of light by a mirror. The planes of atoms will only 'reflect' the incoming

X-rays when their orientation is such that $\sin \theta = n\lambda/2a$. This is a stringent condition. The X-ray wavelength is fixed, so is the spacing of the crystal planes so there are only a few orientations which give Bragg reflection. So-called (Bragg) diffraction patterns therefore can be analysed to reveal details of plane spacing and hence crystal structure. Figure 3.37 shows the arrangement for recording X-ray diffraction patterns. The picture on the photographic plate wrapped around the sample is characteristic of the atomic arrangements in the solid. Figure 3.37 also shows three possibilities—a diffuse halo indicating irregular atom spacing and an amorphous solid, regular dots indicating regular atom spacing and a single crystal, regular rings indicating regular atom spacing and a polycrystalline solid (overlapping dots due to random crystal orientations form rings).

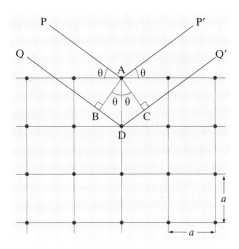

Figure 3.36 X-ray scattering

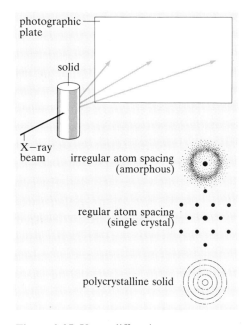

Figure 3.37 X-ray diffraction patterns

amorphous polymer, and at diamond, which in turn can be visualized as an extreme example of a fully crystalline polymer. The respective monomers are $SiO_2$ and C.

To a considerable extent polymer molecules can be regarded as very long thin structures much longer and thinner (in proportion) even than proper Italian spaghetti. How such things pack together will depend on several factors:

- long or short?
- stiff or bendy?
- smooth or lumpy?

- regular or random?
- single or branched?
- slippery or sticky?

You have already seen long, regular and sticky cellulose molecules coiled neatly into rope-like conformations and you may already have come across short, stiff liquid crystal molecules neatly aligned (see ▼**Liquid crystal displays**▲). Now those features can be designed into molecules so that the structures and consequent properties are controlled.

The spaghetti analogy is worth chasing a bit further: if the molecules are long, bendy and smooth like boiled and buttered spaghetti, you can expect a large proportion of them to lie tidily alongside each other and an individual molecule can be so aligned if folded up. Where there are loops, the pattern is spoilt and there can be extensive tangled regions. Within the analogy it is also notable that lifting spaghetti on a fork enhances alignment: so does stretching polymers. Lastly think of the properties of uncooked spaghetti; straight stiff rods can be placed parallel but if they are just dumped a loose criss-cross muddle is inevitable — an amorphous structure. So even without branches or knobs we can envisage distinct ways for polymer molecules to be arrayed in patterns (crystalline), muddles (amorphous) or mixtures of

# ▼Liquid crystal displays (LCD)▲

Liquid crystal molecules of the type

R—O—⬡—⬡—CN

where R is a short hydrocarbon chain, are exploited for the purpose of information display through four effects (two of which are directly attributable to distributed charge).

(a) Weak secondary (van der Waals) bonds can form between molecules lying alongside each other so that they are 'naturally' held almost parallel in large numbers, hence the term 'crystal'.

(b) In a steady electric field, some electrons move along the molecule ($\pi$ bonds and rings!). An electric dipole is thus induced as molecules are positive at one end (electrons lacking) and negative at the other (slight excess of electrons). The induced dipoles then tend to align with the electric field — the 'liquid' nature allows this to happen.

(c) The very high frequency electric fields of light waves pass through the liquid crystal only if the direction of polarization lies parallel to the direction of the molecule axis.

(d) At solid interfaces molecules can be encouraged to adopt a particular orientation, for example molecules preferentially lie parallel to grooves or scratches on a solid surface.

Figure 3.38 shows one type of LCD cell. Can you see how it works?

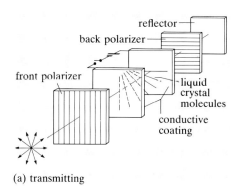

(a) transmitting

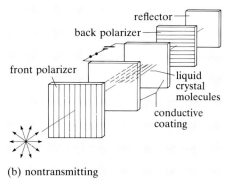

(b) nontransmitting

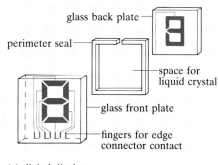

(c) digital display

Figure 3.38 Liquid crystal cell and display

the two. Evidently branches and knobs will generally render pattern-making more difficult, favouring the amorphous condition, while the provision of adhesion between molecules is more a means of preserving the status quo, inhibiting reorganization rather than causing either condition.

## 3.7.2 Long, regular, bendy, single, smooth and slippery

An important application of polyethylene (PE) is as the cup part of hip joint replacements (Figure 3.39). For this purpose it is made to follow the cooked (very long strand) spaghetti analogy rather well.

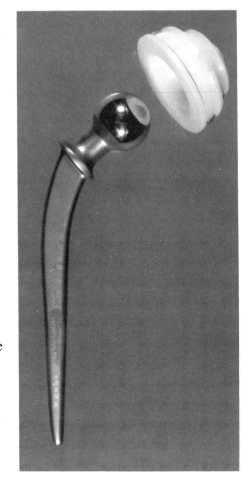

● Long and regular: the basic —CH$_2$— units are almost endlessly repeated; with over a thousand of them per molecule, the molecular mass is 'ultrahigh'.

● Bendy: The C atom is in the sp$^3$ hybrid state so there is freedom for the chain to twist about any C—C bond. A slight twist of successive bonds in the same direction suffices to bend the molecule back on itself.

● Single and smooth: the production process (see 'Producing polyethylene' in Chapter 1) ensures that the chain is unbranched and the tiny H nucleus beds almost unnoticeably into the hybrid lobe of the carbon electron cloud so there are no knobs. The unbranched smooth chains can pack very neatly giving a 'high density' material (HDPE).

● Slippery: C—H covalent bonds are not polar; intermolecular cohesion is by the weak van der Waals force.

Figure 3.39 Hip joint replacement

In consequence most of the material freezes into 'spherulites' giving the microstructure shown in Figure 2.29(a). Spherulites are made up of radially grown ribbons of crystalline polymer. Their formation will be explained in more detail in Chapter 5. The high degree of crystallinity makes for the relatively high density of this PE (960 kg m$^{-3}$) as compared with more amorphous grades (920 kg m$^{-3}$). This ultrahigh molecular mass, high density PE is harder and much more wear resistant than its amorphous cousins. That's why it is used for hip joints. How would you form the cup? In practice it is machined from bulk material or else sintered from powder.

## 3.7.3 With knobs on

There are many artificial polymer species which we can regard as derived from the 'ideal' of HDPE (though historically HDPE is a late-comer). One simple manipulation is to add knobs to the molecule by polymerizing, for example, CH$_2$=CHX where X is anything bigger than a single hydrogen atom. Some examples which just make the polymer molecule 'lumpy' without also making them sticky are drawn in Figure 3.40. The abbreviations used are explained in Appendix 2 at the end of the book. You'll see I've included low density polyethylene, where the 'knobs' are the branches in the chain induced by the method of polymerization.

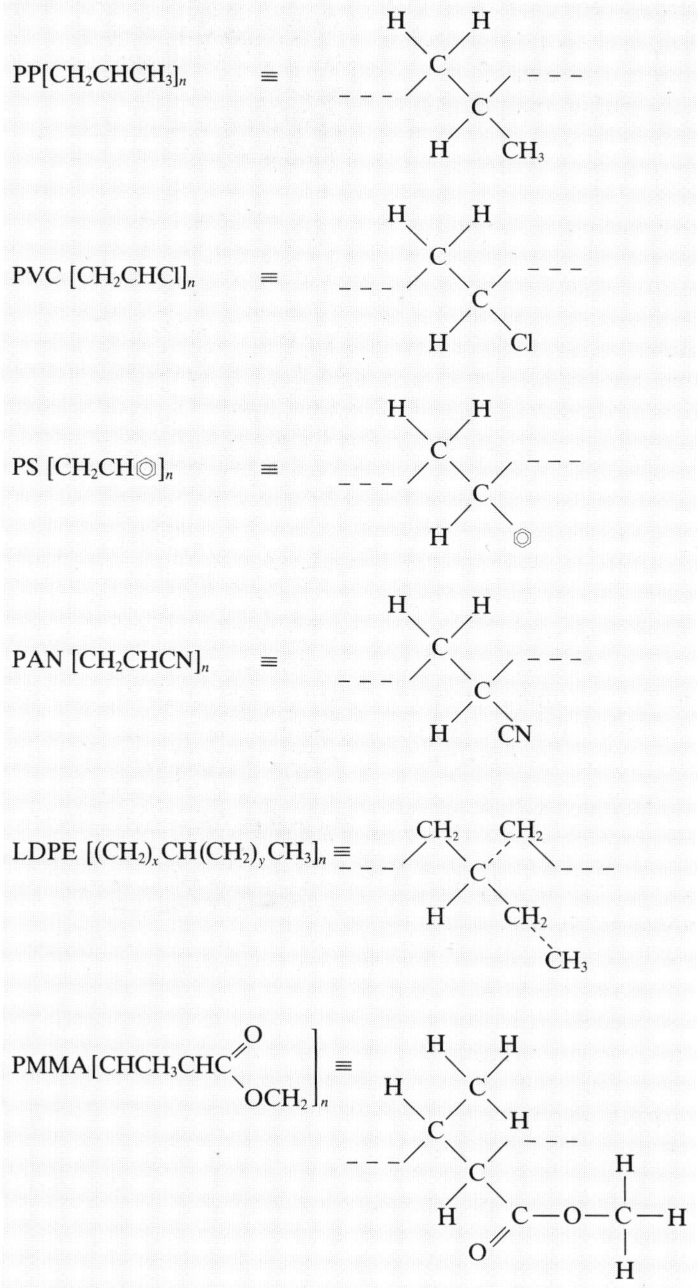

PP[CH₂CHCH₃]ₙ ≡

$PP[CH_2CHCH_3]_n \equiv$

$PVC[CH_2CHCl]_n \equiv$

$PS[CH_2CH\bigcirc]_n \equiv$

$PAN[CH_2CHCN]_n \equiv$

$LDPE[(CH_2)_xCH(CH_2)_yCH_3]_n \equiv$

$PMMA\left[CHCH_3CHC\begin{smallmatrix}O\\OCH_2\end{smallmatrix}\right]_n \equiv$

Figure 3.40 Lumpy and branched polymers

The consequence of putting knobs on the molecules is that it may make it more difficult for them to get into snug alignment to form crystals. Indeed if it turns out that the lumps are extremely bulky or are randomly sited on each site of the —C—C—C— backbone, then the solid is amorphous. Transparent polymeric materials are chiefly amorphous, any crystalline regions being so small that they do not scatter light. Most covalent molecules absorb light only outside the visible part of the spectrum. For example PMMA (traded as Perspex or Lucite) is a high clarity transparent material.

## 3.7.4 Make it stick

Arranging for stronger cohesion between polymer chains may strengthen the crystalline or amorphous state to the benefit of some service properties. For example, in nylon, a crystal structure is formed between parallel chains with regular —NH—CO— (amide) groups. See Figure 3.41(a).

EXERCISE 3.7
(a) What is the source of intermolecular cohesion in nylon?
(b) Would you expect it to melt at a lower or higher temperature than polyethylene?

The extra secondary bonding in nylon gives a high melting temperature and leads to spherulites with a tough combination of crystals and an amorphous phase. One feature of hydrogen linked polymers like nylon is susceptibility to water which 'H-bonds its way in', weakening the crystals and swelling the structure.

A stronger cross-link is formed when one or more covalently bonded atoms join two chains. Vulcanized rubbers use sulphur atom cross-links to form strong covalent bonds between chains at intervals. See Figure 3.41(b). One cross-link every 100 or so carbons stabilizes the natural latex liquid into rubber suitable for elastic bands. A higher density of cross-linking gives a much stiffer material.

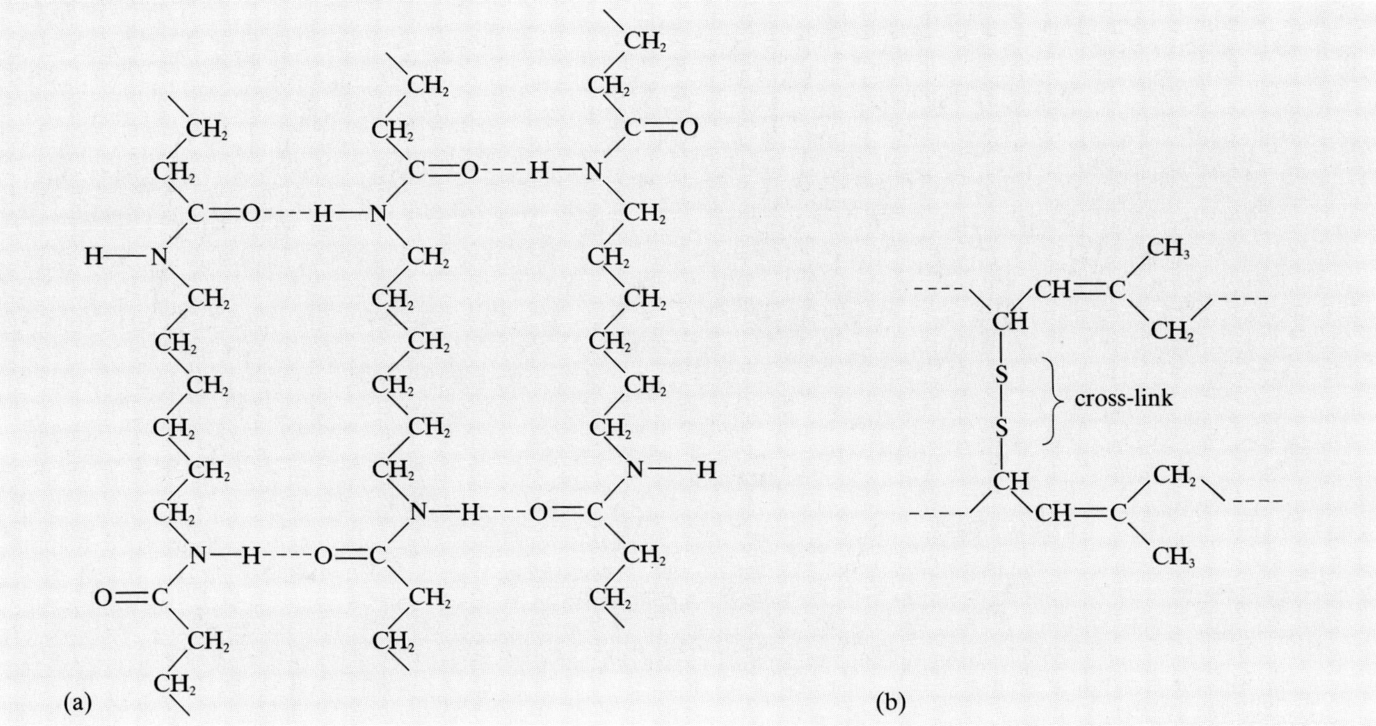

Figure 3.41 Cross-linking in (a) nylon and (b) rubber.

## 3.7.5 Stiffer chains

Polyethylene terephthalate (PET), marketed under several names such as Dacron and Terylene, has the formula

$$[-CH_2-CH_2-O-CO-\langle\bigcirc\rangle-CO-O-]_n$$

It can be formed into a moderate strength fibre and is also blow-mouldable into bottles. What is the function of the six-carbon (benzene) rings? Well, they provide more bonds per unit length of chain than, say, nylon and that immediately means an inherently stronger polymer. Furthermore the ring structures don't twist like the $sp^3$ carbon–carbon bonds so the chains are stiffer. Can you now see why I said that liquid crystal molecules are stiff?

SAQ 3.9   (Objectives 3.2 and 3.7)
(a) Figure 3.42 shows the atomic structure of phenol formaldehyde, which can be moulded as a powder before final polymerization. Will it crystallize? Will it soften on heating?
(b) Say which of the following are thermoplastics — PMMA, LDPE, nylon, PET, vulcanized rubber. Give reasons for your answer.

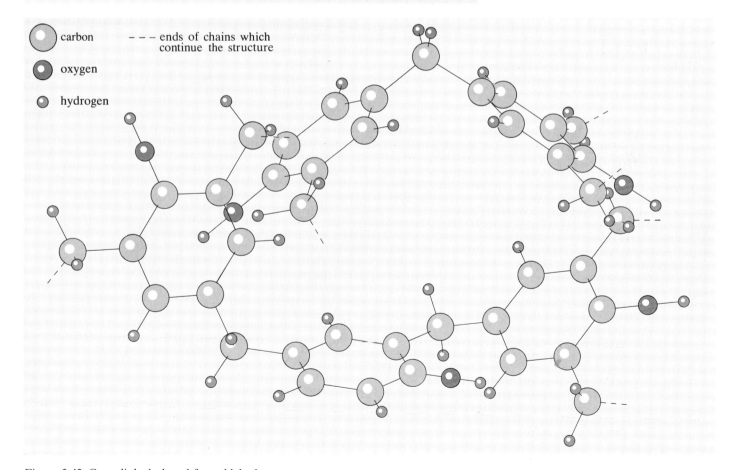

carbon
oxygen
hydrogen
- - - ends of chains which continue the structure

Figure 3.42 Cross-linked phenol formaldehyde

## Summary

Polymers are part crystalline, part amorphous. The more lumpy and branched the polymer molecules, the less crystalline and less dense the material. Many long chain polymers crystallize with a spherulitic microstructure — radial crystalline ribbons separated by amorphous material.

Polymer chains can be 'designed' to give particular properties. The length of the chain, the amount of chain branching, and the degree of cross-linking between chains will affect the density, strength, stiffness and even the optical properties of the material. Polyethylene, for example, can be made with branched or smooth molecules, producing LDPE and HDPE respectively. Transparent polymers such as PMMA are largely amorphous.

The bonds between carbon atoms within a polymer chain are strong covalent bonds. But the bonds causing alignment of the molecules into regular crystalline structures are often weaker secondary bonds. In PE they are due to van der Waals forces. In nylon there is cross-linking through hydrogen bonds, giving higher melting temperatures. In vulcanized rubber, however, the chains are cross-linked by primary covalent bonds provided by sulphur atoms; the more cross-linking, the stiffer the material.

Benzene rings in a polymer chain result in a stiffer (less twistable) molecule. Also, some of the special properties of liquid crystal molecules are a result of the charge distribution in benzene rings and $\pi$ bonds.

# 3.8 Metals

## 3.8.1 Metal crystals

X-ray diffraction reveals metal crystals to be particularly simple structures. Many of them are just the closest packed arrangements which we can make from packing identical spheres together. This means that each atom has twelve other atoms touching it. Several other metals crystallize with each atom surrounded by eight nearest neighbours. Contrast this with the structure of a covalent crystal like diamond, where bond directions dictate a low coordination number (four nearest neighbours). This suggests that metal atoms bond without any directional inhibitions.

Let us see how a close-packed crystal is made. We start with a close-packed plane of spheres and then stack several planes together, as shown in Figure 3.43(a). Any sphere in a close-packed plane is surrounded by six others — there are six hollows into which a sphere of the next layer can nestle. But obviously not all six sites can be used simultaneously. They are too close together. Figure 3.43(b) shows that two alternate sets of nests can hold the next layer of close-packed spheres using hollows 2, 4 and 6, or 1, 3 and 5. (Demonstrate it for yourself using the circles in Figure 3.43(a) and some good tracing

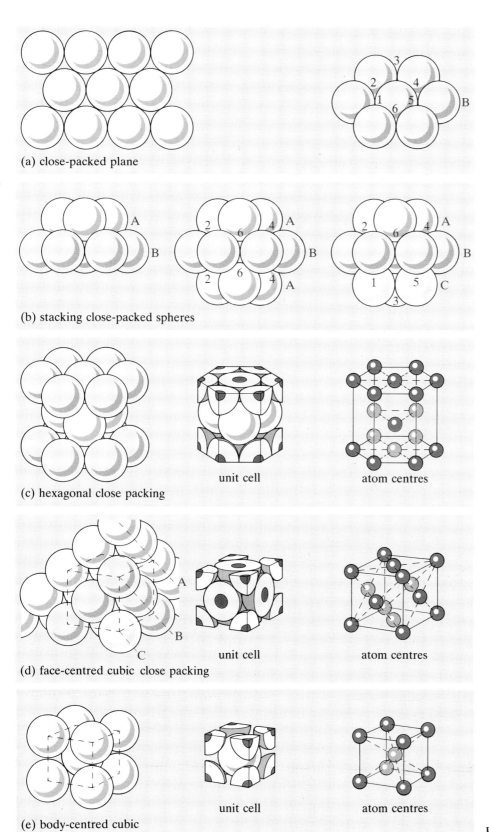

(a) close-packed plane

(b) stacking close-packed spheres

(c) hexagonal close packing    unit cell    atom centres

(d) face-centred cubic close packing    unit cell    atom centres

(e) body-centred cubic    unit cell    atom centres

Figure 3.43  Metal crystal structures

paper.) Where three spheres have been put on the under side of the original seven, the number touching the central sphere is $6 + 3 + 3 = 12$ and there's no way of cramming any more in. We have already seen (in Section 3.4) that in ionic crystals twelve spheres coordinated as a cage which could exactly accommodate one further sphere with the same radius.

In that case we were concerned with cages of negative ions around positive ions. The only difference here is that all the spheres are identical metal atoms.

Notice that Figure 3.43(b) shows two different stacking options. In one case the same set of nests has been chosen on both sides of the original plane. In the other case, the alternative set of nests has been chosen. That subtle difference has a profound effect on the crystal structure and ultimately on the properties of the metal. Let's label the layers A, B or C according to the positions of the spheres in a plan view. If I label my original plane as type B and the one on top as type A, then in the under layer is either another type A, or it's the third type, C. The only rule for stacking planes is that no adjacent planes can be identical. The planes have to 'nestle' not 'perch' on top of each other in order to be close-packed. So in principle, there is an infinite number of close packed crystal structures distinguished by their sequence of planes: ABABABAB, ABCABCABC, ABACABAC and ABABCABABC are a few of the simpler ones. Fortunately metals only adopt the first two of these and the occasional departure from this sequence is known as a 'stacking fault'.

In Figure 3.43(c) and (d), judiciously chosen numbers of close-packed atoms are shown at carefully chosen angles for each of the two simple systems. Carefully chosen cuts reveal building blocks (structure cells) of characteristic shape with which space can be filled. Not surprisingly the two structures are named **Hexagonal close-packed** (HCP) and **cubic close-packed** (CCP). More often we call CCP **face-centred cubic** (FCC). The FCC cell resembles patterns seen in some ionic crystals. For example, in MgO the oxygen ions form an array like the FCC structure cell.

Examples of HCP metals are magnesium, scandium, titanium and zinc. Examples of FCC (or CCP) metals are aluminium, calcium, nickel and copper. See ▼ Voids — the spaces in between ▲

SAQ 3.10  (Objective 3.9)
(a) What sort of experiments can be used to determine the separation between planes of atoms?
(b) It is found that the atoms of solid copper are arranged in a face-centred cubic pattern (Figure 3.43(d)) with cube edge 0.36 nm.
 (i) Estimate the diameter of a single atom in the solid.
(ii) Estimate the density of the solid (the mass of a copper atom is $1.1 \times 10^{-25}$ kg).

# ▼Voids — the spaces in between▲

With ionic structures we began by considering cases of negative ions filling space with positive ions sitting inside. Were any of these negative ion structures close packed? Yes, almost — the oxygen ions in MgO nearly touched and there, we saw the oxygen ion centres as defining octahedra and tetrahedra. Look at Figure 3.17. Can you recognize a face centred cubic array of negative ions? In MgO, magnesium ions sit in all the 'octahedral' spaces of the oxygen ions' FCC structure. The tetrahedral spaces are empty — voids.

You should therefore appreciate that FCC metal structures (with one metal atom where each oxygen ion sits in MgO) have voids, octrahedral and tetrahedral, where small impurity atoms can be accommodated. Such atoms are not part of the regular lattice and are therefore said to be **interstitial**.

The hexagonal close packed structure has similar voids. There is another common metallic structure, body centred cubic (BCC), but this is not close-packed so its void (also octahedral and tetrahedral,

being between 6 and 4 atoms respectively) do not lend themselves to the same easy space filling.

At room temperature pure iron has a BCC structure, but at 910 °C this spontaneously switches to FCC. The voids in FCC structures are a little larger but less numerous than those in BCC ones. It turns out that carbon atoms are just small enough for some to be able to squeeze into the FCC iron structure by pushing iron atoms slightly further apart, so that at 910 °C, about six carbon atoms can be accommodated (interstitially not in the lattice) among about one hundred iron atoms. In contrast, BCC iron can hold only about one hundred times less carbon interstitially. You will see later how the principles underlying these differences can be used to great effect in manipulating some of the properties of steels.

Worked example: Calculate the size of the octahedral void in FCC iron and confirm that a carbon atom must slightly distort the structure if it is to occupy such a site.

The diameters of iron and carbon are 0.254 nm and 0.154 nm respectively.

Answer: In an FCC iron cell, spheres touch along the face diagonals, as shown in Figure 3.43(d). If the cube edge is $a$ and the iron atom diameter is $d_{Fe}$, then the diagonal, which is two iron atom diameters long, is:

$$2d_{Fe} = \sqrt{2}d_{Fe}$$
$$a = \frac{2d_{Fe}}{\sqrt{2}} = \sqrt{2}d_{Fe}$$

The maximum diameter of an interstitial atom which can fit into the iron cell without distorting it is $d_i$ such that:

$$a = d_{Fe} + d_i = \sqrt{2}d_{Fe}$$
$$d_i = d_{Fe}(\sqrt{2} - 1)$$
$$= 0.254\,\text{nm} \times 0.414$$
$$= 0.105\,\text{nm}$$

So the carbon atom, with a diameter of 0.154 nm, is too big and will distort the structure.

## 3.8.2 About the metallic bond

What holds the spheres together? Metal atoms have s state electrons in their outer cloud. Is this common factor the key to metallic bonding? Let's start with some imagination. To make covalent bonds we saw that one way of reducing the electrical potential energy of the negative electrons was to put them nearer to more positive charge. For two atoms close together this was arranged by electrons going into orbits around both nuclei. And if going round two nuclei can be favourable, then orbiting more than two might also be so, though pushing 8 or 12 nuclei close together has an energy cost. Apparently reorganizing the electron clouds can decrease energy enough to outweigh the increase implied for such high coordination number crystals. But why such high numbers? The alkali metals provide a simple case where we can imagine why electron orbits need to overlap several atoms.

EXERCISE 3.8   Which group of the periodic table are the alkali metals? What is their common outer electron configuration?

Alkali metal atoms can form diatomic covalent molecules by overlapping their half-full s orbits. That's what hydrogen ($1s^1$) does to form stable $H_2$ molecules. In fact alkali metals crystallize as body-centred cubic (BCC) structures with eight nuclei as nearest neighbours and another six only 15% further away, as shown in Figure 3.43(e). This is not a close packed structure but it is much tighter than diamond.

So how do the electron orbits behave? One of our images of the s state (Figure 3.6) was of an orbit which penetrated from time to time close to the nucleus so that the time–space map was a sphere. The problem for an electron in such an orbit in trying to establish a low energy two-atom track is that it is only very rarely going to find the second atom! The chances of passing close to another nucleus improve if a lot of nuclei are in close range. A good mental picture for such a set-up would be of very transitory covalent bonds (see Figure 3.20) to one neighbour, then another, then another. What if more electrons per atom and more nuclei could be involved? The metals of second group of the periodic table have two s electrons and they do form close packed crystals (twelve nearest neighbours and six more 41% further off).

---

EXERCISE 3.9  Adjacent pairs of Group Ia and Group IIa metals with their melting temperatures are listed in Table 3.4. Does a stronger bond follow from the involvement of more electrons with more nuclei?

---

Table 3.4

| Element pairs | Melting temperature/K |
| --- | --- |
| Li:Be | 452:1550 |
| Na:Mg | 371:924 |
| K:Ca | 337:1120 |
| Rb:Sr | 312:1042 |
| Cs:Ba | 302:1000 |

As Figure 3.44 suggests, what we now have is a continuum of overlapping s electron zones throughout the crystal. The electrons don't 'know' to which atom they 'belong' so we might expect them to move through this continuum from atom to atom; such delocalization allows the possibility of electrical conduction — a conspicuous property of all metals. See ▼Electrical conduction▲.

But the metallic bond is more than just a transitory covalent bond. If the outer electrons are delocalized from their parent nuclei the crystal lattice is one of like positive ions. It is held together by the 'paste' of delocalized electrons. Although the ions repel each other the fluidity of the electronic glue allows considerable movement of ions without great energy cost. Ductility is another characteristic of metals.

A better image of the metallic bond comes from considering more generally the simultaneous interaction of many atoms with their outer electrons. From this we can develop a view of electrons in solids, whether in a metallic conductor or in a covalent insulator.

## ▼Electrical conduction▲

A metallic conductor is able to carry electric current because some electrons within it are 'free' to accept energy, being accelerated in the externally applied electric field which exists when a voltage source is connected to it. You will be aware that electric currents also heat the metal, so it must be that electrons are 'free' to lose energy as well, transferring their gained kinetic energy into heat energy within the metal. In a metal it is the delocalized electrons which carry the current.

In general, the conductivity of a substance is a measure of both its ability to provide charges to carry the current and its tendency to impede the passage of the current by extracting energy from it as heat. The conductivity depends on both the number and the mobility of the charge carriers. There will be further discussion of these points in the next chapter.

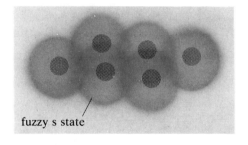
fuzzy s state

Figure 3.44 Overlapping s states

## 3.8.3 Electrons in solids

Using quantum mechanics, solid state physicists have given a much more powerful description of electrons associated with many atoms. This is the so-called 'band theory' which goes beyond metals to all solids and lets us model all properties determined by electrons. The most direct experimental evidence of the nature of the 'disturbed' electron states around densely-packed (solid) atoms comes from X-ray studies. See ▼X-rays and energy bands▲

The principle of the necessary quantum mechanical calculation is to find the electron energy states for one atom with other atoms set around it in the appropriate directions to make a chosen (crystal) structure. The problem is to imagine a way of representing the influence of the surroundings which is both accurate enough for the results to mean something and simple enough to allow us to do the sums. The common output of various methods of approximation is that with many closely spaced atoms the outermost electron states, (both the normally occupied ones and those which would be excited states for isolated atoms), lose

# ▼X-rays and energy bands▲

Figure 3.45 shows part of the X-ray spectra emitted by solid aluminium and aluminium vapour when bombarded by electrons of about 100 eV energy. In the solid spectrum, instead of the spectrum of sharp lines (specific values of energy or wavelength) which would be obtained from single atoms, there is now a band of radiation which is virtually continuous over a range of energy. The origin of the X-ray emission is as follows. First the brute force of the external electron bombardment knocks out an inner shell electron (that is one from the core). Then this 'hole' is filled in by an electron from a much higher energy level. The excess energy which the replacement electron must lose is radiated away as an X-ray of the required energy. The continua exist, it is argued, because there are many nearly equal starting energies for electrons falling to a vacated inner level, that is electrons fall from a band of closely spaced energy levels.

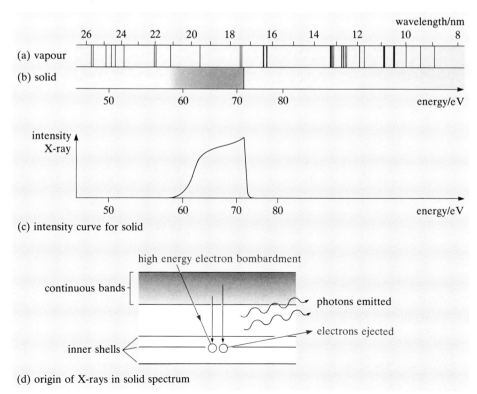

(a) vapour
(b) solid
(c) intensity curve for solid
(d) origin of X-rays in solid spectrum

Figure 3.45 X-ray emission spectra for aluminium

If we suppose that electrons from any one of these levels in the band can drop with equal probability into the hole in the core, then we see two important features of these bands:

(a) There appears to be a fairly sharp maximum energy, characteristic of each material.

(b) There are more levels at higher energies than at lower ones. In other words the 'density of states' (the number of energy levels in a given energy interval) increases with energy.

their atomic precision and merge into bands of very closely spaced energy levels. At some energies bands may overlap, at others there may be gaps between bands (no levels at all). The behaviour of solids depends on the band–gap structure and on the extent to which bands are filled. The core electrons remain bound in energy levels, much the same as in the isolated atoms. Figure 3.46 suggests how the spreading of individual atoms' outer electron energy levels depends on how close the atoms are set. Notice that the spreading provides some lower energy states and so offers the possibility of an overall reduction of energy (bonding) by filling the bands from the bottom up.

It is important to realize that just as with isolated atoms, these are quantum states controlled by a set of quantum numbers and we need a different state (at a different energy) for each electron. Although the outer energy levels of the solid are bunched together into bands there is a tiny energy spacing between states within a band which tells of the quantum distinction between them based on the dynamics of electrons within the whole crystal rather than within one atom. So how many electron states are there? Well, let's consider about a cubic micron $(10^{-18}\,\mathrm{m}^3)$ of a solid alkali metal such as sodium. There will be about $10^{10}$ atoms with overlapping s states merging into a band. Each s state on an isolated atom could contain two electrons, though for alkali metals it is only half full. So for our cubic micron the band must provide $2 \times 10^{10}$ very closely spaced levels — only half of them need to be filled though.

Also note in Figure 3.46 that some of the states which were excited states of the isolated atoms may now be accessible. This accounts for bonding in metals which, as atoms, have all their electrons spin paired. A magnesium atom for example has . . . $3s^2$ as its outer layer so an electron would have to be 'promoted' to higher energy states in order to bond, for example a 3s electron moving to a 3p state. But in our model of the solid some levels derived from the 3p states may have a lower energy than some which were derived from the 3s states, so of course they are occupied preferentially!

Conductors and insulators    This band picture can distinguish readily between conductors (metals) and insulators. First you might like to think of the energy states in a band like rooms in a lake-side hotel. Suppose you (an electron) have been allocated a cheap room at the back and you want to change it. You cannot just move into a 'room with breath-taking views' if the hotel is fully booked — these rooms are not available. Maybe the next hotel, further up the valley (the next band) has suitable rooms . . . but you probably haven't the energy to walk there just now.

Remember that electrical conduction requires the presence of charges which are free to move in response to electric fields, thereby gaining and

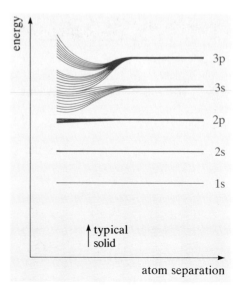

Figure 3.46 Energy levels and atomic separation

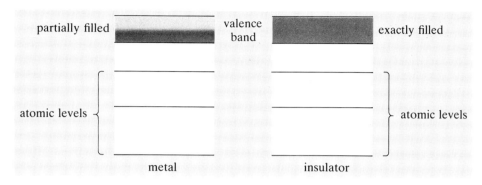

Figure 3.47 Metal and insulator energy bands

losing energy. Electrons in partially filled bands have many empty energy states into which they can hop as they gain and lose energy. In contrast, electrons in a full band have no capacity to contribute to currents unless they can get across the gap into the next unfilled band. It turns out that in insulators the energy gap is wide enough to prevent this. Figure 3.47 illustrates the two extremes. Let's look at some insulators.

Diamond and Silicon   How does the band theory account for the electronic properties of diamond and silicon? We've already described the crystal structure of diamond and silicon in terms of covalent bonds. The band theory provides a model which throws more light on how electrons perceive their crystal environment. Each atom in the solid holds its core tightly — $(1s)^2$ for carbon, $(1s)^2(2s)^2(2p)^6$ for silicon — but as the interatomic distance decreases, the outer states merge into crystal-wide bands of states with gaps on the energy ladder. Both carbon (as diamond) and silicon are insulators at low temperatures (no rattle) so it must be that the four outer electrons per atom exactly fill a crystal-wide band. With no available energy state to hop into, an electron cannot accept energy from an electric field. The full band is termed the **valence band** since it holds outer or valency electrons. The valence band states, when occupied by electrons, give lower overall energy than the corresponding single atom states would. That is why valence band states are responsible for bonding.

If an electron could be given a single lump of energy to kick it across the gap into the next band, it would find itself surrounded by a host of empty states. Thus it would now be able to accept energy from an electric field and participate in conduction, moving within what is termed the **conduction band**. In diamond the size of kick required is several electron volts. In silicon it is around one electron volt, and the rattle of room temperature is just enough (on average) to allow a very few valency electrons into the conduction band. This is why silicon is said to be an (intrinsic) semiconductor, at room temperature. See ▼**Intrinsic and extrinsic semiconductors**▲.

# ▼Intrinsic and extrinsic semiconductors▲

You can think of a conduction electron in silicon as having hopped out of a covalent bond, leaving a vacant state behind — a hole in the valence band. The vacant state could be filled by an electron hopping over from an adjacent covalent bond in which case the hole would move. Note that for a hole to move to the right, an electron must hop to the left. If an electric field causes electrons (with their negative charge) to hop to the left, then holes will thereby be moved to the right just as if they had positive charge. So it is very useful to think of holes as mobile positive charges. Intrinsic semiconductors have one hole in the valence band for each electron in the conduction band and both contribute to current flow (see Figure 3.48). Conduction requires the presence of full and empty electron states with little difference in energy. Electrons in the conduction band are surrounded by empty states and so can conduct and holes in a nearly full valence band are surrounded by occupied states and therefore can also conduct.

Remember that silicon bonds by offering four electrons to covalent partnerships. In an **extrinsic** semiconductor, electrons (or holes) are introduced when a few silicon atoms in the crystal lattice are replaced by elements with either five (or three) bonding electrons respectively. Consider the effect of replacing a silicon atom, $\ldots (3s)^2(3p)^2$, with a single phosphorus atom, $\ldots (3s)^2(3p)^3$. Four electrons settle into covalent bonds leaving a fifth fairly loosely attached. A little rattle kicks the fifth free, that is into the conduction band. Thus, phosphorus 'doped' silicon can conduct with these freed negative charges and is said to be an n-type semiconductor (n denotes a negative charge carrier). Notice that electrons introduced by the phosphorus leave not holes but ions ($P^+$) fixed in the lattice and not eligible for conduction. In a p-type semiconductor positive holes are the effective charge carriers

EXERCISE 3.10 Explain why silicon doped with boron, $(2s)^2(2p)^1$, is a 'p(ositive) type' semiconductor.

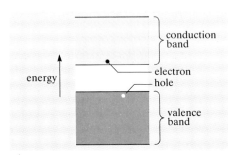

Figure 3.48 Semiconductor energy bands with hole

SAQ 3.11 (Objective 3.4)
Explain the following observations in terms of energy bands:
(a) At very low temperatures, pure silicon is an insulator.
(b) At room temperature, aluminium doped silicon is a p-type semiconductor.
(c) A resistor in part of an integrated electronic circuit can be fabricated by the controlled introduction of impurities such as phosphorus in a thin strip of silicon.
(d) Polymers such as PVC are good insulators.

## Summary

Because metal atoms can bond in any direction, metal crystal structures are often formed by close-packing of identical spheres.

Some metals, such as magnesium and zinc, form hexagonal close-packed crystal structures. Others, such as aluminium and copper, form face-centred cubic (cubic close-packed) structures. Another structure, formed by pure iron at room temperature, is body-centred cubic, which is not close-packed.

The atomic diameter and the packing structure determine the size of the voids left in the structure. These voids can be filled by smaller interstitial atoms, which can change the material properties. Interstitial atoms may distort the crystal lattice.

Metallic bonding involves the sharing of electrons among several atoms, in effect delocalizing them. This results in a more closely packed structure (higher coordination number) than covalent crystals, where the bonds are directional. Alkali metal atoms are held together by the sharing among nearest neighbours of an s orbit electron from each atom, to form a body-centred cubic crystal structure. Metals in the second group in the periodic table share two electrons from each atom, and form a more tightly packed structure.

The quantum mechanical 'band theory' of electron energy levels explains why metals are electrical conductors and other materials, like diamond and silicon, are insulators. In insulators, the outer electron states (those in the valence band) are filled and there are no free electrons to contribute to conduction.

In some insulators, relatively little extra energy is needed to knock a valence band electron up into a delocalized state in the 'conduction band'. Silicon, for example, has enough rattle to do this at room temperature and is known as an intrinsic semiconductor. In extrinsic semiconductors, impurity atoms provide extra electrons or unfilled states (holes) to contribute to conduction.

# 3.9 Transition metals

We conclude this chapter by a look at the very special elements known as transition metals. Electrons in these elements show very complicated behaviour.

## 3.9.1 What are they?

EXERCISE 3.11   Which are the first transition elements in the periodic table? Which electron shell is being filled?

The first group of transition elements is almost a catalogue of the technically important metals. Add some from the later transition blocks of the table, for example molybdenum, niobium and silver from the 4d block, tungsten, platinum and gold of the 5d series, and you'll appreciate that many of the elements with special properties are transition metals. They provide the major and minor ingredients of most of the structural alloys, all the magnetic elements, all the refractory metals and the catalysts for a host of chemical processes. (Which major structural metal is not included?) To comprehend this cornucopia as a whole you need a model for electron behaviour in these metals.

## 3.9.2 Electrons in solid transition metals

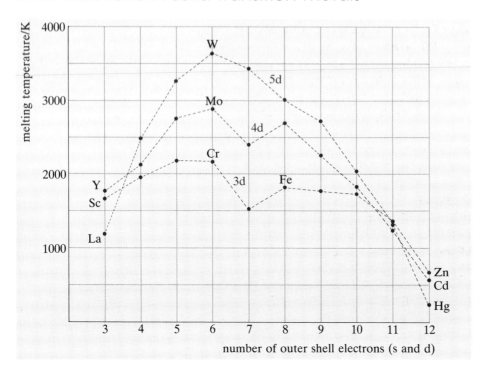

Figure 3.49 Transition element melting temperatures

The evidence of melting temperatures, as shown in Figure 3.49, indicates very strong bonding for transition metals, and that suggests that many electrons are involved. Up to now, the metallic bond has been associated chiefly with s electrons. If extra electrons are to be involved they must come from other atomic states. In isolated atoms the 4s and 3d levels for example are energetically distinct but in the solid metals they have broadened into overlapping bands and that distinction has gone. Even the atomic p states, which were excited states for single atoms, may be available without energy cost. I have already pointed out that X-ray spectra give us evidence for the number of states in a band being unevenly distributed. There are very few states at the bottom of the band and the number increases steadily the further one goes up the band. This distribution of electron states on the energy ladder is referred to as 'the density of states'. If we add to the band diagram the density of states, it must be something like the sketch of Figure 3.50. A very wide band, derived from 4s states and offering two levels per atom, straddles a less broadened band derived from the 3d states and providing ten levels per atom. All the atomic s and d electrons pile into these states working from the bottom up. Real calculations reveal much more complicated diagrams, so it is not worth trying to push the simple version quantitatively.

However, don't suppose that all the levels in this composite band with similar energy will have electrons in them that all behave the same; elephants and rifle bullets might have similar energies! So what can the former s and d electrons get up to? There would seem four possibilities:

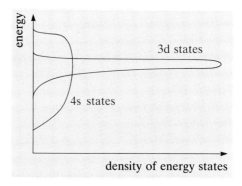

Figure 3.50 Density of states for transition elements

(a) Some electrons can be **completely delocalized**, offering metallic bonds as described in the previous section.

(b) Some electrons may be **tightly localized** onto a single atom and form **spin-pairs** within that environment; this would make them members of the core of electrons playing no part in interactions.

(c) Some electrons may be **tightly localized** onto a single atom and be **not spin-paired**; this gives rise to a magnetic moment (see Section 3.9.5).

(d) Some electrons may be **slightly delocalized** so they can form covalent bonds with neighbours.

The explanations of properties lie in how many electrons have taken up each role. It needs a new answer for every element and every alloy because both the total electrons per atom and the energy ladder they are climbing are different. Furthermore, electrons can divide their time between different roles; for example, the average number of electrons per iron atom in solid iron which contribute to its magnetism is 2.2! The role swapping that this implies is also evidenced by the high resistivity of most transition metals (several times that of silver and copper). When hopping between delocalized and localized states, an electron's ability to contribute to current carrying is somewhat compromised. We will now look at some of the exceptional properties of transition metals.

## 3.9.3 High and ultrahigh melting temperatures

The melting temperatures ($T_m$) of transition metals in Figure 3.49 show a similar pattern for the three blocks of elements. As each set of d states fills the $T_m$ rises to a peak and then falls away. The 4d elements have higher $T_m$ than their 3d analogues and the 5d metals are yet more refractory. These trends are explicable within our model. Moving through one period, say the third, the rise and fall of $T_m$ reflects the numbers of electrons involved in bonding. To begin with (Sc, Ti) there are not many electrons, so bonding is not excessively strong. To end with (Cu, Zn) higher nuclear charge holds more electrons locally onto one atom so fewer form covalent bonds. In the middle, however, there are several electrons and most are making covalent bonds to neighbours, so here the total bonding is strongest. The enhanced effect seen in the later periods suggests that the covalent bonds are stronger for the heavier elements. This is because an orbit embracing two nuclei will give a stronger bond the higher the nuclear charge of the cores penetrated — d orbits do penetrate a little. Thus 4d covalent bonds are a bit stronger than the 3d bonds. The 5d covalent bonds are especially strong because covalent orbitals penetrate inside the atomic 4f shell and see lots more nuclear charge; there are 14 electrons in the full f shell. So for these elements any covalency will considerably tighten up the bonding and the $T_m$ peak is both higher and broader.

EXERCISE 3.12   For the electric light bulb of Chapter 1 (Section 1.8.1) identify those metals chosen for high temperature service and comment.

## 3.9.4 Intermediate compounds

Much of the metallurgical wizardry you will learn about in later chapters depends on the controlled formation and distribution of so-called intermediate compounds, that is, compounds of definite stoichiometric proportions of a host metal and at least one other added component (a metal or carbon). An example occurs in the hardening of titanium by alloying with a few percent of copper. When particles of $Ti_2Cu$ are dispersed through the titanium matrix, they impede the movement of dislocations and increase the yield stress of the metal.

Intermediate compounds are covalent crystal compounds. So a transition metal must be involved to provide the covalent capability. The strongly directional nature of covalent bonds dictates nonmetallic crystal structures of low coordination number, hard and brittle because the distortions necessary to let dislocations flow easily cost too much energy. Yet the transition metal still provides delocalized electrons which can bond the intermediate compound regions into the metal and let it be a conductor. In general the two elements need to be rather different from one another in atom size and electron structure in order to suppress the tendency merely to dissolve in one another in a random array. Thus copper and titanium form a true intermediate compound while copper and nickel are so alike (and both adopt cubic close packed structures) that they make random mixtures in all proportions. Copper and zinc can also form random mixtures up to about 30% Zn, beyond which they form various intermediate compounds.

The best examples of intermediate compounds are where one element provides empty deeply penetrating states and the other has a supply of localized electrons. $CuAl_2$ is such a case with aluminium offering p states to combine with the d states of copper. Carbon is an especially important element in this context with its four vacant (p state) sites. Tungsten carbide embedded in cobalt (used for cutting tools) is another example. In this case carbon p states and transition metal hybrid d and s states form strong covalent links. The behaviour of carbon in steel is the crux of steel metallurgy and $Fe_3C$ plays a crucial role. Controlling the shape of the carbide phase gives some control over the mechanical properties, as we will see in Chapter 5.

EXERCISE 3.13 What property would you expect to be affected by particles of hard iron carbide in a ductile matrix?

Strong hard steels are loaded with carbides of chromium, molybdenum, vanadium, niobium, or tungsten, all of which you should recognize as transition elements.

## 3.9.5  The origins of magnetism

All magnetic fields are associated with electric currents: either macroscopic currents such as those driven through the turns of a transformer or else atomic currents such as those associated with the motion of electrons within and among atoms. We are concerned with the latter here. The dominant source of atomic magnetism in solids is not the orbital motions of electrons (which tend to cancel each other), but that rather elusive property we call spin. Now, an electron spin contributes 'a magnetic moment' to the atom in fixed amounts of one unit, plus or minus according as the spin is 'up' or 'down'. Thus a spin pair contributes nothing. This severely limits magnetic effects in solids (and molecules) because bonding tends to pair opposite spins. Also, isolated spins give rise to only weak effects. However, much stronger effects are evident in ordinary magnets. This needs something a little special; it needs electrons with unpaired spins to interact with each other within and between atoms. Which elements are uniquely placed to offer such electrons?

The transition metal atoms offer unpaired spins in the solid state and some of them are able to bond in such a way as to allow the interaction of spins between atoms. In fact in the first transition series only metallic iron, cobalt and nickel display ferromagnetism. So how do they do it? The following points are prominent.

(a) The isolated atoms of iron, cobalt and nickel have incomplete 3d shells. The lowest energy state has as many electron spins unpaired as possible. For example look back at the electron configuration for iron shown in Figure 3.8.

(b) In the solid state, iron, cobalt and nickel retain some unpaired spins. Some electrons are delocalized, some are confined to local bonds but a few are localized and unpaired. For example solid iron atoms have on average 2.2 electrons locally unpaired.

(c) Localized electrons are 'kept aware' of the spin state of neighbouring atoms through bonding electrons (which visit at least two atoms) and the delocalized electrons. This exchange of spin information between neighbours is termed an **exchange interaction**. In iron, cobalt and nickel it means that the magnetic moments of all bound but unpaired electrons tend to align in the same direction like so many tiny bar magnets (but much more strongly coupled).

(d) In crystals, the regular arrays of atoms together with the exchange interaction produce preferred directions for the alignment of atomic spins. For example in BCC iron, alignment directions along cube edges are preferred.

(e) Thermal rattle is always tending to shake coupled spins out of alignment — at a high enough temperature, the co-operation between atoms is lost and only the weaker interaction remains (refer back to Figure 3.4).

**SAQ 3.12** (Objective 3.6)

In solid iron the maximum (saturation) magnetic flux density is 2.2 tesla. In cobalt it is 1.8 tesla and in nickel 0.6 tesla. Explain why (at room temperature) iron appears to be, atom for atom, more magnetic than cobalt and cobalt more so than nickel. Other transition elements are not ferromagnetic — speculate on the reasons.

That's the atomic view of magnetism. You've already seen, in Chapter 2, a little of what can happen on a larger scale. The magnetic domains described there are made up of very many (billions) of atoms co-aligned. Alloys and compounds involving the ferromagnetic elements can also exhibit magnetism. One example is ▼Ceramic transformer cores▲

## Summary

Each of the transition metals, in which there are unfilled d orbitals in the outer electron shell, has special properties resulting from the complex behaviour of its outer electrons. The amount of delocalization and the number of electrons not spin-paired affects chemical bonding, melting temperatures, electrical properties and magnetic properties.

The variation in transition metal properties can be traced through rows and columns in the periodic table. Melting temperatures are higher in the middle of each row, where about half of the d states are filled, providing a large number of electrons and unfilled states in the outer shell and resulting in stronger bonding between atoms.

In an energy band diagram of a transition metal, the available energy levels are in broad overlapping bands.

Intermediate compounds are covalent crystals involving a transition metal bonded to carbon or another metal. They are characteristically hard, brittle and electrically conducting. Their properties are widely used in metallurgy, for example as iron carbide to harden steel.

Several of the transition metals have unpaired electron spins which interact with those in the surrounding atoms, leading to ferromagnetism.

**SAQ 3.13** (Objective 3.5)

Of the following pure substances, which are brittle, which are ductile and why — magnesia, aluminium, brass (Cu–30%Zn), copper, iron carbide?

**SAQ 3.14** (Objective 3.8)

Compare and contrast the bonding in nylon (Figure 3.41) and glasses based on silica (Figure 3.30). The respective melting temperatures of nylon and glass are 470 K and 1400 K; the tensile strengths are $70 \, MN \, m^{-2}$ and $100 \, MN \, m^{-2}$. Both materials have a high resistivity.

# ▼Ceramic transformer cores▲

Transformer cores are usually made from laminated sheets of silicon steel (Fe–3.25%Si). The laminations and silicon addition are used to limit eddy current losses which arise in the core owing to the changing magnetic field (see 'Choosing iron–silicon alloy for a transformer core' in Chapter 2). The eddy current losses increase rapidly as the operating frequency is increased and neither laminating nor alloying can sufficiently counter the extra loss above a few hundred hertz.

What we require is a magnetic material which is insulating. Ceramics generally have this property and indeed the first magnet material, lodestone, is the natural ceramic oxide of iron, $Fe_3O_4$, now called magnetite. From this has been developed a class of materials called ferrites. They are ionic oxides containing iron and some other divalent metal. For high frequency transformer cores, a mixture of manganese and zinc ferrites is used. It has a resistivity about two million times higher than that of silicon iron. I'll leave you to work out the processing route.

Incidentally, the crystal structure (which is essential for obtaining the magnetic characteristics) can be appreciated using the techniques of Section 3.4 for ionic crystals. Oxygen ions are arranged in a combination of octahedral with tetrahedral cages, just as with MgO. In managanese ferrite, the $Mn^{2+}$ ions and an equal number of $Fe^{3+}$ ions sit in half of the available octahedral cages. The same number again of $Fe^{3+}$ ions sit in one eighth of the available tetrahedral cages. The formula can be written:

$$Fe^{3+}(Mn^{2+}Fe^{3+})O_4^{2-}$$

bracketing the 'octahedral ions'. Zinc ferrite is only slightly different being:

$$Zn^{2+}(Fe^{3+}Fe^{3+})O_4^{2-}$$

I'm not including a drawing because in this case the cage description is quite succinct and the drawing would require thirty-two oxygen, sixteen iron and eight divalent metal ions to define it! Zinc ferrite's structure is in fact the same as that of the mineral 'spinel' which is $Mg^{2+}(Al^{3+}Al^{3+})O_4^{2-}$.

In this chapter we have looked at the various ways atoms can stick together to form molecules and much larger structures — some highly ordered, some rather disordered. We have remarked on certain properties of solids but have left most of the effects of thermal energy (heat) to the next chapter.

## Objectives for chapter 3

You should now be able to do the following.

3.1 Use the energy–atom separation curve to explain the origin of differences in Young's modulus and melting temperatures between materials. (SAQ 3.4)

3.2 Explain the occurrence of specified atomic/molecular arrangements in specified solids in terms of the strength and directionality of the bonding, and the geometry of the molecules and structures involved. (SAQ 3.6, SAQ 3.8, SAQ 3.9)

3.3 For particular materials (for example oxide glasses, ethylene, metals, carbon), relate specified properties to crystal/amorphous architecture. (SAQ 3.7)

3.4 Explain the existence of valence and conduction bands in solids and how they account for conduction, semiconduction, insulation. (SAQ 3.10)

3.5 Relate the ductile or brittle nature of specified materials to their bonding and structure. (SAQ 3.12)

3.6 Describe a model for the electronic origin of ferromagnetism. (SAQ 3.11)

3.7 Explain specified (mechanical) properties of polymers in terms of primary and secondary bonding, molecular geometry. (SAQ 3.9)

3.8 Give a description of bonding mechanisms in a variety of materials which lie between the three pure ionic, covalent and metallic types.

3.9 Describe the X-ray diffraction and other evidence relevant to supporting an atomic view of matter.

3.10 Define or distinguish between:

| | |
|---|---|
| van der Waals bonding | BCC |
| hydrogen bonding | HCP |
| polarization | CCP (or FCC) |
| valence/conduction band | piezoelectricity |
| intrinsic/extrinsic semiconduction | void/vacancy |
| anisotropy | conductor/insulator |
| spherulites | cross-linking |
| chain folding | side-branching |
| ferromagnetism | |

# Answers to exercises

**EXERCISE 3.1**
H–He: first shell s orbital
Li–Ne: second shell s and p orbitals
Na–Ar: third shell s and p orbitals

**EXERCISE 3.2**

$$U = Ar^{-12}$$
$$F = -12Ar^{-13} \; (= dU/dr)$$

Consider forces at distance $r$, and distance $0.95r$, (5% compressive strain)

$$F_1 = -12Ar_1^{-13}$$
$$F_2 = -12A(0.95r_1)^{-13}$$
$$F_2/F_1 = (0.95)^{-13} = 1.95$$

The repulsive force almost doubles. (The attractive force varies much less over the same range, typically as $B/r^2$.)

The steepness of the repulsive energy curve allows us to pretend that the atom cores are hard (virtually incompressible) spheres.

**EXERCISE 3.3** For doubly charged ions at a given spacing the attractive force is four times greater than for singly charged ions. Also the closer spacing gives nearly another factor of 2 from $1/r^2$. The reduction in energy as the doubly charged ions pack is thus much greater. Consideration of the potential energy curve would suggest that MgO has the higher modulus. The potential well of MgO is deeper and is located at a smaller value of $r_0$ so it should have a higher curvature at its base.

**EXERCISE 3.4** The radii are about 0.168 nm for the positive ion and 0.182 nm for the negative ion. The ratio of these is about 0.9, exceeding the minimum of 0.73 for cube cages, so eight $Cl^-$ ions coordinate around a $Cs^+$ but do not touch. The spacing of adjacent $Cs^+$ ions is equal to the length of the cube edge $a$, and the body diagonal of the cube (according to Pythagoras and Figure 3.51) is $\sqrt{3}a$. The body diagonal of the cube is $(0.16 + 0.36 + 0.16)\,nm = 0.68\,nm = \sqrt{3}a$, so the cube edge $a$ is almost 0.40 nm (touching $Cl^-$ ions would give a cube edge of only 0.32 nm).

**EXERCISE 3.5** The principle is the deposition of graphite on a surface by frictional wear. The property is the ability of graphite layers to slide past each other owing to the weak interlayer bonding.

**EXERCISE 3.6** Graphite is in fact a good conductor parallel to the planes of the rings owing to the delocalized $\pi$ bonds (resistivity $\approx 10^{-6}\,\Omega\,m$). Normal to the planes a much lower conductivity is found ($\approx 3 \times 10^{-2}\,\Omega\,m$).

**EXERCISE 3.7**
(a) In nylon, hydrogen bonds link long chains through the $-N-H \cdots O=C-$ parts of amide groups ($-NH-CO-$) in each chain.
(b) In PE the intermolecular links employ relatively weak van der Waals forces. In nylon the somewhat stronger hydrogen bond joins molecular chains. Nylon would therefore be expected to have the higher melting temperature.

**EXERCISE 3.8** Group Ia contains the alkali metals: Li, Na, K, Rb, Cs, Fr. All their atoms end with . . . $ns^1$. (Where $n$ is the number of the period to which the metal belongs).

**EXERCISE 3.9** Yes. The group 2 elements all have higher melting temperatures than the group 1 metals meaning, by the arguments of Section 3.3, that more rattle energy is needed to break their bonds so stronger bonding is confirmed.

**EXERCISE 3.10** Boron is $(1s)^2(2s)^2(2p)^3$ so forms only three covalent bonds. It can only form its fourth bond by taking an electron from a neighbouring bond to become $B^-$, launching a hole into the valence band. The hole behaves like a positive charge — hence p-type conduction.

**EXERCISE 3.11** The first block of transition elements covers these ten metals: scandium, titanium, vanadium, chromium, manganese, iron, cobalt, nickel, copper and zinc. It is the series in which the 3d shell is being filled. You will remember that there are questions as to the order in which energy levels might fill, because there are several with nearly equal energies. The sequence K, Ca, Sc, Ti . . . turns out to be levels filling in the order . . . 4s 3d instead of the more naively expected order . . . 3d 4s.

**EXERCISE 3.12** From Table 1.3 it can be seen that the filament (tungsten) and its support (molybdenum) must be made from refractory metals. The metals used are, not surprisingly, transition metals, tungsten from the 5d series, molybdenum from the 4d. See Figure 3.49.

**EXERCISE 3.13** Hard particles in a ductile matrix will strengthen it, tending to reduce its ductility by impeding dislocations. In losing ductility it is losing toughness and becoming more brittle.

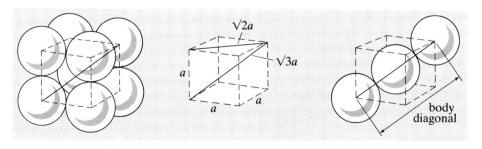

Figure 3.51 Caesium chloride structure

# Answers to self-assessment questions

## SAQ 3.1

(a) Atoms are about $10^{-10}$ m radius with a central nucleus of about $10^{-15}$ m radius.

(b) The nucleus contains protons and neutrons. A cloud of electrons occupies the remaining space.

(c) Almost all of an atom's mass is in the tiny nucleus. The total number of protons plus neutrons, which are of almost equal masses, is the 'mass number' of an atom. Electrons have low mass, only $\approx 1/2000$ that of a proton or neutron.

(d) Protons carry a positive electric charge; electrons an equal and opposite (negative) charge. Thus neutral atoms must contain equal numbers of protons and electrons. The nucleus attracts the electrons but within the cloud electrons repel one another electrostatically, making for very complex dynamics.

(e) Atoms with equal numbers of protons behave identically chemically; this is what defines an element. The number of protons is called the atomic number of the element, symbol $Z$. Every $Z$ from 1 (hydrogen) to 103 (Lawrencium) has been identified in the Earth or made artificially by nuclear reaction. Isotopes of an element have different numbers of neutrons in their nuclei so the atoms have different masses.

## SAQ 3.2

(a) The principal quantum $n$ may take any integral value: 1, 2, 3, . . . This number can be thought of as defining a band of energy within which the allowed energy lies. The lower the value of $n$, the lower the energy of the allowed band.

(b) The azimuthal quantum number $l$ and the letters s p d f refer to orbitals. For any value of $n$, the number $l$ may take any integral value: 0, 1, 2, . . . $(n - 1)$. Thus if $n = 3$, then $l$ may take the values of 0, 1 and 2. When $n$ is fixed, then the lower the value of $l$, the lower the value of the allowed energy. To a first approximation, the energy of any electron in an atom can be defined by specifying the values of $n$ and $l$. Usually the value of $l$ is denoted by a letter:

$l = 0$ is labelled s    $l = 3$ is labelled f
$l = 1$ is labelled p    $l = 4$ is labelled g
$l = 2$ is labelled d

If therefore you wished to refer to an energy level for which $n = 3$ and $l = 2$, it is simply called the 3d level. Similarly 4f refers to the level with $n = 4$ and $l = 3$. The relative energies can be determined by experiment. For each value of $l$ there are $2(2l + 1)$ distinct electron states. An s level, for example, can hold 2 electrons, a p level can hold 6 electrons, a d level can hold 10 electrons and an f level can hold 14.

(c) Carbon:
$(1s)^2$ (2 $\times$ s electrons in $n = 1$)
$(2s)^2$ (2 $\times$ s electrons in $n = 2$)
$(2p)^2$ (2 $\times$ p electrons in $n = 2$)
Oxygen: $(1s)^2(2s)^2(2p)^4$
Sodium: $(1s)^2(2s)^2(2p)^6(3s)^1$

(d) Spin up/spin down: strictly the s orbital is capable of holding two electrons only if they differ in the quantity known as spin — it is either up or down. Similarly the p orbital is made of three spin pairs and the d five spin pairs.

## SAQ 3.3

There are three mechanisms by which atoms bond strongly together, namely ionic, covalent and metallic bonding. Metals obviously use the third of these, in which some electrons are detached from individual atoms and spread throughout the spaces between them. This separation of charge produces a non-directional, cohesive force. The presence of free electrons means that metals are electrical conductors. The interatomic forces in metals are non-directional and the atoms are spheres, so close-packed crystal structures are often formed.

Ionic bonds are typical of ceramics such as magnesium oxide. In this type of bond, electrons are transferred from one atom to another, making oppositely charged ions which attract each other electrostatically to form the bond. These materials are electrical insulators until they melt, when the ions become mobile. The forces between ions are also nondirectional but because ions of like sign repel, ionic crystals are of more open structure than metals.

Polymers are made of distinct molecules which attract each other only weakly but within which much stronger bonds join atoms. In these covalent bonds, electrons are shared between two atoms and the attraction is directional (along the line of atom centres). Because all the electrons are locked into bonds, polymers are electrical insulators. Polymer molecules are generally long chains and so get tangled and only partially achieve the organization needed to be crystalline.

## SAQ 3.4

See Table 3.5.

(a) High modulus and high melting temperature do roughly correspond. You should have plotted a graph to reveal this.

(b) The observed and theoretical strengths differ drastically in magnitude but trends remain.

Comment: Carefully made, defect-free specimens can get close to the theoretical limit (glass and diamond), but even the best fall well short. Bulk engineering materials are even further from their theoretical maximum strength.

## SAQ 3.5

Structural defects include point defects such as impurities and vacancies; line defects such as stacking faults (a

Table 3.5

| Substance | Tensile strength $\sigma/\mathrm{GN\,m}^{-2}$ | |
|---|---|---|
| | calculated | observed |
| Alumina | 56 | 0.7 |
| Copper | 55 | 0.4* |
| Diamond | 285 | 50 |
| Glass | 14 | 3.6 |
| Graphite | 9 | 0.1 |
| Iron | 52 | 0.20* |
| Salt (NaCl) | 6 | 0.1 |
| Silicon | 46 | 3.3 |
| Tungsten | 116 | 1.5* |

* substantial yield before fracture.

'wrong' atomic plane between two 'right' ones); and gross defects such as cracks. Clearly atom–atom images of bulk solids fail if many defects make the overall picture very different from the simple atomic pair or if local weaknesses develop as a result. Dislocations allow planes to slip past each other at a fraction of the stress required for defect-free material. Cracks harbour stress concentrations and propagate through a brittle material at low stress.

SAQ 3.6 The radii are $Ti^{4+}$ = 0.068 nm, $Ba^{2+}$ = 0.135 nm and $O^{2-}$ = 0.145 nm. The ratios of ion size are Ti:O = 0.47 and Ba:O = 0.93. For the titanium ion the ratio lies a little above the critical value for octahedral cages. Six oxygen ions could therefore, standing slightly apart, provide an octahedral cage.

For the barium ion, the ratio is well above the critical value for a cubic cage but only just below that for a cuboctahedral cage. Now we've somehow got to fit the titanium and barium cages together so although the cube with oxygen ions standing well apart sounds good on its own, we can't fit it in with the octahedron. We can put the octahedron with a cuboctahedron though. The consequence of this is that the barium ion will have to rattle around the cuboctahedral cage — it's too small.

The perovskite structure is indeed just that prediction; eight octahedral cages joined corner to corner generate a cuboctahedral

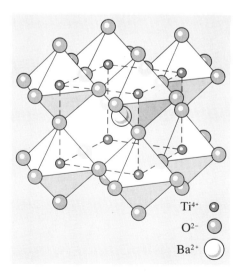

Figure 3.52 Barium titanate structure

cage in their midst, see Figure 3.52. The diagram highlights cages. Figure 3.19 showed the cubic structure cell which can be cut out of this array.

SAQ 3.7 Diamond is covalently bonded. Covalent bonds are strongly directional and so limit the slip which is characteristic of softer materials. The bonds share electrons, localizing them to the region between and around just two atoms, so it is an insulator. Graphite is also nominally covalently bonded but with a high degree of delocalization of electrons among a network of π bond rings. Large sheets of six-carbon rings locked by σ bonds are internally strong but each sheet is somewhat free to slide relative to its neighbours. Graphite conducts electricity and is soft as a consequence of the π bonds.

SAQ 3.8
(a) Radius of $Si^{4+}$ ≈ 0.04 nm
Radius of $O^{2-}$ ≈ 0.14 nm
The ratio of the two radii is about 0.3 and lies between the lower limits for an octahedron (0.41) and a tetrahedron (0.22) so we could have a tetrahedral cage with oxygen ions standing a little apart.
(b) $sp^3$ hybrids of silicon (analogous to those for carbon in diamond) would give a tetrahedral configuration for silica.

SAQ 3.9
(a) It is heavily cross-linked by random interconnections which prevent ordered crystallization. There are no secondary bonds, the breaking of which would allow softening. Phenol formaldehyde is a thermoset.
(b) PMMA and LDPE (in fact all polyethylenes) are van der Waals bonded thermoplastics, as is PET. Nylon is an H bonded thermoplastic. Vulcanized rubber is a thermoset, being covalently cross-linked.

SAQ 3.10
(a) X-ray Bragg diffraction experiments produce data which depend on atom spacing. Waves (X-rays) scattered by planes of atoms can constructively interfere when the separation of parallel planes causes one wave to lag behind another by an exact number of wavelengths — the extra distance arising from the path length between planes,

which depends on the angle and spacing. For a particular wavelength, only certain combinations of angle and spacing will give this constructive scattering.
(b) (i) The atoms in a face-centred cubic array may touch along the diagonal of the cube face (see Figure 3.44(d)) this is $\sqrt{2}$ times the length of the cube edge $l$ by Pythagoras. But the cube-face diagonal is also equal to $(\frac{1}{2} + 1 + \frac{1}{2})$ times the atom diameter $d$. So

$$2d = \sqrt{2}l$$
$$d = \frac{l}{\sqrt{2}} = \frac{0.36\,\text{nm}}{1.4} = 0.26\,\text{nm}$$

(ii) The cubic structure cell has a volume $l^3$ and contains four atoms made up as follows. Each corner shares an atom with eight similar cells, so each corner contributes an eighth of an atom, making one atom. Each face centre atom is shared with two cells, so the six faces contribute three more atoms between them. The density (mass/volume) ρ follows:

$$\rho = \frac{4m}{l^3} \text{ where } m \text{ is the atomic}$$
mass
$$\rho = \frac{4 \times 1.1 \times 10^{-25}\,\text{kg}}{(0.36 \times 10^{-9})^3\,\text{m}^3}$$
$$= 9000\,\text{kg m}^{-3}$$

(Note that using only two significant figures in the data gives an answer no better than one significant figure.)

SAQ 3.11
(a) At low temperatures, only a small amount of 'rattle' energy is available. The valence band in cold silicon is exactly full. No electrons are free to accept (or lose) a small amount of energy from an electric field and so they cannot take part in conduction.

(b) Aluminium offers three electrons for bonding. In a silicon lattice, four electrons are required from each atom so when aluminium joins the lattice in place of silicon it is one short. It may 'borrow' an electron rattled (thermally) free from some other silicon–silicon bond. This deficiency may in turn be made up from another bond. The wandering deficiency is a 'hole' which can conduct electricity (in the valence band).

(c) Resistance depends upon resisivity (or reciprocal conductivity) and the dimensions of the material. A semiconductor such as that in (b) has a resistivity which can be controlled by adjusting the impurity concentration. The dimensions of a resistive strip can also be adjusted to give the required resistance.

(d) The carbon–carbon bonds in polymers like PVC are covalent σ bonds and are strongly directional, just as in diamond. The valence band for a chain in say PVC must be full and the gap between it and the conduction band must be a few electron volts.

SAQ 3.12 The 3d shell of iron atoms contains six electrons. In the isolated atom, four of these would be unpaired. In the solid state, the evidence of magnetic saturation is that about 2.2 remain unpaired (point (b) in Section 3.9.5). By happy coincidence, in SI units this just happens to lead to saturation at about 2.2 tesla.

Cobalt has one fewer and nickel two fewer unpaired 3d shell electrons in the isolated atom, so it is not surprising to find the solid state exhibiting magnetic saturations similarly reduced below that of iron. In fact Ni has about 1.7 unpaired electrons on average while cobalt has only about 0.6.

The nonmagnetic transition metals may either require all their electrons to be delocalized, or else spin paired, or else may find a minimum energy condition in which the magnetic effect of the electron one atom is cancelled by that of a neighbouring one. Manganese, for example, does this.

SAQ 3.13 Magnesia and iron carbide are respectively ionic and covalent — they are brittle. MgO is resistant to slip because electrostatic repulsions between like ions restrict slip systems. $Fe_3C$ is resistant to slip because of the directionality of the covalent bond. Without the ability to slip, crystal structures are brittle. Al, Cu–30%Zn and Cu are single phase metals. The metallic bond allows dislocations to move easily thereby bestowing ductility.

SAQ 3.14 Glass is brittle and moderately strong whereas nylon is tough. Both are good electrical insulators.

Nylon consists of regular polymeric covalently bonded chains containing carbon and nitrogen in the backbone, with hydrogen and oxygen side groups. Hydrogen bonds are formed between aligned chains, via the amide (nitrogen-containing) group, giving crystalline features. As a consequence nylon melts when these secondary bonds are thermally ruptured, but retains its molecular integrity to much higher temperatures. Resistivity is high because the electrons are localized to the bonds.

Glass also has distinct molecular units (silica tetrahedra) but in a much more tangled fashion than the nylon chain. Glasses are not crystalline. The inclusion of an ionic component such as $Na^+$ or $Mg^{2+}$ in glass inhibits the formation of long rigid chains and tends to lower the softening temperature.

Both liquids have a high viscosity owing to the persistence of large molecular structures into the liquid state. In silica-based glass electrons are localized in covalent bonds or in the oxygen ions. In both materials the bonding is a mixture of covalent and electrostatic.

# Chapter 4 Temperature as an agent of change

## 4.1 Thermal effects in outline

This chapter is about the 'rattle' half of the 'stick-and-rattle' imagery used in Chapter 3 for modelling materials at the atomic scale. Essentially we shall see two things in this chapter:

1 That the conflict of atom sticking forces with the muddle-making effects of atomic motion determines whether any particular change can take place *at all* at a given temperature. (Perhaps the word 'entropy' is in your mind already. We'll come to that later.)

2 That the *rate* of a change, once allowed, is dominated by thermal influence.

Clearly these are two factors of great technical interest. Between them they determine the effects of rattle as it impinges on processing conditions and on the suitability of materials for particular service.

Temperature is, of course, the measure of 'thermal' conditions. Nowadays it is measured by thermometers and expressed as a number on an agreed scale. Some features of thermometers and of their use are discussed in ▼**Thermometers and process control**▲ but it is worth noticing that the use of instruments to measure temperature in materials processing is a comparatively recent innovation in the history of materials technology. For thousands of years craftspeople have been 'getting the fire right' (or wrong) without any instruments at all. Judgements by eye or of how your spit fizzes can be surprisingly sensitive. Tching te Tchen was the centre of the Chinese porcelain industry for centuries. Not only were the raw materials all local, but the rock used as flux happened make the porcelain tolerant to imprecision in the firing temperature. A kilnmaster had to judge 1300 °C ± 30 °C by eye. A big kiln held tonnes of pots (Figure 4.1). The slow thermal response of such a monster probably eased the control of temperature but the stakes were high. Father d'Entrecolles, a seventeenth-century Jesuit missionary in the city wrote, 'It seldom happens that a baking succeeds altogether well, and sometimes the whole is lost, and when the oven is opened they find the ware and the cases reduced to a mass as hard as stone. A fire too fierce, or cases in a bad condition, may ruin the whole. And it is not easy to regulate the degree of fire, for the nature of the weather changes in an instant the action of the fire, the quality of the subject on which it acts, and that of the wood which serves for fuel. Thus for one workman that grows rich there are a hundred that are ruined, and yet they tried their fortune with expectation of success, and the hopes of setting up a merchant's shop.'

(Duhalde P., *The General History of China*, 1741)

Figure 4.1 Chinese porcelain kiln
(Courtesy of Benteli Verlag)

Even today there are craftspeople who can do without thermometers. For example in steel, hardness and toughness are mutually compromising properties. A plain carbon steel quenched from high temperature becomes extremely hard but very brittle. Reheating to some chosen temperature encourages new crystals to grow giving a tougher, fine-grained matrix holding the ultra-hard component. This is 'tempering' and non-colourblind blacksmiths were taught to get the right degree of temper without using a thermometer. Figure 4.2 (right) is what they learned to interpret.

| Colour of object to be tempered | Temp./°C |
|---|---|
| very pale yellow | 430 |
| light yellow | 440 |
| pale straw-yellow | 450 |
| straw-yellow | 460 |
| deep straw-yellow | 470 |
| dark yellow | 480 |
| yellow-brown | 490 |
| brown-yellow | 500 |
| spotted red-brown | 510 |
| brown-purple | 520 |
| light purple | 530 |
| full purple | 540 |
| dark purple | 550 |
| full blue | 560 |
| dark blue | 570 |

Figure 4.2 (above) Chart of colours and temperatures for tempering steel. The colour is caused by interference in the surface oxide. From Machinery's Handbook, Industrial Press Inc. 10th ed., 1939

# ▼Thermometers and process control▲

Measuring temperature is not like measuring length with a ruler, which itself has some of that same physical quality. Thermometers are transducers of the condition of temperature into other physical things, such as resistance, voltage, pressure. The meter reading is really a graduated indication of one of those other physical quantities. It is meaningless until the instrument has been calibrated against a defined numerical scale of temperature.

Here are some effects by which temperature is sensed and given number.

• Expansion of a liquid in a tube, usually mercury in glass.
• Resistance of a metal wire, platinum in good instruments.
• Voltage output from a thermocouple, that is, the junction between wires of two different metals.
• Pressure of a constant mass of gas kept at constant volume.
• Brightness or colour of radiation emitted by a furnace.

Processes may require steady or changing temperature. Changes may be spatial or temporal. Temperature may be tightly controlled by using a thermometer as the sensor in a feedback loop. On the other hand, if it is simply necessary to know whether the temperature is high enough for a process to work, or low enough for safety, the thermometer may merely monitor conditions or be coupled to an alarm to indicate when a threshold has been crossed. With such a diversity of uses, accuracy may not be the paramount virtue of a thermometer. Any of the following features may be of particular relevance for certain applications.

• Accuracy. A thermometer will agree with the International Temperature Scale to a specified precision and have been calibrated at some standard temperature.

• Range. The instrument must be able to give readings at the process temperature. For example, below $-38\,°C$ and above $360\,°C$ a mercury-in-glass thermometer is useless. Radiation pyrometers are used to measure high temperatures.

• Sensitivity. A process may need close control of temperature or need to be tolerant of some deviation. Close control needs a sensitive thermometer, that is one giving noticeable response to a small change in temperature.

• Response time. A long response time means that short, sharp fluctuations of temperature are averaged out.

• Repeatability. Actual values may be less important than that successive batches of a process get the same treatment.

• Reliability. A malfunctional instrument might close down the plant.

• Corrosion-proof. Furnace atmospheres can be very reactive chemically; the thermometer sensor may have to be protected by a ceramic sheath, which will slow its response.

• Output. Telemetry of signals to a central control point favours thermometers with electrical output.

• Compatibility with other instrumentation.

• Cost.

SAQ 4.1 (Objective 4.1)
Find a thermometer, preferably in a processing application and enquire which attributes are relevant to its application.

**EXERCISE 4.1** Think of, or otherwise discover, another example of non-thermometric temperature judgement.

To say that temperature is 'that which is measured by a thermometer' is not the definition which helps for modelling at the atomic scale. The *theoretical construct* of temperature, relating it to the energies of atoms, gives much better insights into the way temperature regulates processing. We will come to that in Section 4.2 and then make use of it for the rest of the chapter. But first let us review the kinds of property changes caused by temperature variation.

## 4.1.1 How things change with temperature

The temperature-varying effects used in thermometers have a fairly steady change over a good range of temperature (Figure 4.3a). By contrast, phase changes, of which melting and boiling are the common examples, happen at sharply critical temperatures; $+1\,°C$ and your dahlias are alright, $-1\,°C$ and they are dead of frostbite. Between these extremes of gradual and sudden change are temperature-sensitive phenomena which accelerate from insignificance to drama over a narrow but finite temperature range. Examples you have already met are the increasing flexibility of thermoplastics beyond their glass transition temperatures, the rates of chemical reactions, and the collapse of performance of semiconductors as temperature goes up. Figure 4.3(b) illustrates one of these and although it will not always be obvious whether to regard a particular change as 'gradual', 'accelerating' or 'sudden' we shall find the theoretical models for each category to be sharply distinct.

• Gradual changes are modelled in terms of the *average* thermal energy of the particles of the system.
• Accelerating changes are modelled in terms of the fraction of the particles which have *much higher than average* thermal energy.
• Sudden changes are modelled in terms of the balance between *chaos* generated by 'rattle' and *order* induced by interparticle forces.

Because the models differ, our ideas about influencing each sort of behaviour differ. We wish to know how to influence these behaviours

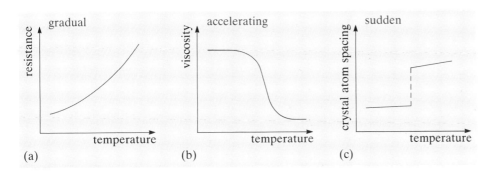

Figure 4.3 Changes. (a) Gradual. (b) Accelerating. (c) Sudden

because, within each category, there are 'good things' and 'bad things', from both the service and process points of view.

My classification into 'gradual', 'accelerating' and 'sudden' is in terms of scientific principles. In Table 4.1 examples of those other aspects of PPPP, namely products, processes and properties, are related to these principles. The classification will give us a structure for the chapter.

Table 4.1  Examples of gradual, accelerating and sudden changes

|  |  | Gradual | Accelerating | Sudden |
|---|---|---|---|---|
| Service | Good | Electrical resistance thermometer<br>Bimetal switches<br><br>**A** | Single-component heat curable glues<br>Enhanced performance of detergents in hotter water<br>**B** | Onset of ferroelectric properties<br>Can leave cooking to simmer<br>**C** |
|  | Bad | Crazed glaze on crockery<br><br>Need to compensate timer in clocks and watches<br><br>**D** | High temperature limit on semiconductor performance<br><br>Excessive corrosion of high temp. components<br>**E** | Some permanent magnet materials demagnetize at modest temperature<br><br>Burst water pipes by freezing<br>**F** |
| Process | Good | Density variation provides convective mixing in castings<br><br>Metal tyres and bearing sleeves can be shrink fitted to wheels/shafts<br>**G** | Temperature control of workability of glass<br><br>Sintering of powders to solid (a route to near net shape forming)<br>**H** | Chemical reaction selection by use of critical temperature<br><br>Melting allows casting processes<br>**I** |
|  | Bad | Viscosity gradually decreases well beyond $T_m$<br>Convection in liquid zone of 'float' zone purification remixes impurities (do it in space!)<br>**J** | Overageing of precipitation hardened alloys<br>Continued diffusion of previously implanted dopants in subsequent processing of microcircuit chips<br>**K** | Upset metallurgy in the heat affected zones of welds<br>Cracks in porcelain due to crystal transition in quartz<br>**L** |

EXERCISE 4.2   Here is a list of twelve items to fit into Table 4.1 as **A** to **L**. Can you place them? You may not succeed with all of them yet; return to this exercise as you go through the chapter.

1  Creep deformation.
2  Homogenization of castings.
3  Glass shattering under thermal shock.
4  Convection in domestic water system.
5  Quenching and tempering of steel.
6  Temperature increases of resistivity demands ultra uniformity of lamp filament cross section.
7  p–n junctions are stable against homogenization at room temperature.
8  Fuses in electrical circuits.
9  Superconducting transitions all at inconveniently low temperatures (so far!)
10  Provision of surface compressive stress enables 'safety glass' for car windows.
11  Onset of unwanted reactions in chemical processes.
12  Jam–toffee transition makes boiling temperature critical in jam-making.

## Summary

Thermometers *measure* temperature. They are transducers providing observable and quantifiable signals in variables other than temperature. Thermometers are calibrated to give numbers in accord with an internationally agreed scale. Various attributes influence the selection of an instrument for a task.

Temperature can determine whether a change can occur at all and if so at what rate.

Distinct modelling patterns characterize gradual, accelerating or sudden property variation with temperature.

# 4.2 Temperature and energy

The intimate connection between temperature and the kinetic energy of rattling atoms is now developed. The proposition is that the temperature of an object is a measure proportional to the *average kinetic* energy of its atoms. Both the emphases, on average and kinetic, are important. The first implies that not all atoms have the same energy, a fact we shall make much of in due course. The second warns that there are energies in matter which are not kinetic, and in a moment we shall debate whether or when they are part of the thermal energy. Within this model the concept of a **perfect gas** gives the simplest understanding and that is the starting point.

## 4.2.1 Random energy of a perfect gas

The definition of a perfect gas is that the only energy its particles possess is the translational kinetic energy of their constant random motions. 'Perfection' requires the molecules to be true points and not to exert any forces on one another! A monatomic, chemically unreactive gas, such as helium at low pressure, approaches the ideal well enough for theoretical analysis and experimental observation to be compared, and that provides the connection between unseen atoms and measured temperature. Experiments conducted even before temperature was well defined suggested that the pressure, volume and temperature ($p$, $V$, $T$) of a mass of gas were related by

$$p \times V = \text{constant} \times T$$

The familiar calculation for the pressure, in which Newton's laws are applied to atoms of mass $m$ bouncing off the walls of a container, gives

$$p \times V = \frac{Nm\bar{u}^2}{3}$$

where there are $N$ atoms and $\bar{u}$ is their root mean square speed. So $p \times V$ is proportional to both $T$ and to the average kinetic energy of these atoms, $m\bar{u}^2/2$. Therefore $T$ is proportional to the average kinetic energy of the atoms of a perfect gas.

The magnitude of the constant of proportionality is discussed in ▼**Boltzmann's constant**▲. It follows from the above that the pressure reading of a constant-volume gas thermometer (Figure 4.5) is a direct measure of the theoretical construct, that is it measures in an 'absolute' temperature scale. Moreover, in that zero kinetic energy is a conceivable condition, so is zero temperature. This defines the **absolute zero** of temperature, that is 0 K (meaning 'nought kelvins'), otherwise recognized as $-273\,°C$. (Strictly there is a small quantum correction which dictates a small residue of kinetic energy at the absolute zero of temperature.)

## 4.2.2 Which energies are random?

The **internal energy** $U$ of a piece of matter is all the energy which is randomly distributed among the particles it is made of. For a perfect gas it is all translational kinetic energy but in real matter it will include contributions from other energy forms. Heating raises the temperature and hence the average kinetic energy. But, with variable efficiency, the mechanism of random collisions converts random kinetic energy into other forms. Such conversions will drain away kinetic energy so more energy will have to be added for a given rise of temperature. For example Figure 4.4 shows the specific heat of hydrogen gas rising from the 'perfect gas' value as the random energy content comes to include molecular rotation and vibration.

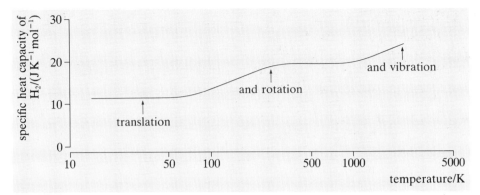

Figure 4.4 Variation of specific heat of hydrogen gas

# ▼Boltzmann's constant▲

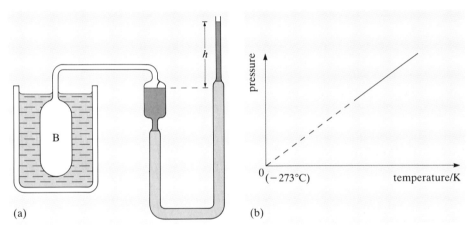

Figure 4.5 (a) Pressure in bulb B varies with temperature and is measured by the height $h$ of the mercury column. (b) Extrapolation gives absolute zero

The sum of all the kinetic energies of the $N$ atoms of gas is

$$U = N \times \frac{m\bar{u}^2}{2}$$
$$= N \times \text{constant} \times T$$

The constant is set as $3k/2$ with $k$ known as Boltzmann's constant. The $\frac{3}{2}$ is there for convenience in theory, which does not concern us here. So

$$U = \frac{3NkT}{2}$$

and the average energy of a single atom (or molecule in a molecular gas) is $3kT/2$.

To find Boltzmann's constant, the specific heat capacity (at constant volume to avoid a big $PdV$ work term) of an approximately ideal gas is measured using a known number of atoms. For one mole of monatomic gas there are Avogadro's number of atoms ($N_0 = 6 \times 10^{23}$) and

$$c_v = \frac{dU}{dT} = \frac{3N_0k}{2}$$

The experimental value of $c_v$ is $12.5 \, \text{J mol}^{-1} \, \text{K}^{-1}$, so

$$k = \frac{2c_v}{3N_0}$$
$$= \frac{2 \times 12.5}{3 \times 6 \times 10^{23}} \, \text{J K}^{-1}$$
$$\approx 1.4 \times 10^{-23} \, \text{J K}^{-1}$$

or, and usefully remembered,

$$k \approx 86 \, \mu\text{eV K}^{-1}$$

163

The basic criterion which determines the coupling of an energy form into the thermal (random) energy content is whether sufficient kinetic energy can be transferred in a collision to effect a jump between the quantum levels characteristic of that form of energy. A collision cannot convert more than the total kinetic energy of the colliding atoms. Where this is smaller than what is needed to induce a quantum jump, no jump can occur. Where the average kinetic energy, approximately $kT$, is enough, jumps become frequent and that energy form becomes part of the randomly held energy.

EXERCISE 4.3   Use the results of 'Boltzmann's constant' to work out in electron volts (eV) the average kinetic energy of a perfect gas atom at 300 K. Is this big or small compared to the quantum gaps between the atoms' electron energy levels ($\approx 1\,\text{eV}$)? Will electron energies become random under the influence of atom collisions at room temperature?

In solids the atoms are vibrating. Figure 3.10(b) showed the potential energy of an atom under the influence of a nearby atom and the kinetic 'rattle' energy added on. Any vibrating system has two forms of energy in constant interchange and for atoms in solids the exchange is between kinetic energy and the potential energy manifested as interatomic forces. With many atoms all at different stages of oscillation both these energies should be random.

The molar heat capacity for most simple atomic solids is about $25\,\text{J}\,\text{K}^{-1}\,\text{mol}^{-1}$ at room temperature (Figure 4.6). This observation is **Dulong and Petit's law**. The value is twice that of a perfect gas. Assuming that the kinetic energy is still changing by $\frac{3}{2}N_0 k\,\text{J}\,\text{K}^{-1}$ per mole, the change of potential energy seems to be contributing just as much to the specific heat capacity as the change of kinetic energy. (For a continuous exchange of energy between kinetic and potential there would be, on average, equal amounts of kinetic and potential energy.) At low temperatures where $kT$ is not as big as the gaps between quantum levels for vibrations, the specific heat falls away towards zero. Only a small fraction of the atoms are in action.

Pure metal elements aside, most engineering materials are structurally more complicated than atomic solids and these simple ideas do not hold up well. One trouble is that displacing any one atom affects the forces acting on *several* others so the thermal vibrations are coupled into waves running through the solid. It is the energy of these waves which is quantized rather than that of individual atoms, and you can imagine that with atoms of different masses joined by bonds of various strengths this gets too complicated long before reality is in sight. It is easier to regard the waves as 'packets' of elastic energy behaving independently of the atoms, rather as the quanta or energy packets of radiation are supposed to. Those are called 'pho*t*ons' after the Greek for light: the packets of atomic vibration wave energy are 'pho*n*ons' after the Greek

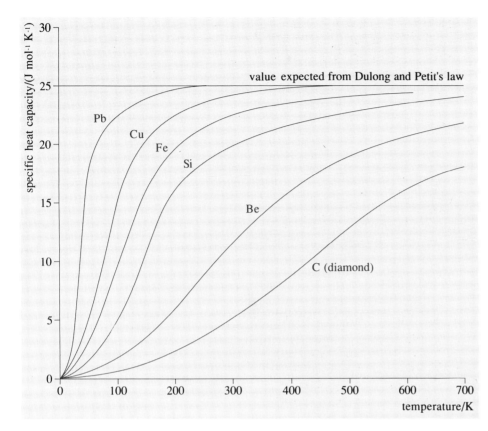

Figure 4.6 Variation of specific heat capacity with temperature for selected solids. Note trend towards Dulong and Petit's predicted value. The stronger the bonding, the more gradual the trend

for sound. To model the effect of temperature on electrical resistivity it is necessary to talk in terms of quantum interactions between delocalized phonons and delocalized electrons instead of collisions between particulate atoms and electrons. More about this later.

In molecular solids, motions other than vibrations are important. The rotation about the C—C bond in long polymer molecules allows them to fold and wriggle. Entanglement and cross-linking profoundly affect their temperature-sensitive technical qualities, such as stiffness for service or fluidity for processing. Another useful distinction arises from the fact that the cohesion between atoms within molecules is much stronger than the cohesion between whole molecules – stiff springs versus sloppy springs as it were. That ensures that much of the vibrational energy of the atoms does not couple out to neighbouring molecules to form phonons. Vibrations within molecules have frequencies characteristic of the atom species and the bonding. This is the basis of chemical analysis by infra-red spectroscopy.

But it doesn't take much imagination to think of lots more ways for a solid to hold energy other than the whole-atom kinetic energy and the potential energy due to electron cloud overlap. For example, is it not surprising that the specific heat capacity of iron fits Dulong and Petit's law? Iron can conduct electricity so what about the kinetic energy of its free electrons? Does heating it change its magnetic domain pattern so that there is a work term for the influence of the Earth's magnetic field?

EXERCISE 4.4   List another half-dozen energy forms which might exist in a solid.

What matters in describing the changes of properties with temperature is whether there is a mechanism connecting the atomic kinetic energy to the particular form of energy responsible for the property in question. Thus there is a very direct coupling between kinetic energy and potential energy due to electron cloud overlap. Evidently, as the lamp filament story of Chapter 1 tells, the coupling to electromagnetic radiation, never mind *how* at the moment, rapidly grows stronger as $T$ rises ($E = \sigma T^4$) and astrophysics tells us that at a few million kelvins even nuclear energies are exchangeable with thermal rattle.

SAQ 4.2   (Objective 4.2)
The diagram Figure 4.7 indicates some possible energy exchanges (A to F) between atomic kinetic energy (which is proportional to temperature) and various other forms of energy in a solid. Identify items A to F as 'strongly coupled', 'slightly coupled' or 'weak/not coupled' at room temperature. Comment on this as an indicator of energy forms to be included as part of the thermal energy of a solid.

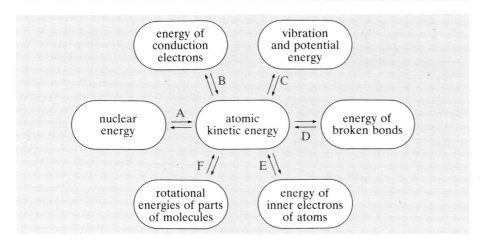

Figure 4.7

## Summary

Temperature is proportional to the average kinetic energy of the atoms of an object. Absolute zero of temperature is, for our purposes, when the atoms are at rest.

For perfect gases the average translational kinetic energy of an atom is $3kT/2$, is random, and is the *only* internal energy. The value of $k$ is $86\,\mu\text{eV}\,\text{K}^{-1}$ and $kT$ is used as a rough estimate of all atomic kinetic energies.

For all real matter there are other random internal energies which exchange between atoms by the agency of atomic collisions. The amount of internal energy for any of these forms depends on the effectiveness of the exchange mechanism and this will vary with temperature. Vibrations of atoms in solids strongly couple random kinetic and potential energies (due to interatomic forces).

Atomic vibrations are coupled along lines of atoms and may be regarded as quantum wave packets covering several atoms. These are phonons: they are the quanta of energy that the lattice of atoms can exchange in quantum jumps.

## 4.3 Average energy effects 1: thermal expansion

This section and the next provide models for properties interpreted in terms of the average thermal energy of all the atoms. Since $T$ measures the average atomic kinetic energy we shall expect to be looking at properties which change gradually with $T$, roughly proportionally, over a wide range. In terms of the classification introduced in Figure 4.3, we shall be looking at changes like Figure 4.3(a). The experimental descriptions of such changes usually use power series, setting the property $X$ as a function of $T$ with

$$X = X_0(1 + aT + bT^2 + \ldots)$$

Often the constant $b$ is much smaller than $a$ so that $X$ is almost proportional to $T$. Then $a$ is called the **temperature coefficient** of the property, and $a = (1/X_0)(\mathrm{d}X/\mathrm{d}T)$.

In particular, models will be set up for thermal expansion in Section 4.3 and for the electrical resistivity of metals in Section 4.4.

EXERCISE 4.5   Do you know of any other properties with this sort of temperature dependence?

## 4.3.1 Room to rattle: modelling thermal expansion

As the temperature of a piece of solid is raised its volume increases. Evidently the average spacing of its atoms has increased. The model needs to link more atomic kinetic energy (temperature) with bigger spacing. How? Expansion is a property well described by the two-atom model we started to use in Chapter 3. There the cohesion of a solid was expressed by the atoms being in a potential well. Figure 3.10 described how the potential energy of atoms varies with their spacing, and it was noted that the very strong repulsion induced when the atom cores try to overlap gives a steep inside edge to the curve, while the attractive forces are much less distance-critical so the outer face of the potential well is less steep. This asymmetry is the crux of the model of expansion.

At absolute zero temperature, with no kinetic energy, all the atoms will be spaced at the separation giving minimum potential energy so the length of a piece of solid is the sum of a vast number of these equal distances (I am supposing an element, with all atoms the same). At any higher temperature the atoms will have some higher total energies (quantum levels) and will be vibrating over some amplitude characterized by the shape of the potential well.

What is the kinetic energy of the atoms at those spacings where the total energy line crosses the potential energy curve?

Zero. The span of the well is the allowed range of motion.

The length of a line of atoms is now the sum of many different atom spacings, best added by trying to establish a time-average value. The average for any pair must be somewhere near the mid-point of their oscillation and because the well is asymmetrical that average is greater than the bottom-of-the-well separation. Also as the temperature rises the average kinetic energy goes up and takes atoms to higher levels in the well so the effect continues as Figure 4.8 shows. Thus thermal expansion fits into the model. Did you notice the phrase 'somewhere near the mid point' just now? Why not exactly? Because the well is asymmetrical the oscillations are not exactly simple harmonic, the atoms spend a little bit more time on the 'soft force' distant part of their motion than on the 'hard force' close approach part. This effect emphasizes the effect of the well shape. In Chapter 3 you saw various well shapes for various types of bond, so now you should be able to correlate types of bonding with an expectation of expansion coefficient.

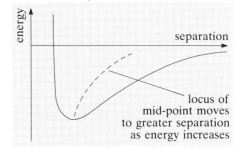

Figure 4.8

SAQ 4.3   (Objective 4.3)
Figure 4.9 shows potential curves A, B and C. Here also is a list of three substances, three values of melting temperature and three expansion coefficients. Can you match them all up and explain your reasons? Key: Magnesia (MgO), aluminium alloy, nylon; 250 °C, 2600 °C, 500 °C; $15 \times 10^{-6} K^{-1}$, $23 \times 10^{-6} K^{-1}$, $100 \times 10^{-6} K^{-1}$.

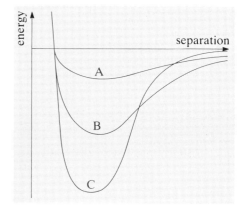

Figure 4.9

The model based on average spacing of atoms works best for metals with their close-packed atom structures. In the relatively open covalent crystals and glasses, angular distortions of the structure can accommodate some of the increase in average bond length. Consequently these materials show rather lower expansion coefficients $\alpha$; compare soda glass, porcelain and silica with the metals in Table 4.2. And there are other ways of compensating for thermal expansion. See ▼Invariable metals▲.

Table 4.2

| Material | $\alpha$ | $\mathscr{S}_v$ | $E$ | $\kappa$ | $K_{1c}$ | M = ? |
|---|---|---|---|---|---|---|
| | $10^{-6}\,K^{-1}$ | $10^6\,J\,m^{-3}\,K^{-1}$ | $10^9\,N\,m^{-2}$ | $W\,m^{-1}$ | $10^6\,N\,m^{-3/2}$ | |
| porcelain | 2.2 | 2.6 | 70 | 1.4 | 1 | |
| strong steel | 11 | 3.5 | 210 | 50 | 60 | |
| fused silica | 0.5 | 1.85 | 80 | 1.3 | 0.7 | |
| alumina | 8 | 4.0 | 345 | 21 | 4 | |
| nylon | 100 | 1.8 | 1.5 | 0.3 | 3 | |
| dural | 23 | 2.5 | 150 | 70 | 40 | |
| soda glass | 7 | 2.25 | 70 | 0.7 | 0.7 | |

## 4.3.2 Thermal stresses

What if there is a temperature change but some constraint prevents the proper thermal size changes? The constraint has to exert force to prevent the change of size so a thermal stress will be induced. There are many technical situations where thermal stress is made use of (for example, thermostat switches, toughened glass) or has to be worried about (cracked welds, bent rails, gas-tight seals).

# ▼Invariable metals▲

Invar, the 64% iron 36% nickel alloy, has a minute coefficient of expansion at room temperature, $0.9 \times 10^{-6}\,K^{-1}$, less than 8% of that of either constituent. It is *invariable*. What could be compensating for the thermal effect? The answer is in the temperature dependence of the magnetic repulsion of the atoms. For this particular alloy the rate at which the magnetic coupling slackens with temperature and allows the chemical bonding to pull the atoms closer just balances the normal thermal expansion effect. The alloy was discovered by Guillaume in 1899 and immediately replaced the awkward bimetal methods of expansion compensation in clock pendula, Figure 4.10. Guillaume went on to discover Elinvar in 1912. This alloy, 35% Ni, 10% Cr, 55% Fe, has zero temperature coefficient of Young's modulus implying not only a symmetrical potential well but a parabolic one.

Elinvar – *el*astic *invari*ant – solved the worst problem of inconstancy of spring driven balance wheels in watches, which was that the elasticity of the spring decreased with rising temperature and slowed the watch. A combination of beryllium bronze wheel and an elinvar spring with a *slight* thermo-elastic effect allows a simple structure whose oscillation is temperature independent, with expansion and elastic effects compensating one another. Here we have examples of newly discovered *properties* being taken up in *products* which could benefit from it. The *process* of clockmaking was affected in that assembly of the complex bimetal structures was no longer necessary. If the new alloys had been more expensive they might not have been favoured. It would be nice to think that the *principle* described led to the discovery, but the explanation comes after the discovery.

Figure 4.10 Clock pendulum using bimetallic compensation (courtesy of British Museum)

Every case is a problem of stress analysis in itself, but there are essentially three kinds of situation:

• A homogeneous body (all the same stuff) subject to external constraint (fixed in some way) suffers a uniform change of temperature (it all gets hot together). That's railway lines and cracking welds. (See ▼Homogeneous body, uniform $\Delta T$, external constraint▲.)

• A composite body (at least two different materials) whose constituents have different properties generates internal constraint (each material interferes locally with the other) when $T$ changes. Glass-to-metal seals needed in vacuum technology have to beat this problem; so do pottery glazes. (See ▼Composite body, uniform $\Delta T$, internal constraint▲.)

• Homogeneous body again but with non-uniform temperature change (different bits at different temperatures). When the temperature change is rapid the body is said to suffer thermal shock. This is what cracks bottles when you try to sterilize them by pouring in boiling water. How does your ceramic hob get on when you pour cold water on it? (See ▼Homogeneous body, non-uniform form $\Delta T$▲ and ▼Ceramic hob materials▲.)

# ▼Homogeneous body, uniform $\Delta T$, external constraint▲

(a)

Figure 4.11 (a) Buckled rail (Press Association). (b) Calculating the strain

The classic example of the buckled railway line, Figure 4.11(a), is an easy geometry to calculate. Suppose the length of a rail is $L$ when its ends just touch adjacent rails and then the temperature goes up by $\Delta T$, Figure 4.11(b). The length ought to become $L(1 + \alpha\Delta T)$ but is constrained to $L$. It is therefore as if the rail had expanded and then suffered a compressive strain $\alpha\Delta T$ which would have needed a stress (elastic) $E\alpha\Delta T$. This could bend the rail. The way out of the problem is to make provision for expansion and so remove the constraint: rail gaps, bridge roller bearings, and so on.

Some cracked welds belong to this class of problem. Here thermal stress has added to service stress intolerably, and Figure 4.12 is the result. It is good practice to anneal out any thermal stresses before the piece gets cracked.

$L$

ought to be $L(1 + \alpha\Delta T)$

compressed back to $L$ by $\sigma = E\alpha\Delta T$

(b)

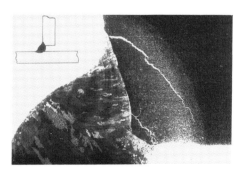

Figure 4.12 Cracked weld

# ▼Composite body, uniform $\Delta T$, internal constraint▲

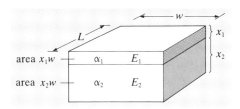

Figure 4.13 Glaze (upper part) on ceramic slab (lower part)

A relatively simple example to model is the glaze on a pot. In Figure 4.13, a thin glassy sheet ($x_1$), the glaze, is stuck onto a thick slab ($x_2$). Let $\alpha_1 > \alpha_2$. When, as the pot cools from its firing, the glaze gets below its glass transition temperature, it can no longer accommodate to the body. A thermal stress develops. The glaze should contract more than the pot, but both must move together. So the glaze is in tension and the pot in compression. At the interface the forces must balance. If naively we imagine different but uniform stresses $\sigma_1$ and $\sigma_2$ in the two layers (and treat just one direction)

$$\text{force in glaze} = \sigma_1 \times \text{area}_1$$
$$= + \sigma_1 x_1 w$$
$$\text{force in slab} = -\sigma_2 \times \text{area}_2$$
$$= -\sigma_2 x_2 w.$$

These forces balance as vectors when

$$(+\sigma_1 x_1) + (-\sigma_2 x_2) = 0 \quad (4.1)$$

The effective strains in the two parts are the differences between what they should have done ($L\alpha\Delta T$) and what they actually did ($\Delta L$):

$$\varepsilon_1 = \frac{(L\alpha_1\Delta T - \Delta L)}{L}$$

$$\varepsilon_2 = \frac{(L\alpha_2\Delta T - \Delta L)}{L}$$

The resulting stresses, assumed elastic with $E_1$ and $E_2$ the respective Young's moduli, would be $\sigma_1 = E_1\varepsilon_1$ and $\sigma_2 = E_2\varepsilon_2$. A short algebraic thrash gives the following stresses:

$$\sigma_1 = + \frac{x_2 E_1 E_2}{x_1 E_1 + x_2 E_2}(\alpha_1 - \alpha_2)\Delta T$$
$$\text{(glaze)}$$

$$\sigma_2 = - \frac{x_1 E_1 E_2}{x_1 E_1 + x_2 E_2}(\alpha_1 - \alpha_2)\Delta T$$
$$\text{(pot)}$$

Notice that the positive tension, $\sigma_1$, is much bigger than the negative compression, $\sigma_2$, because $x_2 \gg x_1$. That could crack the glaze, which might produce a pretty pot (Figure 1.14) but it is not the way to make a *strong* one.

Figures 4.14 and 4.15 show where similar thinking wins the prize. Gas tight glass-to-metal seals (Figure 4.14) are vital in vacuum technology and may have to survive thermal cycling. The seal can only stay gas tight if the expansion coefficients match accurately, crafted by informed composition choice. Figure 4.15 shows a bimetal switch. This is a cheap thermal switch. Equation 4.1 does not apply because the forces cause acceleration, that is the parts move.

Figure 4.14 Glass-to-metal seal

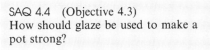

SAQ 4.4 (Objective 4.3)
How should glaze be used to make a pot strong?

Figure 4.15 Bimetal switch

# ▼Homogeneous body, non-uniform $\Delta T$▲

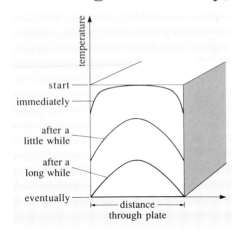

Figure 4.16 Variation of temperature with time within body subject to rapid cooling

Thermal shock arises because heat flow is not instantaneous. Suppose the surface of a slab is suddenly cooled, energy being extracted at some rate $\mathcal{Q}$ (measured in watts per square metre). The temperature inside cannot be uniform and will vary with time somewhat in the manner Figure 4.16 shows. At time $t = 0$, when cooling

begins, and after an infinite time, there are no internal stresses. At all other times every slice of the object should be a slightly different size from its adjacent slice, but is constrained by its neighbour. Complex non-uniform stresses are the result. We can, however, see what factors influence the magnitude of these stresses. Clearly very steep temperature gradients make for big stress and the gradient depends on how much energy has to flow from the inside to the surface, how quickly it can do so and the cooling rate at the surface ($\mathcal{Q}$). So, fast cooling of a thick, hot slab made of a material which has poor thermal conductivity, high volume specific heat capacity and high expansion coefficient, will induce intense stresses. If, additionally, the material efficiently converts the 'effective strain' to high stress via a high Young's modulus and if it is crack sensitive, you've got a problem on hand. Consider the chances of a ceramic turbine component when the engine is shut down; or a more homely example, how a ceramic hob survives having cold water split on it. The trick in that case is rather neat, as we shall see.

SAQ 4.5 (Objective 4.3)
The materials properties mentioned in the above analysis were:

- Thermal conductivity $\kappa$ ($W\,m^{-1}\,K^{-1}$).
- Volume specific heat $\mathscr{S}_v$ ($J\,m^{-3}\,K^{-1}$).
- Expansion coefficient $\alpha$ ($K^{-1}$).
- Young's modulus $E$ ($N\,m^{-2}$).
- Crack sensitivity (toughness) $K_{1c}$ ($N\,m^{-3/2}$).

(Don't worry if you haven't encountered the last one and its symbol before.)

We can combine this suite of properties into a 'merit index' $M$, expressing the ability of a material to survive thermal shock. If we set

$$M = \frac{\alpha \mathscr{S}_v E}{\kappa K_{1c}}$$

do we want $M$ to be big or small to express good performance? Appraise the materials cited in Table 4.2.

# ▼Ceramic hob material▲

The photograph of a domestic ceramic hob, Figure 4.17, suggests a puzzling suite of properties. The hot spots must be localized, so the material should not be a good thermal conductor or that will spread the heat sideways. If the hob is a bad conductor it must be thin to allow sufficient power transfer for its function. So how can it be strong enough to survive the following?

- Thermal shock if cold water is poured onto a hot ring.
- Thermal stress due to steep temperature gradients at the edges of the hot spots.
- Having a full saucepan dropped on it from one metre height – the acceptance test.

What is your recommendation for thermal expansion coefficient?

It must be as small as possible; thermal shock resistance can be achieved by low thermal strain and by reducing temperature gradients (by having good conduction). The latter is denied us so low expansion is imperative.

Fused silica glass has a very low expansion coefficient and a low enough thermal conductivity. What is unacceptable about a glass for this purpose? (Think of Chapter 2.)

Glasses are very untough as soon as they are scratched. They are bound to be scratched in service.

The solution is a ceramic based on lithium aluminium silicate, LAS, which crystallizes in a hexagonal form and is therefore anisotropic. Its expansion coefficients are $+6.5 \times 10^{-6}\,K^{-1}$ across the hexagon and $-2.0 \times 10^{-6}\,K^{-1}$ along the hexagon. Yes, a negative expansion. As the atoms spread sideways the layers can pack down into their hollows a bit more. This curious behaviour confers near zero coefficient upon polycrystalline LAS and chemical substitutions can control the exact value.

To make a ceramic hob in the required dimensions would be impossible by ordinary ceramic techniques, but plate glass is easily made. So the material is first

prepared as a glass and sheets are made by hot rolling (it is very viscous). Then a heat treatment encourages crystallization with the hexagon axes randomly oriented. If the crystals are very fine they can accommodate anisotropic shape changes without inducing large thermal stresses between differently oriented crystals. Also fine grain size confers strength on the ceramic body by limiting the size of intrinsic flaws which could act as cracks. How the nucleation of crystals is controlled is discussed in Section 4.7.

Figure 4.17 Ceramic hob

# 4.4 Average energy effects 2: resistivity of metals

The temperature variation of materials' electrical resistivity displays considerable diversity, Figure 4.18. This section will deal with metals' resistivity, indicated by curve A on the figure, which fits the classification as a 'gradual' temperature effect and is therefore a splendid thermometric property. Platinum resistance thermometers are amenable to precise calibration in terms of the international thermometric scale and are favoured where accuracy is the paramount requirement. However some more general comments now will help put this restricted topic into a wider context.

Electrical conduction is easily pictured as a flow of charged particles, and resistance as the flow being impeded somehow. Two energy conversions are involved: electric force times distance is work done on the charges to produce the current; and, since a current causes heating, the obvious impediment is by 'losses' to atoms (deliberate vagueness for later development). The identities of the charge carriers are also diverse. In electrolytes ions move, and in ceramics charge is carried by electrons hopping from ion to ion or by the diffusion of oxygen vacancies. In metals the agents are the electrons of the metallic bond. Whatever the particles transporting the charge may be, conductivity must depend on how many there are, $n$ per unit volume, what charge $q$ each holds, and some description of how easily they can move around. The memorable equation for conductivity is

$$\sigma = nq\mu$$

The last factor $\mu$ of this equation is called the **mobility** of the particles and is a measurable property of the material. Just as tractive force is needed to keep a car running at steady speed against wind friction, so the heating effect of an electric current warns of some 'friction' dissipating the work done by the electric force. By analogy we could suppose that the carriers move at a steady speed, known as the 'drift velocity' $v_D$ under the influence of the driving force. Mobility $\mu$ is then defined as the drift velocity per unit electric field strength $E$,

$$\mu = \frac{v_D}{E}$$

The value of the equation for conductivity is that it separates internal features of materials so that temperature variations of conductivity can be explored; either the number of carriers might change with $T$ or it may be the processes by which their flow is impeded. We shall find examples of both as the subject develops. The other term, the charge per particle does not change. In a metal the number of electrons available and their charge is constant so we must look to variations in mobility to account for the temperature effects on resistance.

Electrons ought to be dealt with by quantum mechanics, so the idea of energy being continually drawn from the applied electric field in a

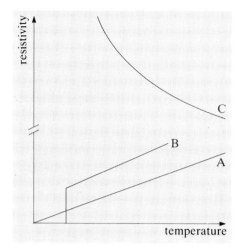

Figure 4.18 Variation of resistivity with temperature. Curve A is a metal; B is a superconductor; C is an intrinsic semiconductor

conductor and continually drained off by friction should not satisfy us. Rather, each electron must take in and lose energy in jumps. As it turns out the separation of energy levels for electrons in solids is so very small that the uptake of energy from the source can indeed be treated by non-quantum methods.

The quantum-level separation in a metal is estimated as $10^{-11}$ eV. How does this compare with $kT$ at 300 K?

At 300 K, $kT \approx \frac{1}{40}$ eV, 2.5 billion times the free electron quantum-level spacing.

The jumps by which energy is lost are much bigger, $\approx kT$, so these are really quantum events rather than continuous. One consequence, deducible from observed drift velocities, is that electrons apparently only 'collide' with about one in a hundred atoms. That is difficult to imagine for charged 'particles' being driven through a jungle of other charges. The electrons responsible for metallic bonding are delocalized (Chapter 3) and are better thought of as wave packets than as point particles. Quantum treatment then shows that a perfectly regular lattice is essentially transparent to electrons. But phonons, that is, spread out wave packets of vibration energy, spoil this regularity. By thinking of electrons and phonons as wave packets quantum theorists have built good models of their mutual interactions through which electrons absorb energy *from* the lattice or deliver energy *to* it. That theory is not presented here. For now it suffices to regard the electrons as particles for their energy take-up from the source of electric power, and to imagine that somehow they manage to interact infrequently with the atoms. Then some quite graphic interpretations of resistance and mobility can be drawn.

Think first of a metal *not* carrying current. Since it neither heats up nor cools spontaneously, yet the free electrons are buzzing around in the crystal, the interactions with phonons must go both ways with equal facility. Every now and then an electron takes up the energy of a phonon and jumps a few billion levels to a higher quantum state. At other times it drops down a few billion levels and launches a phonon in the lattice. Exact dynamic balance, with equal (*not zero*) amounts of energy transferred in each direction, expresses 'thermal equilibrium' between the electrons and the atoms.

With a voltage across the metal, electrical energy is taken up with non-quantum mechanical smoothness, as each electron accelerates in response to the electric field.

If atoms are $2 \times 10^{-10}$ m in diameter and an electron accelerates past 100 atoms in a field of 0.5 V m$^{-1}$, how much kinetic energy will it gain? Is this still small compared with $kT$ at 300 K?

The distance travelled is $100 \times 2 \times 10^{-10}$ m so the potential drop is only $10^{-8}$ V and the electron has gained $10^{-8}$ eV of kinetic energy; still very much less than $kT$.

Now, because the electrons have taken in energy from the field, they are a bit 'hotter' than they should be. So, to re-establish thermal equilibrium, a bit more energy is transferred *from* electrons *to* the lattice than *vice versa*. More phonons are launched than are absorbed. What our thermometers respond to (atomic kinetic energy) increases, and we have a picture of the electric heating effect. Because there is an interaction, the orderly energy of the flowing electrons is dissipated into the randomly held energy. The memory of the energy which had just been gained from the field is lost and, on average the electron starts from rest again. Figure 4.19 shows the energy changes for an electron over a short time interval.

Because *all* phonon energies greatly exceed the ordered electrical energy of the electrons, all interactions between the flowing electrons and the vibrating atoms disrupt the smooth take up of energy from the electric field. Evidently the resistance of the lattice to the electrons' progress depends on how often these interruptions occur. That must be to do with how many phonons the lattice is supporting, which is roughly measured by the average atom energy ($\approx kT$). So we find metals' resistances obeying the law

$$R = R_0(1 + aT + bT^2 + \ldots)$$

and can regard the temperature effect as gradual.

The idea of a steady drift velocity, got by analogy with friction, does not square with this quantum-jump model. More realistically $v_D$ is to be seen as an *average* speed as electrons are repeatedly accelerated in an orderly direction by the field and scattered back to randomness by the phonon interactions. If, as in Figure 4.20(a), just the orderly speed for an electron were plotted against time there would be an irregular sawtooth progress to average. Figure 4.20(b) shows that if phonon encounters are more frequent, which is the effect of higher temperature, the sawteeth get smaller and the average is reduced. Smaller drift velocity, lower mobility, reduced conductivity is the sequence which accounts for the temperature effect.

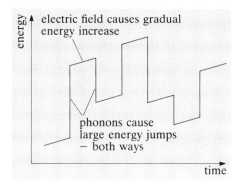

Figure 4.19 Energy changes of electron subject to electric field in metal at temperature above 0 K

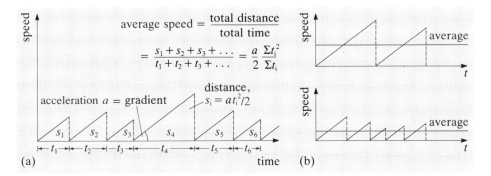

Figure 4.20 (a) Finding the average speed of the electron. (b) The average of many small sawteeth is less than that of fewer large ones

But phonon scattering is not the only way of impeding electrons. Property control is what we are interested in; what other ways can we find? See ▼**Resistivities of transition metals**▲

EXERCISE 4.6   Between two phonon interactions an electron accelerates past 1000 atoms of $2 \times 10^{-10}$ m diameter in a field of $1.4\,\mathrm{V\,m^{-1}}$. What maximum speed will it reach? What is the equivalent steady speed?
Electron mass $\approx 10^{-30}$ kg, electron charge $\approx 1.6 \times 10^{-19}$ C.

## 4.4.1  A glimpse at superconductors

Item B on Figure 4.18 showed a resistivity which jumps to zero at a critical temperature; not a 'gradual' phenomenon but one it is convenient to speak of here. This is superconductivity, which hit the headlines in 1986–7 when new ceramic materials were discovered with their critical temperature above liquid nitrogen temperature and therefore accessible by cheap refrigeration. Hitherto the record high had been only 23 K and the effect could only be used in conjunction with expensive liquid helium refrigeration.

In those earlier superconductors the charge carriers were still electrons of the metallic bond but they were somehow immune to interruption by phonons. How? The answer is again only really adequate in quantum modelling but, again, a caricature which is at least a starting point can be given. If electrons behaved by 'big world' mechanics, they would momentarily attract the nearest positive ions as they moved through a metallic crystal lattice, and a sort of bow wave of elastic displacement would diverge from the electron's track. Another electron moving at just the right speed and place on a parallel track could find the crystal being opened by this wave to let it through. A weak coupling between pairs of electrons can be envisaged in quantum mechanics using this effect. It doesn't produce the continuous loss of energy which bow waves demand of ships! At low enough temperatures when random phonon energies are small they are unlikely to disrupt the pairs. Just a few of these, with ultra-high mobility, confer virtually infinite conductivity upon the metal. As you might expect, a strong magnetic field, which will try to bend the track of the electrons, can break up the pairs so the critical temperature falls as magnetic field is applied, Figure 4.21. The proper quantum version of this tale won a Nobel physics prize but the explanations are apparently not instantly transferable to the newer superconducting materials.

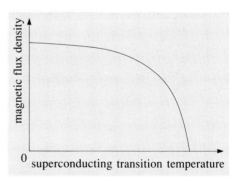

Figure 4.21 Superconducting transition temperature falls as magnetic intensity increases

# ▼ Resistivities of transition metals ▲

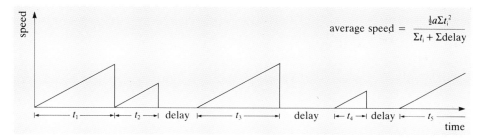

$$\text{average speed} = \frac{\frac{1}{2}a\Sigma t_i^2}{\Sigma t_i + \Sigma\text{delay}}$$

Figure 4.22 Delays reduce average speed

The bonding in transition metals is partly by delocalized electrons and partly by covalency, and individual electrons switch roles frequently (Chapter 3). An electron caught in a covalent bond stops doing any conduction until it jumps again into a mobile state. Just as a break in a journey brings your average speed down so do these delays reduce conductivity, Figure 4.22. By how much depends on how many traps there are, how well they catch electrons and how long the delay is. Many traps can produce high resistivity. So titanium has lots of traps because few of its covalent orbitals are full. It has a resistivity of $\approx 60 \times 10^{-8}\,\Omega\,\text{m}$. Copper has all its covalent states in action so trapping is infrequent and a lower resistivity of $1.6 \times 10^{-8}\,\Omega\,\text{m}$ results. Apart from manganese, an element which consistently defies simple explanations of its properties, there is a downward trend of resistivity through the 3d transition group, see Figure 4.23.

In alloys the effects are more confused. The law of mixtures (Chapter 2) does not apply to alloys' resistivity, as Figure 4.24 shows for copper–nickel alloy. Some

alloys are technically useful because the temperature coefficient of resistivity can be controlled by composition. You can guess that trapping delays might get shorter if there's more rattle in the solid, because increased phonon activity may break covalent bonds. So an increase in conductivity with rising temperature from this effect may compensate for the straight thermal decrease. The 50/50 Cu–Ni alloy, known as constantan gets the balance just right (see the red $d\rho/dT$ curve on Figure 4.24) and is used for temperature-stable resistors. Compare this compensation effect with that for thermal expansion in invar; transition metals offer nice possibilities in all sorts of property fiddling.

The concept of trapping is vital to explanations of many other phenomena, the luminescence of the materials used to coat television screens for example.

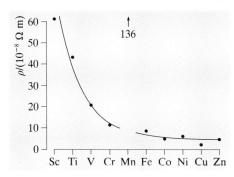

Figure 4.23 Downward trend of resistivity through the 3d series of transition metals

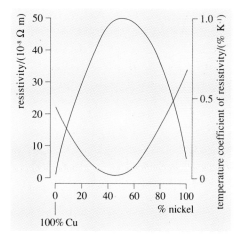

Figure 4.24 Resistivity and temperature coefficient of resistivity for copper–nickel alloys

## Summary of Sections 4.3 and 4.4

Gradual changes may be described by power series. When the change is reasonably linear it is useful to define a temperature coefficient for the effect.

Thermal expansion is caused by the average spacing of atoms increasing with temperature. It can be modelled by noting the asymmetry of the potential well describing interatomic forces. It is therefore related to bonding characteristics such as $T_m$.

Compensating effects can produce near zero expansion, for example in crystalline invar or lithium aluminium silicate.

Thermal stress arises when temperature change is accompanied by mechanical constraint. Constraint can be external, internal due to different materials fixed together or internal due to temperature gradients. There are benefits and problems from thermal stresses.

Mobile electrons in metal are the sources of electrical conductivity. They appear to absorb energy from an applied electric field with 'classical' smoothness because quantum levels are very very closely spaced in energy. However, the interactions with the lattice are much bigger energy jumps, and are best thought of as absorption or launching of phonons rather than individual electron–atom impacts. This model takes care of the delocalized natures of both electrons and lattice wave packets.

Conductivity $\sigma = nq\mu$. It enables temperature effects to be separated into dependence on $n$ or dependence of $\mu$ on $T$. In metals, resistive heating is the surrendering of drift energy to the lattice by 'collisions' launching phonons. How often this happens affects the drift velocity and hence $\mu$.

Mobile electrons may also become trapped in a local chemical bond. Trapping accounts for higher resistivity of transition elements with d-shell states empty.

Temperature changes in localized and delocalized bonding can compensate for temperature changes in phonon resistance. Constantan has almost zero temperature coefficient of resistivity because of this effect.

Superconductivity depends on weakly coupled pairs of electrons carrying each other through 'collisions'.

# 4.5 Extreme energy effects 1: atoms and molecules

The models of the last section used the *average*-energy atoms ($\approx kT$) and had no concern for what distribution of energies might give that average. The effects which 'accelerate' as $T$ rises depend on interactions of atoms which chance to have much higher than average energies. Because, as we shall see, the population of such atoms is strongly temperature dependent, so are the rates of the consequent thermally activated events. Hence adjusting temperature to control chemical processes and shaping processes may be a matter of some delicacy. For the same reason chemical or deformational service performances (corrosion rate, Chapter 7, and creep resistance, Chapter 6) can be strongly temperature dependent. Diffusion, where atoms spontaneously intermingle under the influence of their thermal energy is, in solids, another thermally activated phenomenon. It is used to case harden steel (Figure 4.25) by getting extra carbon into its surface. The carbon is quickly diffused in at $\approx 950\,°C$; in service at $\approx 100\,°C$ in the crankshaft of your car, diffusion is millions of times slower so the carbon stays just where it was put. The steep temperature rise of conductivity in semiconductors, explained by the few most energetic electrons getting to an empty band of energy levels, is thermal activation again, though here we have to find out how *electron* energies are distributed, and that is quite a different story from what we shall see for atoms and is treated in the following section. Thermal activation is the key idea for explaining accelerating temperature sensitivity of properties or processes. So what are the models for this behaviour?

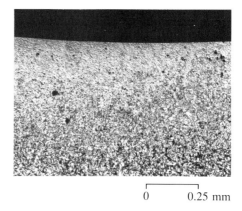

Figure 4.25 Case hardened steel. Carbon has been diffused in from the edge, producing a lighter coloured microstructure. The carbon concentration diminishes with distance from the edge, as can be seen by the darkening of the microstructure away from the edge

## 4.5.1 The energy distribution of perfect gas atoms

As ever that simplest state of matter, the perfect gas offers the easiest insight. A chemical reaction in a gaseous mixture will require some particular bonds in a molecule to be broken. The rate of such a reaction will depend on how many molecules per second suffer broken bonds. One way bonds get broken is by crashes between sufficiently thermally energetic molecules. In that there are strong bonds and weak bonds, one question is: how much energy is sufficient? Many bonds need $\approx 1\,eV$ to rupture them, which you'll remember is forty times the average at 300 K, so we really are thinking of the few molecules with exceptionally high kinetic energy as useful agents. The next questions are: how many molecules have at least sufficient energy and how does that number depend on temperature? Figure 4.26 depicts the energy distribution of atoms in a perfect gas and ▼**How do you know the molecular kinetic energy distribution?**▲ describes an experiment by which this can be measured. The number of atoms in a narrow energy range $E$ to $E + dE$ is represented by the area of a strip under the curve. This area is the height of the strip times its width, or $N(E) \times dE$. According to high powered analysis and in agreement with experiment,

$$N(E) = A(E/kT)^{1/2} \exp(-E/kT)$$

in which $A$ is a constant and where both $E$ and $T$ are important. For this case the low-energy end of the curve is controlled by the $E^{1/2}$ term, which rises with $E$ but flattens out. But the exponential term *falls* as $E$ rises and beyond the hump dominates the function, especially the part we are interested in, where $E$ is much greater than the average $kT$. Up

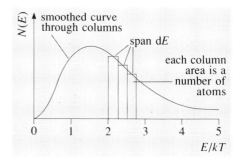

Figure 4.26 Energy distribution of atoms in perfect gas

179

# ▼How do you know the molecular kinetic energy distribution?▲

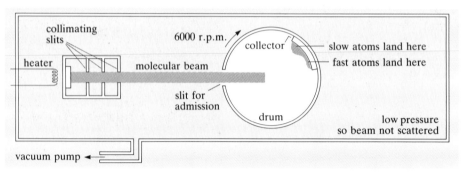

Figure 4.27 Apparatus for determining velocity distribution of bismuth atoms

The ideas of energy distribution presented are part of the theoretical foundation of classical physics, so experimental confirmation was in its day a matter of some importance. Figure 4.27 shows the set up used in experiments in the 1930s. A small furnace provided vapour of the heavy metal bismuth which escaped into a vacuum chamber through a series of slits to form a collimated beam of atoms.

Bismuth was a wise choice having a low boiling temperature; and its atoms, being heavy, carried the kinetic energy appropriate to the furnace temperature at modest speeds. In front of the hole a drum rotated at 6000 r.p.m., admitting the beam through a slit. On the far side of the drum the atoms condensed on a chilled glass plate. You can see that the atoms in a burst will cross the drum in different

times depending on their speed and so be spread out on the moving plate. The furnace temperature and rotor speed were held constant for long enough to build up a deposit on the plate. To compare the numbers of atoms at each speed the varying thickness of deposit was judged by its transparency. So the distribution for that temperature was discovered, and found to agree with the theory.

here we can say, roughly, that each successive strip of the distribution $kT$ wide has only $1/e$ as many atoms as the preceding one.

For the purpose of modelling thermal activation we really want to know the fraction of atoms having any energy greater than the threshold amount, the activation energy. It is easy to integrate the theoretical function from $E$ to infinity to reveal the fraction of atoms having kinetic energy greater than $E$ as $(E/kT)^{1/2}\exp(-E/kT)$. For virtually all practical purposes, activation energies are many $kT$ and the exponent dominates the fraction. Now try Exercise 4.7.

> EXERCISE 4.7 In a molecule one bond breaks in collisions giving 1 eV of energy and another breaks more easily, needing only 0.8 eV. What are the relative chances of each kind of break at 300 K? How much effect does the $(E/kT)^{1/2}$ term have?

The answer to the exercise may be quite a surprise! A 20% change in the fragility of a bond renders it enormously more likely to be broken. Notice though that we do not yet have any estimate of the absolute chance of either break; they might both be very small at 300 K (petrol is safe in the tank).

Now let us see the effects of temperature on the curve of Figure 4.26. As $T$ goes up $E/kT$ gets smaller, so $\exp(-E/kT)$ gets bigger. A family of energy distribution curves for different temperatures is shown in

Figure 4.28. The important feature is that higher $T$ produces more atoms with extra high energies. But again what matters is the fraction of atoms with energy exceeding the threshold value. *In fact this rise is also exponential.* This is just what we need in order to explain the accelerating effect. Such effects are triggered by just a small proportion of the atomic collisions, but that proportion rises rapidly with temperature. In the hot gas around the spark in an engine, the reaction between petrol and air becomes fast.

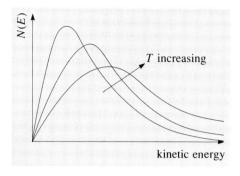

Figure 4.28 Energy distribution changing with temperature

EXERCISE 4.8 Compare the number of atoms in a gas having $E > 1\,\text{eV}$ at 1000 K and 300 K.

## 4.5.2 Arrhenius's rate equation

Exercise 4.8 took a chemical reaction in the gas which required 1 eV bonds to be broken, and compared the rate of the reaction at 300 K and the rate at 1000 K. To add perspective it implies that what would take 30 000 years at room temperature is done in 1 second at 1000 K. Although a bond will not be broken in every collision involving 1 eV of kinetic energy, the rate of bond breaking must be proportional to the number of such events per second and in turn that must be proportional to the population of these high-energy molecules. So the **Arrhenius rate equation** for gas reactions is an automatic deduction from the kinetic energy distribution.

$$R = A \exp(-E_a/kT)$$

is Arrhenius's equation with $R$ the rate, $E_a$ the activation energy and $A$ a constant for the effect which contains the probability that activation will result from a collision of sufficient energy. The term $A$ is relatively insensitive to temperature. The equation proclaims that fast rates need low activation energy and/or high temperature. The role of catalysts in chemical processes is to provide low-activation-energy routes, which allow reactions to go fast at feasible temperatures.

By good fortune, the same exponential factor dominates the distribution of all those other energies mentioned in Section 4.2.2 among atoms and molecules of real matter at the interesting high energy end of their spectra. In particular, the distribution of energy in vibration is exponential. Arrhenius's equation is therefore also valid for the many mechanisms in solids which depend on small proportions of atoms shaking free from the constraints of bondage. The thermally activated phenomena cited at the start of this section are examples.

A further matter crucial to decisions about process temperatures is to understand how the rates of *competing* effects with different values of $E_a$ are differently sensitive to temperature. Then things can be organized so that what we want to happen does happen, and what is to be avoided is avoided.

The dependence of rate on temperature is quickly shown. If

$$R = A \exp(-E_a/kT)$$

then

$$\frac{dR}{dT} = \left(+\frac{E_a}{kT^2}\right) A \exp(-E_a/kT)$$

or

$$\frac{1}{R}\frac{dR}{dT} = +\frac{E_a}{kT^2}$$

Thus the temperature coefficients of all rates are positive – increasing $T$ increases $R$. But the $E_a$ which comes to the front by differentiation is important. It ensures that proportionally the rates of effects with bigger $E_a$ terms increase faster with temperature than those with lower $E_a$ terms. This is a general theorem, so be aware of its consequences when trying to hasten a process by increasing temperature. Although the process may be speeded – the flow index of a molten polymer improved enabling easier extrusion say – some undesirable effect, the oxidation rate of the material perhaps, may accelerate even more. An engineering example of careful temperature control of a process is cited in
▼Temperature control of a chemical process▲

# ▼Temperature control of a chemical process▲

Ammonia, $NH_3$, is an essential substance in the production of several polymers, for example nylon, Acrilan, superglue and so on, to say nothing of fertilizer, dyes and explosives. The reaction is reversible and throws out heat (*exo*thermic) as it goes forward,

$$N_2 + 3H_2 \rightleftharpoons 2NH_3 + heat$$

It follows, Figure 4.29, that the activation energy for the forward reaction is less than that for the reverse route. As soon as the gases enter the reaction chamber and meet the catalyst, they begin to react and the heat given out starts to raise the temperature. How long can this be allowed? On the one hand it makes the reaction yet faster, a 'good' thing, on the other hand we are headed for an explosion. Not only that, but the reactor designer has to worry about the rate of the reverse reaction. As the concentration of ammonia rises, the reverse reaction becomes more active; there are more $NH_3$ molecules to crash into one another. If the reaction comes to equilibrium, ammonia being created and destroyed at the same rate, the expensive hardware is achieving nothing of value. But our theorem says that the rate of the reverse reaction, with the larger activation energy, is the more

temperature sensitive. So if the temperature of the gases is brought down, the reverse reaction will be slowed more than the forward one, and ammonia will continue to form. The temperature profile along the reactor, Figure 4.30, is controlled by judicious extraction of heat (used elsewhere). But that can't go on indefinitely.

Why not?

Because the reactor gets less and less effective. The rate of the forward reaction is also suppressed by falling temperature. Some fancy modelling is needed to design the best compromise.

What financial factors would enter the 'best compromise' model?

Return on capital; that depends on how much more capital is needed to build a bigger reactor, what the cost of using that capital is, and what extra revenue would accrue from the plant. The last two items are time dependent while the plant is running because interest rates and the value of ammonia may change. The best design will be insensitive to these factors over a range of production rates so that demand fluctuations can be inexpensively followed. Storage of stocks would be expensive.

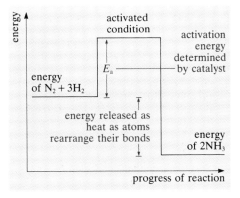

Figure 4.29 Energy versus progress of reaction for production of ammonia

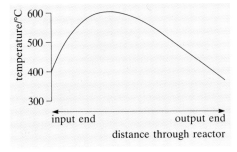

Figure 4.30 Temperature versus distance through reactor for ammonia production

The final trick I want to show you with Arrhenius's equation is how to extract the constant $A$ and $E_a$ from experimental data. If the equation is 'inverted' by taking natural logarithms of both sides it becomes:

$$\ln R = \left(-\frac{E_a}{k}\right)\frac{1}{T} + \ln A$$

$$y \quad = \quad m \quad x + \quad c$$

Comparing the logarithmic equation with the standard equation for a straight line you can see how to plot rates $R$ obtained at several temperatures $T$ to get a straight line. If the Arrhenius model is valid, $\ln R$ against $1/T$ is what should be plotted.

How do you get the $A$ and $E_a$ from the graph?

$A$ comes from the intercept at $1/T = 0$ and $-E_a/k$ is the gradient. ▼Using Arrhenius plots▲ shows how to do this.

# ▼Using Arrhenius plots▲

A hot-curing single-part epoxy glue called, say, Ticky Tacky sets in 6 minutes at 137 °C, 10 minutes at 127 °C, or 30 minutes at 100 °C. That is sufficient data for an Arrhenius plot to be drawn. The rate of reaction is the reciprocal of the setting time, and temperatures need to be in kelvins. Table 4.3 converts the data to the appropriate form and Figure 4.31 is the corresponding graph.

To find the activation energy for the curing reaction we must calculate the gradient of this logarithmic line. Using the arrowed points

$$\frac{-E_a}{k} = \frac{-3.5 - (-1.5)}{(2.7 - 2.35) \times 10^{-3}}$$

$$= \frac{-2000}{0.35} \text{ K}$$

Taking $k = 86 \times 10^{-6}\,\text{eV K}^{-1}$ gives

$$E_a = \frac{2000 \times 86 \times 10^{-6}}{0.35}\,\text{eV}$$

$$= 0.49\,\text{eV}$$

The graph shows a typical awkward feature. Because measurements are over a narrow range of $T$, the $T^{-1}$ axis is very stretched so $T^{-1} = 0$ is not immediately readable. By calculation instead, taking the data for $T = 400\,\text{K}$ (that is, $T^{-1} = 2.50 \times 10^{-3}\,\text{K}^{-1}$ and $\ln R = -2.30$)

$$\ln A = \ln R + \frac{E_a}{kT}$$

$$= -2.30$$

$$+ \frac{2000}{0.35} \times (2.50 \times 10^{-3})$$

$$= 11.99$$

So $A = 160\,000\,\text{mins.}^{-1}$

SAQ 4.6 (Objectives 4.5 and 4.6) You have to choose between Ticky Tacky and a similar glue, Bangiton, for your home-based toy production line. You expect to use about 1 kg of glue per week and there's a good discount for orders of at least 10 kg. Bangiton sets in 5 minutes at 127 °C or 30 minutes at 100 °C. You have a domestic refrigerator which runs at 0 °C for storage. Unfortunately the blurb divulges nothing about shelf life for either glue. Plot an Arrhenius graph for Bangiton using a wide enough $T$ axis to reach 0 °C ($T^{-1} = 3.66 \times 10^{-3}\,\text{K}^{-1}$), transfer the line of Figure 4.31 onto your graph and use it to decide whether bulk purchase of either glue is wise. Evaluate the activation energy for the curing reaction of Bangiton.

Table 4.3 Data for Ticky Tacky

| Temperature $T$ | | $T^{-1}$ | Setting time | Rate $R$ | $\ln R$ |
|---|---|---|---|---|---|
| °C | K | $(10^{-3}\,\text{K}^{-1})$ | (mins.) | $(\text{mins.}^{-1})$ | |
| 137 | 410 | 2.41 | 6 | 0.17 | $-1.80$ |
| 127 | 400 | 2.50 | 10 | 0.10 | $-2.30$ |
| 100 | 373 | 2.68 | 30 | 0.033 | $-3.40$ |

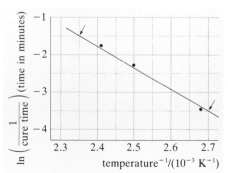

Figure 4.31 Curing curve for Ticky Tacky

## 4.5.3 The glass transition of armophous polymers

A good example of thermal activation is found at the glass transition of amorphous polymers. You will remember that polymer molecules are primary bonded along their length but weaker (secondary) bonds hold the chains together. When sufficient of the between-chain bonds are thermally disrupted, lengths of molecule become much freer to move. Since the chances of rattle overcoming stick go up exponentially with temperature, we should expect the onset of relatively free movement to occur at a fairly well determined temperature. This temperature is known as the glass transition temperature, $T_g$.

Several properties all of which depend on the ease of molecular movement show accelerating change at or around the same temperature. Notably the stiffness of the material alters markedly from rigid below $T_g$ to floppy above. This softening effect has been noted for silicate glasses in Chapter 3 and is the origin of the name. For those materials it marks the change from the hard brittle condition to the ductile workable state used by glass blowers.

Young's modulus of some organic polymers is plotted as a function of temperature in Figure 4.32(a). Some polymers find service above $T_g$ as rather floppy or flexible materials. Polyethylene and rubbers are examples. Others, polystyrene, polycarbonate, poly(methyl methacrylate), are used below their glass transition temperatures and are recognized as harder, often brittle, materials. PVC is interesting. Its $T_g$ is widely adjusted down from 80 °C by blending with much short chain plasticizers. UPVC, with no such addition, is now popular as a stiff and durable window frame material requiring no painting. Plasticized PVC provide the ubiquitous simulated leather for such applications as luggage. $T_g$ for this material is some 100 K lower.

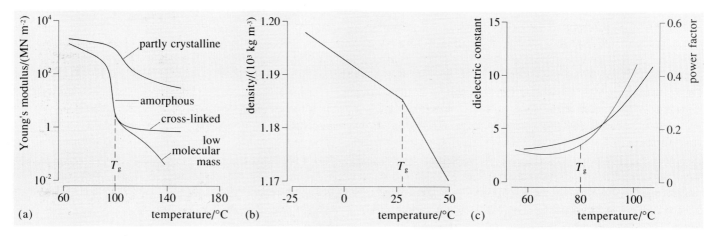

Figure 4.32 Temperature-variation of properties of organic polymers. (a) Young's modulus for polystyrene. (b) Density of poly(vinyl acetate). (c) Dielectric constant of unplasticized poly(vinyl chloride)

Figures 4.32(b) and (c) show the changes in some other properties around $T_g$. There is a change in density; freer molecules rattle around in more space than locked ones. Above $T_g$ the distributed charge on polymer molecules can move in response to an applied electric field. So the dielectric constant increases. Unfortunately the loss factor also goes up; that is, more energy is converted from electrical to thermal forms by the movable charges as they vibrate in response to an alternating voltage. For this reason polymer dielectrics for capacitors have to be used below $T_g$ and rely on electronic rather than molecular displacement for their polarization. Polystyrene, with phenyl rings full of delocalized electrons and with $T_g \approx 100\,°C$, is a favourite.

▼**Polymer molecular structure and the glass transition**▲ relates $T_g$ to molecular structural features for several polymers.

Metals can be given the amorphous glassy structure by very rapid cooling from the melt. Can these be processed by glass techniques?

No. Raising the temperature to their $T_g$ would induce crystallization. A very fine grain structure with good mechanical properties results. But they lose their peculiar and useful magnetic isotropy.

## 4.5.4 Diffusion in solids

Diffusion is spontaneous movement of matter caused by atoms' thermal motion. Management of diffusion as a means of altering local composition is vital in many fields of engineering. Figure 4.25 showed an example where carbon had been diffused in iron to harden it. Dopants are manoeuvred in semiconductors by diffusion, and the same phenomenon is responsible for the sintering processes for joining powders into solid masses. (Sintering is used for high-performance ceramics, allows near-net shape forming of metals and enables awkward polymers such as PTFE to be moulded.) That matter really moves is shown in Figure 4.33, where three fine nickel wires have been sintered into a single piece.

Diffusion in solids is a slow process. In solids, the kinetic energy is in the form of vibration, so on the whole nothing moves anywhere – except 'occasionally' when an atom manages to jump from one lattice site to another. This possibility is thermally activated, but few mechanisms for such movement have low enough activation energy to allow any reasonable transport rate. (We shall see what the mechanisms are later.) Even when atoms do jump, they jump in random directions so the overall travel of mass is very small. For example in carburizing iron at 950 °C for three hours, a typical treatment, it is estimated that each carbon atom has travelled about 1 km by making $10^{13}$ jumps, each of about $10^{-10}$ m. Yet in Figure 4.25 the carbon has penetrated less than 1 mm; there are nearly as many steps back as forward. (These numbers imply that 'occasionally' means a thousand million times per second, but since atoms vibrate at about $10^{12}$ Hz, a jump every 1000 vibration cycles is fairly occasional.)

(a) 5-1/4 HOURS

(b) 168 HOURS

(c) 241 HOURS

(d) 241 HOURS

Figure 4.33 Successive stages in the sintering of wires (c) and (d) show different places in the twisted wires

# ▼Polymer molecular structure and the glass transition▲

Two kinds of factor affect the temperature of the glass transition in polymers. First, the strength of bonding *between* the long molecules. Second, their ability to 'wriggle' once a section has become free. There are few, if any, instances where changes of molecular structure unequivocally affect one of these factors without affecting the other. Table 4.5 provides examples and comments.

High density polyethylene with very long unbranched chains of

$$(-CH_2-CH_2-)_n$$

and hydrocarbon rubbers

$$(-CH_2-CH=CH-CH_2-)_n$$

(polybutadiene) have minimal interchain bonding (van der Waals bonding only) and are flexible by virtue of easy rotation about the C—C bonds. Waves or 'skipping rope' motions of segments of chain become possible as soon as a few repeat units of length are free. Low $T_g$ is found.

Significantly strong interchain bonding acts between nylon molecules by virtue of hydrogen bonds from the —NH— group to the

$$-\overset{\|}{\underset{O}{C}}-$$

group. This is the main reason for the higher $T_g$ values of all nylons since there are single (twistable) bonds all along the backbone and most of it is —CH$_2$—CH$_2$—.

Much less flexibility is available if bulky side groups hang onto the main chain or if rings are incorporated into the backbone. The table gives instances of both. Impedance of chain flexure by side groups is called **steric hindrance**. Obviously the effectiveness of a group in hindering flexure depends on its size and how often

it appears in the chain. Thus the side group —CH$_3$ in polypropylene is fairly small compared with the C$_6$H$_5$ ring in polystyrene and $T_g$ values of 5 °C and 100 °C respectively are recorded. PTFE, which has two fluorine atoms on every C atom, is very stiff since the fluorine atoms interfere strongly with one another if rotation is attempted. This substance shows a high $T_g$. Another stiff backbone is provided by the rings-within-the-chain of polycarbonate and again $T_g$ is high.

Thermosetting polymers do not show any glass transition below their decomposition temperatures since chains have been crosslinked together by strong primary bonds.

Vulcanized rubbers, however, offer a trend from thermoplastic to thermoset as the degree of crosslinking can be controlled. $T_g$ rises as crosslinking is increased (see Table 4.4).

Table 4.4  $T_g$ of vulcanized natural rubber

| Sulphur content/% (degree of cross-linking) | $T_g$/°C | Application |
|---|---|---|
| 0 | −65 | |
| 2 | −50 | tyres |
| 10 | −40 | shock absorbing bushes |
| 20 | −24 | |
| 32 | +60 | ebonite |

SAQ 4.7  (Objective 4.7)
Here are three more polymer repeat units. By comparing their structures with examples in Table 4.5 speculate what $T_g$ values they might display.

(a) Poly(chlorobutadiene)  (Neoprene rubber)

$$-CH_2-\overset{\underset{|}{Cl}}{C}=CH-CH_2-$$

(b) Poly(methyl methacrylate) (Perspex)

$$-CH_2-\overset{\overset{CH_3}{|}}{\underset{|}{C}}-$$
$$COOCH_3$$

(c) Poly(phenyl ethersulphone) (plastics for computer casing)

$$-\bigcirc-O-\bigcirc-\overset{O}{\underset{O}{\overset{\|}{S}}}{\underset{\|}{}}-$$

Table 4.5

| Polymer | $T_g/°C$ | Repeat unit of chain | Method of raising $T_g$ |
|---|---|---|---|
| Polybutadiene | −107 | —CH₂—C=CH—CH₂—<br>    H | Lowest $T_g$. The double bond makes adjacent C—C easier to rotate. |
| Polyethylene | −20 | —CH₂—CH₂— | C–C bonds only |
| Polypropylene | +5 | —CH₂—CH—<br>    CH₃ | Small neutral side group: weak steric hindrance (SH). |
| Poly(vinyl acetate) | +28 | —CH₂—CH—<br>    O<br>    O=C—CH₃ | Larger, weakly charged side group. SH not too strong because side group is flexible. |
| Nylon 6,6 | +60 | —N—(CH₂)₆—N—C—(CH₂)₄—C—<br> H    H O    O | Hydrogen bonding cross links between chains. |
| Poly(ethylene terephthalate) | +65 | —O—(CH₂)₂—O—C—⬡—C—<br>       O   O | Stiffer backbone. |
| Poly(vinyl chloride) | +80 | —CH₂—CH—<br>    Cl | Strong SH by very frequent charged side group. Weak polar bonding between chains. |
| Polystyrene | +100 | —CH₂—CH—<br>    ⬡ | Very bulky and frequent side groups give strong SH. |
| Poly(tetrafluoro ethylene) | +115 | —CF₂—CF₂— | Ultra strong SH by very frequent charged side atoms. |
| Poly(bisphenol-A-carbonate) | +142 |    CH₃<br>—⬡—C—⬡—O—C—O—<br>   CH₃   O | Very stiff backbone –O–C–O– has some flexibility. |
| Polyimide | +330 |   O   O<br>  C   C<br>—N⟨ ⬡ ⟩N—⬡—O—⬡—<br>  C   C<br>  O   O | Virtually rigid backbone |

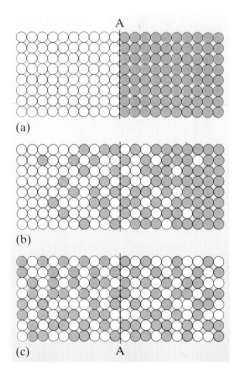

Figure 4.34 Red atoms diffuse across the boundary A–A

**Self diffusion**, identical atoms changing sites, goes on all the time. But technologists are interested in diffusion in non-uniform compositions, where the process is one of slowly changing concentrations at all places. Starting with an abrupt interface between two solids, the concentration profile after some time of diffusion must follow the trends of Figure 4.34. It is easy to see that the gradient of concentration encourages the change of composition. In Figure 4.34 random jumping will move more red atoms across the plane A–A from right to left than the other way simply because there are more of that species on the right of the plane. That reduces higher concentrations and raises lower ones. When uniformity is reached (no gradient) the jump rates across the plane are the same in each direction so there is no further change of concentrations. During the interesting transient phase, differential equations (proposed by Adolf Fick in 1855) can relate concentrations to time and place.

**Fick's first law** asserts the flux of diffusing atoms across a plane to be proportional to their concentration gradient, with the constant of proportionality being a temperature-dependent material property known as the **diffusion coefficient $D$**. Working in one dimension to keep the maths simple, we have

$$J = -D\frac{\mathrm{d}C}{\mathrm{d}x}$$

where $J$ is the flux, usually measured in atoms $\mathrm{m}^{-2}\mathrm{s}^{-1}$, and $\mathrm{d}C/\mathrm{d}x$ is the concentration gradient, usually measured as $(\mathrm{atoms\,m}^{-3})\,\mathrm{m}^{-1}$. The minus sign ensures that a *positive* flow of atoms goes in the direction of a falling concentration, that is $(\mathrm{d}C/\mathrm{d}x)$ *negative*. The wide use of the concept of fluxes being 'driven' by gradients is reviewed in ▼**Fluxes and gradients**▲.

The flux is a rate of flow, so for a thermally activated phenomenon should contain the Arrhenius factor. That is embedded in the diffusion coefficient,

$$D = D_0\exp(-E_a/kT)$$

in which $E_a$ is the **activation energy** for the mechanism by which the atoms get to another site in the crystal and $D_0$ is the material property which converts the mere possibility of jumps (the exponential term) into an effect. It contains the atom vibration frequency and the distance per jump.

To design a technical diffusion process an engineer has to determine how the concentration profile varies with time from some initial distribution. That involves solving the differential equation which is ▼**Fick's second law**▲. In the simplest case of a one-dimensional flux from a source which is laid down and not replenished during the process, the solution is

$$C(x,t) = \frac{A}{\sqrt{(Dt)}}\exp\left(-\frac{x^2}{4Dt}\right)$$

where $C(x,t)$ means the concentration at place $x$ at time $t$, and $A$ is a constant related to how much source was provided. You can confirm this as a solution if you wish! A useful rule of thumb coming from this solution is that the diffusing material has travelled distance $\sqrt{(Dt)}$ after time $t$. At $x = \sqrt{(Dt)}$ the concentration is $\exp(-1.4)$ or 78% of its $x = 0$ value, and at twice that range the concentration has dropped away to 37%. But to use any of these results requires knowledge of $D$ and of its temperature dependence for each system. ▼**Measurement of materials' diffusion parameters**▲ describes how this information is gained.

# ▼Fluxes and gradients▲

Flow, current, conduction are all words which suggest something moving, for example fluid flow, electric current, thermal conduction. Flux is another such word but has the specialized meaning of quantity per second per unit area of the channel in which flow occurs. The familiar observation that mountain torrents move faster than meandering streams suggests that flow rates (fluxes) are determined by *gradients* – of what is yet to be considered.

Several models of flow are formalized, with varying accuracy, by equations of the general shape

$$\text{flux} = \frac{1}{\text{area}} \times \frac{d(\text{quantity})}{d(\text{time})}$$

$$= (\text{materials property})$$

$$\times (\text{driving gradient})$$

Let's think about three fluxes which might flow in a metal bar: a flux of atoms (diffusion); a flux of phonons (thermal conduction) and a flux of electrons (electrical conduction). Refer to Figure 4.35. In general any of these flows is the result of a 'drift' added on to random motions, so the net transport across a plane (such as A–A in Figure 4.34) is the difference between a bit more going one way than the other. You saw this idea in action in Section 4.4, where the drift velocity of electrons was conveniently separated from their other motions (Figure 4.20). So the fluxes are all to be expressed as *net* numbers of carriers of the property (thermal energy, electric charge and so on) crossing unit area of channel per second. We have

$$\text{charge flux } j = \frac{1}{\text{area}} \times \frac{dq}{dt}$$

$$= \frac{1}{\text{area}} \times \frac{dNe}{dt}$$

with $e$ the electron charge and $N$ again the *net* number crossing the plane.

$$\text{phonon flux } \mathscr{Q} = \frac{1}{\text{area}} \times \frac{dH}{dt}$$

$$= \frac{1}{\text{area}} \times \frac{dN\varepsilon}{dt}$$

with $H$ a quantity of 'heat energy' made up of $N$ phonons of average energy $\varepsilon$.

$$\text{atomic flux } J = \frac{1}{\text{area}} \times \frac{dN}{dt}$$

with $N$ the *net* number of atoms jumping across a plane.

So the defining equations for all the fluxes are formally the same. What of their driving gradients? The easiest one to pin down is the electric one because Ohm's law is a familiar experimental fact.

Since electric current is $dq/dt$ the electric flux is a current density

$$j = \frac{I}{A}$$

where $A$ represents the cross-sectional area of the conductor. But Ohm's law advises

$$I = \frac{V_0}{R} = \frac{V_0 \sigma A}{l}$$

whence

$$j = \frac{I}{A} = \sigma \frac{V_0}{l}$$

The potential drop $V_0$ over length $l$ is uniform and for a small region can be replaced by the potential gradient, $(-dV/dx)$, the minus sign ensuring that the flux is in the direction of falling potential. Thus the flux is the product of a material property and a driving gradient:

$$j = -\sigma \frac{dV}{dx}$$

For thermal conduction the heat flux $\mathscr{Q}$ is the rate of transport of heat energy per unit area:

$$\mathscr{Q} = \frac{1}{A} \frac{dH}{dt}$$

In this case the temperature difference $T_1 - T_2$ across the length $l$ drives the heat flow. So, defining the materials property of thermal conductivity $\kappa$, we can mimic the charge flux and say

$$\mathscr{Q} = -\kappa \frac{dT}{dx}$$

The final case of diffusion feels different because the things drifting (atoms) are driven by themselves. Concentration gradients merely change the probabilities of the different species of atoms moving one way or the other as they *all* experience jumps between sites. Fick's first law therefore has the same form:

$$J = -D \frac{dC}{dx}$$

Notice also that more than one type of gradient may drive a single type of flux. For instance diffusion of ions in ceramics subject to an electric field can cause charge flow. Then

$$J_{\text{ions}} = -D_{\text{statistical}} \frac{dC}{dx} - D_{\text{electrical}} \frac{dV}{dx}$$

Thermal gradients and stress gradients can also drive diffusion.

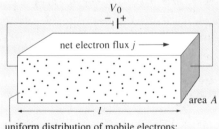

uniform distribution of mobile electrons; conductivity $\sigma$

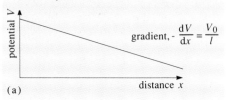

(a) gradient, $-\dfrac{dV}{dx} = \dfrac{V_0}{l}$

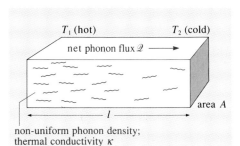

non-uniform phonon density; thermal conductivity $\kappa$

(b) gradient, $-\dfrac{dT}{dx} = \dfrac{T_1 - T_2}{l}$

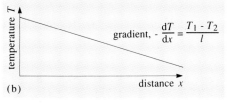

non-uniform concentration of one species of atom; diffusion coefficient $D$

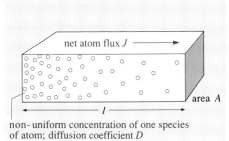

(c) gradient, $-\dfrac{dC}{dx}$

Figure 4.35 Fluxes and gradients. (a) Electrical. (b) Thermal. (c) Atomic

# ▼Fick's second law▲

The particle flux down concentration gradients will itself change the concentration gradient. (The same is true for heat flow which changes temperature gradients, the very things which drive the flow.) Fick considered this as well. Figure 4.36 shows a small region into which particles flow at a rate $J_1$ and out of which they flow at a rate $J_2$. If $J_1$ is bigger than $J_2$, the particles must accumulate in the region. Now $J_2$ will be different from $J_1$ if there is a gradient of flux $\mathrm{d}J/\mathrm{d}x$.

$$J_2 = J_1 + \left(\frac{\mathrm{d}J}{\mathrm{d}x}\right)\Delta x$$

$J_2$ and $J_1$ are numbers of particles flowing per unit area per second. So in time $\Delta t$ the number of particles in the region increases by

$$(J_1 - J_2)A\Delta t$$

The concentration (particles m$^{-3}$) increases by $\Delta C$ in the same interval of time, so

$$\Delta C A \Delta x = (J_1 - J_2)A\Delta t$$

So the rate of increase of concentration is

$$\frac{\Delta C}{\Delta t} = \frac{J_1 - J_2}{\Delta x} = -\frac{\mathrm{d}J}{\mathrm{d}x}$$

This is written using proper calculus as

$$\frac{\partial C}{\partial t} = -\frac{\partial J}{\partial x}$$

Fick's first law has already told us what the flux $J$ is in terms of concentration gradient so

$$\frac{\partial C}{\partial t} = +\frac{\mathrm{d}}{\mathrm{d}x}\left(D\frac{\partial c}{\partial x}\right)$$

If we are lucky and $D$ is constant

$$\frac{\partial c}{\partial t} = D\frac{\partial^2 C}{\partial x^2}$$

This is Fick's second law.

This equation, or three-dimensional versions of it, is what has to be numerically solved in computers to guide such complex processes as diffusion regimes for building integrated circuits. The same mathematics applies to the other flux equations. Transient heat flow is often technically important; thermal shock is one example and heat flow from a casting

as the latent heat of fusion is released is another. The latter is important for getting microstructure right and for getting cycle times for moulds economically short. The equivalent equation is

$$\frac{\partial T}{\partial t} = \frac{\kappa}{\mathcal{S}_\mathrm{v}}\frac{\partial^2 T}{\partial x^2}$$

with $\kappa$ as the thermal conductivity and $\mathcal{S}_\mathrm{v}$ the volume specific heat capacity of the material (that is, heat required per unit volume of material to give 1 K temperature rise). Exercise 4.9 prompts you to derive this for yourself.

EXERCISE 4.9 (Objective 4.8)
Follow the steps used to set up Fick's second law for the case of heat flow using these hints:

(a) Heat energy, $\Delta H$, accumulating in the volume $A\Delta x$ causes temperature change $\Delta T$ given by

$$\Delta H = (\mathcal{S}_\mathrm{v}A\Delta x)\Delta T$$

(b) With $\mathcal{Q}_1$ and $\mathcal{Q}_2$ as the heat fluxes into and out of the region the net inflow in time $\Delta t$ is

$$\Delta H = (\mathcal{Q}_1 - \mathcal{Q}_2)A\Delta t$$

(c) The gradient of the heat flux is $\mathrm{d}\mathcal{Q}/\mathrm{d}x$ and is negative since $\mathcal{Q}$ gets smaller as $x$ increases so for the interval $\Delta x$

$$\mathcal{Q}_1 = \mathcal{Q}_2 - \frac{\mathrm{d}\mathcal{Q}}{\mathrm{d}x}\Delta x$$

(d) Heat flux is related to temperature gradient by

$$\mathcal{Q} = -\kappa\frac{\mathrm{d}T}{\mathrm{d}x}$$

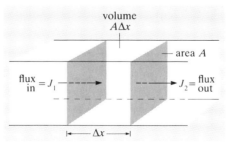

volume $A\Delta x$

Figure 4.36 Flux in, $J_1$. Flux out $J_2$

# ▼Measurement of materials' diffusion parameters▲

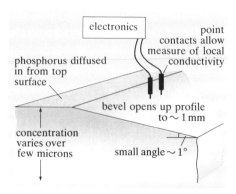

Figure 4.37 Measuring conductivity to ascertain concentration profile

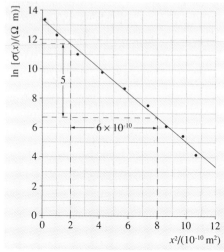

Figure 4.38 Determination of diffusion coefficient $D$

Experimental determination of diffusion parameters consists of concentration measurements in specimens where controlled diffusion has occurred in a geometry for which Fick's second law is solvable. An example taken from the semiconductor world has a known amount of phosphorus deposited on the surface of silicon wafers and diffused in for known time $\tau$ at known temperatures. The electrical conductivity of the resulting $n$-type semiconductor is proportional to the phosphorus concentration (barring some small corrections), so point measurements of conductivity down a bevel, Figure 4.37 establishes the concentration profile. The solution of Fick's second law provides:

$$\sigma(x) = \sigma_0 \exp(-x^2/4D\tau)$$

with $\sigma_0$ a constant of the experimental set up.

Taking logs of this equation gives

$$\ln \sigma(x) = -\frac{x^2}{4D\tau} + \ln \sigma_0$$

$$Y = mX + c$$

What variables of the experiment would plot to give a straight line? What gradient would it have?

Variable $Y$ maps to $\ln \sigma(x)$ and $X$ to $x^2$ to fit the straight line equation. The gradient $m$ is $-1/(4D\tau)$.

Since the time for the diffusion is known the diffusion coefficient can be found. Figure 4.38 shows a set of results found after diffusion for 16 hours at 1520 K. $D$ is found as follows.

$$\ln \sigma(x) = \ln \sigma_0 - \frac{x^2}{4D\tau}$$

$$\Delta \ln \sigma(x) = -\frac{\Delta x^2}{4D\tau}$$

$$D = -\frac{\Delta x^2}{4\tau \,\Delta \ln \sigma(x)}$$

$$= -\frac{6 \times 10^{-10}}{4 \times 16 \times 3600 \times (-5)} \, \mathrm{m^2\,s^{-1}}$$

$$= 5.2 \times 10^{-16} \, \mathrm{m^2\,s^{-1}}$$

The results of a series of measurements of this sort for diffusions at different temperatures are plotted in Figure 4.39. The excellent straight line relation between $\ln D$ and $1/T$ shows the validity of Arrhenius's law and, allows $E_a$ to be calculated.

EXERCISE 4.10   Measure the gradient of the line on Figure 4.39 and evaluate the activation energy for phosphorus diffusion in silicon. Use $k = 86 \times 10^{-6}\,\mathrm{eV\,K^{-1}}$ so that $E_a$ comes out in eV units.

Substitute the result and data taken from any chosen point on the graph to show that $D_0$ is in the range $10^{-4}$ to $10^{-5}\,\mathrm{m^2\,s^{-1}}$.

Similar experiments produced the information of Figure 4.40 (p. 194). Self diffusion was made measurable by electroplating the specimens with radioactive silver. After diffusion for known times at various temperatures the radioactivity was measured as successive thin slices were machined from the surface. This established the concentration profile of the radioactive silver. Since isotopes are chemically identical they diffuse at the same rate.

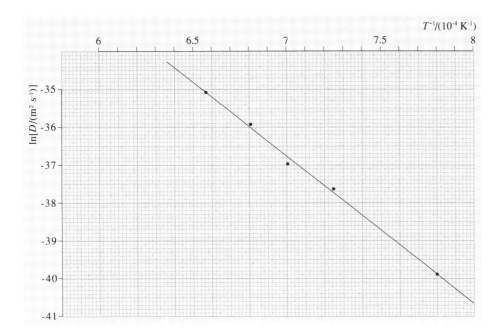

Figure 4.39 Arrhenius graph for calculation of $E_a$

EXERCISE 4.11    Phosphorus has to be diffused into silicon to make transistors. At 125°C, $D$ for this system is $5.6 \times 10^{-16}\,\mathrm{m^2\,s^{-1}}$. Carbon is diffused into iron at 950°C to harden it and for this $D$ is $1.4 \times 10^{-11}\,\mathrm{m^2\,s^{-1}}$. How far will the phosphorus go while the carbon goes 1 mm?

## 4.5.5 Diffusion mechanisms in solids

Diffusion in a perfect crystal would be virtually non-existent because atoms swapping places, even if they moved round in rings, would need high activation energy. The common mechanisms of diffusion rely on defects in crystal structures to lower $E_a$. There are three significant types of imperfection to consider: dislocations, vacancies and interstitially sited atoms.

But remember first that crystalline materials are mostly used as polycrystals, so the grain boundaries provide a continuous network of poorly packed atoms along which diffusion is particularly easy and which reaches right through the piece. A major reason for building electronic circuits on single crystal silicon rather than on cheaper polycrystal substrates is to avoid uneven diffusion. Many of the metallurgical transformations you will study in Chapter 5 depend on atom movement along grain boundaries to achieve practical rates. These are the trunk lines of the atomic freight. Transport to and from them, the short distances *across* crystal grains, is much slower because less extensive defects have higher activation energy.

The nearest thing to a grain boundary *inside* a grain is a dislocation. It is a line defect and there is considerable lattice distortion around its core. Dislocations provide low $E_a$ tracks and are responsible for such diffusion as there is at low temperatures. But there are not many dislocations within a crystal and they anneal out as $T$ goes up. So at higher temperatures another mechanism grows to dominance.

Crystals necessarily contain single-atom vacancies and these provide the next, more difficult, mechanism. An atom adjacent to a vacancy can jump into the gap without too large an $E_a$. Generally there is no bias in the direction of the jumps of atoms into a vacancy, so it is easier to see the effect as diffusion of the vacancy itself. The discussion of diffusion around Figure 4.34 appealed only to concentration gradient without involving a mechanism. The argument would not have been affected had the diagram included some vacancies. On the whole a vacancy would have been more likely to have had a red atom on its right-hand side, so red atoms would have diffused to the left as we said. It may also be, because of differences of bonding, that one kind of atom can jump more easily than another, so it stands a better chance of getting the vacancy when one comes by. The diffusion of two species past each other can go at different rates (the **Kirkendall effect**). The dominating temperature effect is usually the formation of vacancies themselves, a process which, having bigger $E_a$ than the actual diffusion jumps, is more sensitive to rising $T$. Diffusion by vacancy is therefore more important at higher temperatures, say above $0.5\,T_m$ in metals. Figure 4.40 shows Arrhenius plots for vacancy diffusion and grain boundary diffusion in polycrystalline silver. Vacancy diffusion dominates above 700 °C.

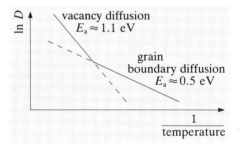

Figure 4.40 Arrhenius graph of self diffusion in silver

The discussion of crystal structures in Chapter 3 described the 'interstitial' spaces. An atom in such a site may be able to jump into a neighbouring space if the 'walls of the cage' open enough to let it through. Small atoms, such as carbon, occupy these sites in both close-packed and in the slightly more open BCC metal crystals. Also the small metal ions are in the interstitial spaces of a lattice of bigger anions in ionic crystals. The activation energies for any of these jumps, and hence the diffusion coefficients, vary widely because so many factors affect them. They depend on the size of the interstitial atom, on the openness of the host lattice, and on the population of the interstitials. Thus, for example, carbon dissolved interstitially in iron to a concentration of a few atomic percent moves quickly, and a hundred times faster in BCC iron that in FCC iron at the same temperature. Uranium diffuses easily in the open cubic $UO_2$ crystal with half the cages empty (Figure 4.41); but $Mg^{2+}$ ions occupy all the octahedral sites in MgO and have to squeeze through triangular gaps, so the mechanism is useless.

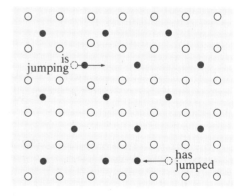

Figure 4.41 Conceptual diagram of interstitial diffusion

You'd be amazed at the ingenuity brought to bear in imagining how atoms can move, but the more erudite the explanations become the less broad their application, so we will leave this subject here.

---

SAQ 4.8   (Objectives 4.8 and 4.9)

Table 4.6 compares concentrations at depth $x$ after diffusions of times $t$ and $2t$ for phosphorus diffused into silicon. The wafer contained boron at a uniform concentration of 0.10 expressed in the same units. Where the boron and phosphorus have equal concentrations, there is said to be a 'p–n junction'. Where is the junction after each diffusion time?

Sketch graphs of $C$, $dC/dx$ and $d^2 C/dx^2$ for both boron and phosphorus at time $t$ and use them with Fick's second law to explain why the junction moves deeper into the wafer as the diffusion time is increased.

Table 4.6

| Depth $x$ | Concentration after time $t$ $A\exp(-x^2/4Dt)$ | Concentration after time $2t$ $(A/\sqrt{2})\exp(-x^2/8Dt)$ |
|---|---|---|
| 0 | 1 | 0.71 |
| $\sqrt{(Dt)}$ | 0.78 | 0.62 |
| $2\sqrt{(Dt)}$ | 0.37 | 0.43 |
| $3\sqrt{(Dt)}$ | 0.105 | 0.23 |
| $4\sqrt{(Dt)}$ | 0.018 | 0.095 |
| $5\sqrt{(Dt)}$ | 0.002 | 0.031 |

---

## 4.5.6 Electrical conduction in ceramics

Ceramics, being ionic and/or covalently bonded solids, without delocalized electrons, are supposed to be insulators aren't they? The bonding leaves no free electrons and the ions can't move. Diffusion upsets all such cheerful illusions and makes electrical conductivity a controllable variable. Technical applications exploit the complete range of conductivities possible. For example:

• The transparent cover slip over the numbers on a liquid crystal display must be conductive. Tin oxide, $SnO_2$, doped with indium oxide, $In_2O_3$, is the answer.
• The body of a spark plug must be an insulator under high voltage. Alumina has all its ions held tight.
• The ferroelectric derivatives of $BaTiO_3$ used in capacitors must have low conductivity — unless a charge leak has to be provided as in some impact detonators for armour-piercing shells.
• Magnetic ceramics for high-frequency transformers need high resistivity to suppress eddy currents. Manganese doping kills one troublesome conductive mechanism.
• Gas analysis in automobile exhausts relies on diffusion of oxygen to and from a sliver of $ZrO_2$ doped with $Y_2O_3$ to generate a voltage which depends on the oxygen concentration of the gas.

Basically in oxide ceramics two modes of conduction cover most of the practical cases:

1 Electron hopping between ions of variable valency.
2 Vacancy diffusion of oxygen ions.

The first of these is a diffusion of electrons rather than of atoms. Consider the example (Figure 4.42) of a magnetic ceramic containing both $Fe^{2+}$ and $Fe^{3+}$ ions. The extra electron on $Fe^{2+}$ needs only a little inducement ($\approx 0.09\,eV$) to hop to a nearby $Fe^{3+}$ and in homogenous material such jumps will be randomly going on all the time. But if there is an electromotive force across the specimen, say an induced e.m.f. if it is being used in a fluctuating magnetic field, jumps down the electric potential gradient will be easier than jumps in the opposite direction, conferring some conductivity on the material. When high resistivity is demanded by the application, means of suppressing this tendency are needed. Firing the ceramic in oxygen to ensure that all the iron is in the $Fe^{3+}$ state is one route. Alternatively small additions of manganese as $Mn^{3+}$ help because when the hopping electrons land on one of these ions they are trapped; the activation energy for $Mn^{2+}$ to lose the electron again is higher.

The second mechanism of conduction is exemplified by tin oxide doped with about 1% indium oxide. The crystal structure will be riddled with vacancies where $O^{2-}$ ions should be. That's because the structure is right for $SnO_2$, but some $In^{3+}$ ions have been substituted for $Sn^{4+}$. A hundred positive ions need 200 oxygen ions to build the crystal; but if 98 'molecules' of $SnO_2$ and one of $In_2O_3$ are provided, the 100 positive ions only have 199 negative ions to live among. So somewhere in the crystal there's an empty oxygen site (Figure 4.43). These vacancies should stay near to $In^{3+}$, where they cause less electrostatic distress, but of course some of them, with rattle to help, will diffuse away. If there's an electric field the $O^{2-}$, which would jump towards the positive end of the piece, will have a better chance of moving into the vacancy than any other neighbour, so there is a net conduction of charge again.

The mechanisms discussed here are examples of diffusion under the influence of a potential energy gradient (a force) rather than of a concentration gradient. But the processes are still diffusion because in each case the charge carrier has to be thermally activated before it can jump and it has to have an adjacent site to jump to — an electron 'vacancy' on $Fe^{3+}$ or an empty atom site for the diffusing $O^{2-}$ ions. When diffusion is driven by a potential energy gradient *and* there is a concentration gradient, it may be that the force effect overwhelms the statistical effect, making it look as if the diffusion is going the wrong way. Another example is provided by creep deformation where a mechanical stress can drive diffusion. This is dealt with in Chapter 6.

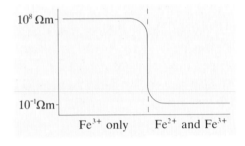

Figure 4.42 Resistivities for ferrites with $Fe^{3+}$ only and ferrites with $Fe^{3+}$ and $Fe^{2+}$

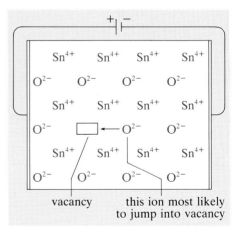

Figure 4.43 Vacancy movement in $SnO_2$ doped with $In_2O_5$

## Summary

The few atoms that have *much* higher random energy than average are responsible for 'thermally activated' events. The fraction with energy greater than $E$ is $\exp(-E/kT)$. This fraction

(a) falls to very low fractions as $E$ increases beyond a few $kT$ (that is, it is larger for lower energies);
(b) rises quickly with $T$ for a given $E$;
(c) rises *more* quickly with $T$ for bigger $E$ than for smaller $E$.

Arrhenius's equation describes the *rates* of thermally activated phenomena. Plots of ln(rate) against $T^{-1}$ are used to extract the constants of the Arrhenius equation. Control of processes using these phenomena is sensitively achieved by temperature choice.

At $T_g$ in polymers the long molecules become free enough to wriggle. Several factors influence $T_g$; the number and strength of bonds between chains; steric hindrance to rotation due to side groups; stiffness of molecular backbone.

Diffusion in solids involves atoms jumping into vacant sites (interstitial or substitutional) when opportunity and sufficient energy coincide. Grain boundaries and dislocations provide fast diffusion paths. Fick's laws govern the rates of movement and the diffusion coefficient $D = D_0 \exp(-E_a/kT)$ expresses the thermal activation.

Diffusion in an electric potential gradient accounts for the noticeable electrical conductivity of ionic solids.

# 4.6 Extreme energy effects 2: electrons

In solids, the bands of energy levels for the outer electrons of atoms span several electron volts. They are filled from the lowest energy upwards in accord with Pauli's exclusion principle. The lowest energy electrons are thus not vulnerable to any thermal disturbance save by the most energetic collision or at extremely high temperatures. That *some* thermal influence is felt is evidenced by thermionic emission of electrons, where electrons escape from a heated surface. This is used as the source of electrons in cathode-ray tubes. In this, and other temperature-sensitive electronic properties of solids, the population of electrons in the upper part of the energy spectrum is the crucial factor of the model. This section describes the distribution of electrons in that part of the spectrum, and uses the distribution to model thermocouple and semiconductor behaviour.

## 4.6.1 Energy distribution of electrons at 0 K

The distribution of available energy states for electrons is called the **density of states function**. This function describes the pattern of rungs on an energy ladder. For the electrons of the outer zones of atoms in solids, Chapter 3 revealed the pattern to be a series of bands of closely spaced levels interspersed by gaps — energies which no electrons could have. The density of states functions, $g(E)$, for electrons in solids are generally complex as they are expressions of the overlap or separation of the bands of energy levels; and there are, of course, many more states than electrons because states derived from excited states of isolated atoms contribute to the conduction band.

The function $g(E)$ might, for example, be like Figure 4.44. The energy distribution of the *electrons* (as opposed to the available states) is decided by how they are arranged among the states as well as by the energies of those states. To make counting simple let us define a single energy level which can be 'filled' by two electrons (Pauli's principle, with allowance for opposite spins) as two 'states'. Let us also say that the probability of occupation of a state at energy $E$ is $p(E)$ (some mathematical expression having a value between 0 and 1), and that the number of states in a narrow strip of energy of width $dE$ at energy $E$ is $g(E) \times dE$. Then the number of electrons in that energy range is

$$dN(E) = p(E) \times g(E) \times dE.$$

The probability of occupation of a state will depend in some way upon temperature since 'rattle' will give some electrons more energy. We'll come to that in a moment but first let us consider the energy spectrum at 0 K. At 0 K all states are filled from the lowest up, just as far as is necessary to accommodate all the electrons, Figure 4.45. I will call the highest energy used the **Pauli level** because Pauli's principle dictates what energy $E_P$ it is at. So at 0 K the electron energy spectrum of these electrons, how many have what energy, is entirely fixed by $g(E)$, the

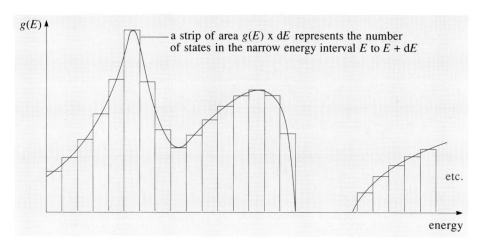

$g(E)$

a strip of area $g(E) \times dE$ represents the number of states in the narrow energy interval $E$ to $E + dE$

etc.

energy

Figure 4.44 Sketch of density of states, $g(E)$

energy distribution of the quantum states. Up to $E_P$ the probability of occupation is 1, but immediately above it falls stepwise to zero.

Figure 4.46 shows some simplified examples of density of states distributions with Pauli levels marked. Such variety gives plenty of scope for modelling! Here is one example.

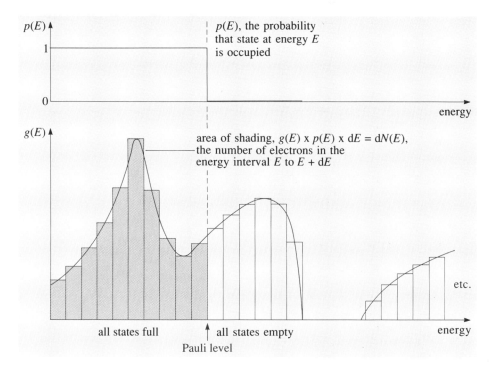

Figure 4.45 Putting the $p(E)$ curve onto $g(E)$ at 0 K

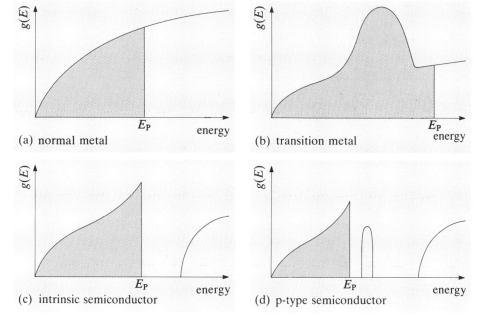

Figure 4.46 Sketches of $g(E)$ for different classes of material

Imagine, with Figure 4.47(a), a tank of water filling two inter-connected buckets of different capacities at different depths below the potential energy datum set by the water level in the tank. The connection ensures that the level of the water surface is the same in both buckets. Obvious? Yes, and analogous to how electrons behave in two metals in contact (Figure 4.47b).

The zero datum of electrostatic potential energy is with charges separated to infinity. If the electrons of a block made of two metal elements could be removed to this datum the energy levels of the metals would be like the two buckets waiting to be filled. They are interconnected because metals are conductors. When all the electrons have been poured in, the levels will have filled to some common energy ($E_P$). But because the densities of states are not the same, there are not the right number of electrons in each metal to match the atoms' positive charges. There are more electrons in one metal than there should be, so it has a net negative charge; fewer than proper in the other, so that's left positive. A **contact potential** exists. You can see why an ionic solution, such as sea water, sitting on the junction can be chemically nasty. The resulting electrolytic reactions are the origin of corrosion.

Note. The contact potential must affect the positions of the energy bands relative to the charges-at-infinity datum, but of course the electrons know what they are doing without our artificial method of placing them in their quantum states. The water bucket analogy can be pushed to include this effect; we could hang the buckets on springs so that the wider one (heavier) sank a bit further as the water was added.

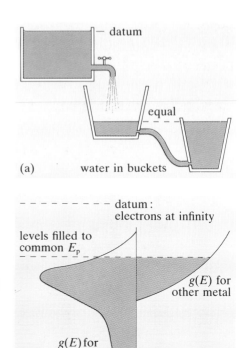

(a)      water in buckets

(b)

Figure 4.47

## 4.6.2 How $kT$ gets at electrons

So much for models at $0\,\text{K}$. Let us turn our attention to the effect of temperature. At any temperature above $0\,\text{K}$ the step of $p(E)$ is smeared out, with the probability of occupation increasing for levels above the step and decreasing for states below the step. That must be so because the total number of electrons to be fitted in hasn't changed; it's just that a few electrons have adopted higher energies. Figure 4.48 shows the effect on the imaginary model used in Figure 4.45. The relation for $p(E)$ is known as the **Fermi–Dirac probability function** after the physicists who proposed it, and Figure 4.48(b) shows how the smearing out changes with temperature. Notice that these curves are symmetrical about the energy at which the probability is exactly a half. This level is called the **Fermi level** $E_F$ and is a characterizing parameter of the curves. ▼**What's what in the Fermi–Dirac probability function**▲ outlines the algebra of the curves, and Exercise 4.12 shows you that the energy span over which $p(E)$ drops from 1 to 0 is only a few $kT$, rather narrow compared to the whole energy spectrum of the electrons from the bottom of a band up to $E_P$. So only a small fraction of the electrons, typically 1%, are subject to thermal influences. Appreciate this result as a correction to Exercise 4.3. It means that *most* of the electrons' energies are not random rather than *none* of them. The total electron energy is therefore

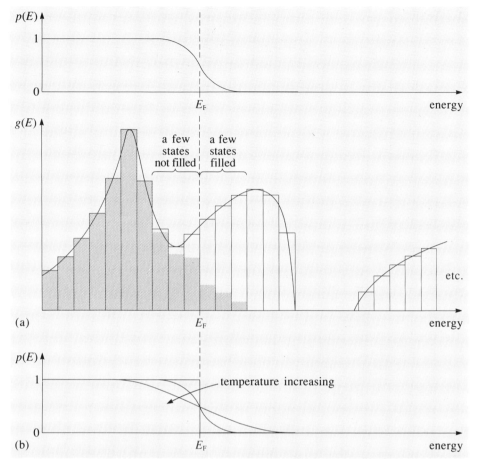

Figure 4.48 Above 0 K the shape of $p(E)$ changes. States around $E_F$ are only partially filled

# ▼What's what in the Fermi–Dirac probability function▲

Algebraically the probability of occupation of an electron energy state at energy $E$ is

$$p(E) = \frac{1}{1 + \exp[(E - E_F)/kT]}$$

You need to know your way around this function; in particular, you need to know what happens to it at various values of $E$ and $T$.

First recognize that at absolute zero it describes a step as required. To see this note that $(E - E_F)$ switches from negative to positive as $E$ passes $E_F$, so the exponential, with $T = 0$, goes from $\exp(-\infty) = 0$ to $\exp(+\infty) = \infty$, making $p(E)$ drop suddenly from 1 to 0.

For all temperatures above zero the step is less abrupt. Look at three cases: $E \ll E_F$, $E = E_F$ and $E \gg E_F$, always with $E_F \gg kT$. A level at the Fermi energy has

$E = E_F$, which makes $E - E_F = 0$ and, since $\exp(0) = 1$,

$$p(E) = \frac{1}{1 + 1} = 0.5,$$

as defined.

At the extremes, $E \ll E_F$ makes

$$p(E) = \frac{1}{1 + \exp(-\text{big number})}$$

$$\approx \frac{1}{1 + 0} = 1$$

and $E \gg E_F$ makes

$$p(E) = \frac{1}{1 + \exp(+\text{big number})}$$

$$\approx \frac{1}{1 + \infty} = 0$$

What is interesting, however, is to appreciate the energy span over which the probability of occupation of levels falls

from near enough certain to virtually nothing.

EXERCISE 4.12 How big is $p(E)$ at $E = (E_F - 2.5\,kT)$ and $E = (E_F + 2.5\,kT)$ for $T > 0$ but $E_F \gg kT$?

Exercise 4.12 has shown that the narrow range of $5\,kT$ about $E_F$ contains almost all the change.

Finally it is also worth noticing that at the high-energy tail of the distribution, where $E \gg E_F$, the exponential term is approximately $\exp(+E/kT)$, and because this is much greater than the 1 in the denominator, that 1 can be neglected. So $p(E)$ degenerates to our simple and familiar $\exp(-E/kT)$ *which is the same as for atoms*. This fact becomes useful in semiconductor analysis.

almost insensitive to temperature and makes but a tiny contribution to a solid's specific heat capacity.

Virtually all explanations of electronic effects and the devices made to exploit them arrive, sooner or later, at debates about how the Fermi level lies in relation to steps or gaps or peaks or flat parts of the $g(E)$ curve. The details of $g(E)$ close to the Pauli level dictate the position of the Fermi level, and hence how electron energy spectra and electronic properties vary with temperature.

In Figure 4.49(a) I have sketched a graph of the 'smeared out' probability function $p(E)$. The curve is symmetrical about the Fermi level $E_F$ and within the space of $5kT$ about $E_F$ $p(E)$ goes from $\approx 1$ to $\approx 0$. I want to look at the way electrons distribute themselves amongst the available states when the occupancy of those states is governed by this probability function. Because I am going to use a highly idealized model, it will suit me to use the approximation sketched in Figure 4.49(b), in which I consider $p(E)$ to be constant within bands of energy $1kT$ wide.

Suppose $g(E)$ is constant, as in Figure 4.50(b). At 0 K all the available levels up to the Pauli level $E_P$ are filled. At temperatures above 0 K, $p(E)$ is approximately modelled by Figure 4.50(a). I have arranged for $E_F$ to coincide with $E_P$. Figure 4.51(a) and (b) together show us that, for example, the strip of energies $1kT$ wide just below $E_P$ (or $E_F$) can be filled to 62% full. The remaining 38% of the electrons that would be there at 0 K are thermally excited to higher energies above $E_P$. Figure 4.50(b) shows they can be accommodated in the strip just above $E_P$, where 38% of states are available. The same argument applies for other strips below $E_P$. The thermally excited electrons can always be accommodated in the corresponding strip of energy symmetrically above $E_P$.

But $g(E)$ is not usually constant; suppose it slopes down, as shown in Figure 4.50(c). Suppose further that $E_F$ still coincides with $E_P$. Now the electrons promoted from below the Pauli level cannot find sufficient accommodation in the corresponding strips above $E_P$ because those strips do not provide enough states. Let us see what happens if I adjust the Fermi level relative to the Pauli level. In Figure 4.51(b) it has been adjusted upwards by a half $kT$ relative to the Pauli level. Now, as Figures 4.51(b) and (c) show, there is sufficient room to accept the promoted electrons. In fact we could move $E_F$ by less than a half $kT$ and still have ample space to receive the electrons. In reality, of course, all the promoted electrons are accommodated, and $E_F$ exceeds $E_P$ by just as much as is required to represent this physical fact.

The above explanation, it must be stressed, is highly idealized, being closer to geometry than physics. However, you should be able to see that for any substance where $g(E)$ is not constant near $E_P$, the Fermi level must be displaced relative to the Pauli level at all non-zero temperatures if the $p(E)$ curve is to model the distribution of thermally excited electrons.

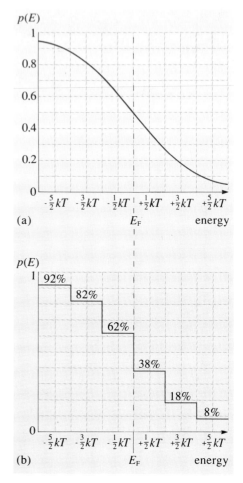

Figure 4.49 (a) $p(E)$ curve above 0 K. (b) Approximation

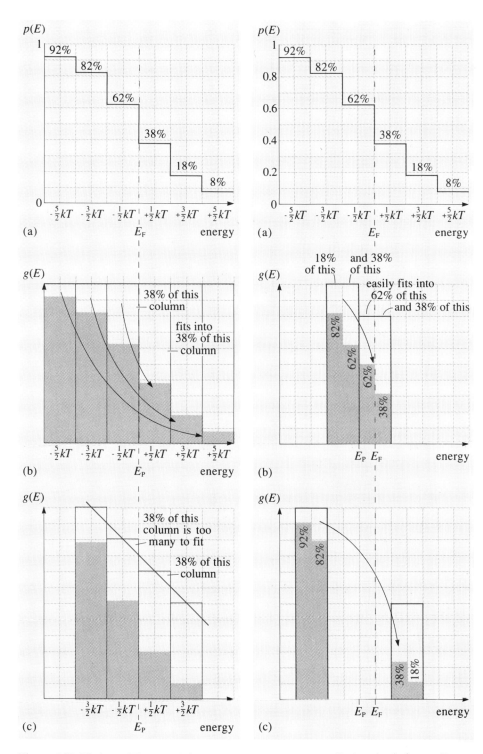

Figure 4.50 Fitting $p(E)$ to $g(E)$ for (b) constant $g(E)$ and (c) falling $g(E)$

Figure 4.51 $E_F$ displaced relative to $E_P$. Electrons can now be accommodated

Now let's apply these ideas.

The first interesting application concerns the change of contact potential with temperature. If the two materials of Figure 4.47 were metals in contact at $T > 0\,\mathrm{K}$, the Fermi level of one would apparently move up while the other went down. But being metals, and hence conductors, the materials allow surplus electrons from the falling $g(E)$ side to move to the other side of the join to find vacant states. The transfer of charge further adjusts both Fermi levels until equilibrium is established with $E_\mathrm{F}$, and hence the whole $p(E)$ curves, matching in the two metals. In consequence of the transfer of charge, the contact potential is changed, which is of course the basis of action of thermocouples. A sufficient condition for a thermoelectric effect is that the gradients of the $g(E)$ curves of the two metals are different at $E_\mathrm{F}$. Thermocouples are widely used as thermometers. ▼Choosing metals for thermocouples▲ looks at some practical considerations.

Explanations of semiconductor behaviour are also heavily adorned with Fermi level modelling so a basic understanding of their electron energy spectra is the other example I will explore. **Intrinsic semiconduction** occurs when a few electrons are kicked upstairs from their valence band to their conduction band. They are *semi*conductors because this excitation is a jump of many $kT$ in useful semiconductors, so the number of carriers $n$ in $\sigma = ne\mu$ is much smaller than in metallic conductors. By placing the Fermi level correctly on their $g(E)$ distributions we shall see how temperature changes $n$ and thereby affects conductivity. Figure 4.46(c) shows $g(E)$ for an intrinsic semiconductor and the Pauli level placed exactly at the top of a band. In this condition it cannot conduct at all because there are no quantum states for electrons to move into if they took up energy from an applied electric field. Now let's put on the probability curve. Figure 4.52(a) suggests putting $E_\mathrm{F}$ at the top of the valence band. It must be wrong because the levels at the top of the valence band lose electrons and there are no levels for them to enter at the energy where $p(E)$ is not zero. According to our discoveries on Figure 4.51, the fact $g(E)$ falls, in this case very precipitately, implies that $E_\mathrm{F}$ must be moved to a higher energy. Try putting it half way up the gap, Figure 4.52(b). Now the tail of $p(E)$ reaches into the conduction band and the curve just falls away from 1 at the top of the valence band. That's better. It models a few electrons leaving states in the lower band and taking up states in the upper band where they are free to move under an applied voltage. Actually even this can't be exactly right because $g(E)$ is large at the top of the valence band and small at the foot of the conduction band, so to get the same number of states vacated and newly occupied $E_\mathrm{F}$ has to be a touch over half way up the gap. You can also see, Figure 4.52(c), that as the temperature goes up the population of the conduction band goes up exponentially because the exponential tail of the Fermi–Dirac curve is the only part which reaches the conduction band. With a gap as big as that in silicon, $E_\mathrm{gap} = 1.1\,\mathrm{eV}$, and with $E_\mathrm{F}$ at the middle of the gap, the

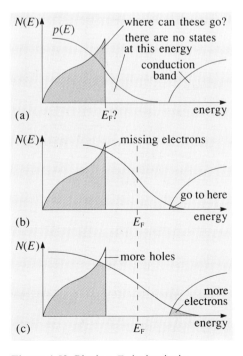

Figure 4.52 Placing $E_\mathrm{F}$ in intrinsic semiconductor

# ▼Choosing metals for thermocouples▲

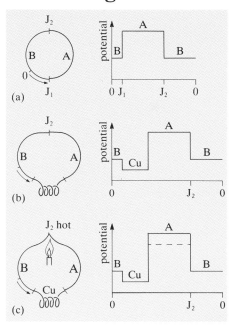

Figure 4.53 Changes of potential going round thermocouple loop

A thermocouple is a loop of two metals joined in two places. In Figure 4.53(a) the loop is all at the same temperature. The potential graph shows a step up and a step down if we trace round the loop. Now break one junction and put in a voltmeter, Figure 4.53(b). As far as the materials of the circuit are concerned, the meter is just a length of copper wire. The steps up and down still cancel out. But heat up one junction, Figure 4.53(c), and its contact potential changes. That puts an electromotive force, e.m.f., in the circuit, which is recorded by the meter. If you know how the e.m.f. varies with $T$, you have a thermometer. If you know about densities of states near the Pauli energy in metals, you can choose pairs of metals which give a big effect!

The density of states functions for transition metals sketched in Figure 4.46(b) shows sharp $g(E)$ variations for the d-band. Those elements or alloys which have this band nearly full will have their Pauli energy lying in the steeply falling part. A good example is nickel which has nine of the ten d-states filled.

The 'normal' metals, Figure 4.46(a), have much flatter $g(E)$; copper will do, because its d-states are full and its Pauli level lies in the flatter s-state region of the valence band. So a combination of these metals would give a big thermoelectric e.m.f. An even bigger difference is obtained for copper–constantan (the latter has the d-states very nearly full). This is the schoolteachers' favourite for easy demonstrations, and is interestingly non-linear over a range of only some 300 K. For serious thermometric purposes an e.m.f. which changes linearly over a wide temperature range is more important than the actual magnitude of the voltage; minute voltages are relatively easy to measure accurately. Linearity is achieved by using a pair of fairly similar transition metal alloys. Process thermocouples are usually either Pt with Pt/10% Rh or the much cheaper pair Chromel (NiCr) and Alumel (NiAl) giving about 40 μV/K over about 1200 K range.

EXERCISE 4.13   Look up Pt and Rh in the periodic table. What will be the effect on the Pauli level of a small addition of rhodium to platinum?

Theory shows the 'absolute thermoelectric power' of transition metal alloys, that is the e.m.f. which would be produced against a metal with constant $g(E)$, to be given by

$$e \approx \frac{-1.2 \times 10^{-2} T}{E_0 - E_F}$$

(units $\mu V\,K^{-1}$)

over a wide range, with $E_0$ being the energy of the top of the d-states. $E_0 - E_F$ measures how close to the top of that steep bit the alloy is working. Since this gap can be delicately controlled by alloying, the performance of the couples can be chosen, as with the platinum–rhodium system. However, the calibration will be sensitive to pollution, and a small-volume junction working at high temperature could soak up atoms from its surroundings by diffusion. Thermocouple junctions for high-temperature use are customarily encased to protect them from this hazard.

tail of $p(E)$ becomes $\exp(-E_{gap}/2kT)$. In $\sigma = ne\mu$, this increase of $n$ swamps any change in $\mu$ due to phonon scattering and the conductivity follows an exponential trend, as sketched in the Arrhenius plot of Figure 4.54. Extrinsic semiconductors are a bit trickier to understand, but as these are the practical materials they are worth attention. See ▼ **Placing the Fermi level in extrinsic silicon** ▲.

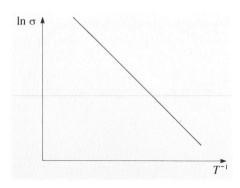

Figure 4.54 Arrhenius plot of conductivity and temperature for intrinsic semiconductor

SAQ 4.9   (Objectives 4.5 and 4.10)
At the time of writing there have been recent reports of polycrystalline diamond being deposited as a thin film. If the process could be improved to make single-crystal film, it might be possible to plant dopants in it to make p–n junctions which would remain extrinsic to much higher temperatures than silicon, which becomes intrinsic at about 500 K. The energy gap in silicon is about 1.1 eV, and that of diamond is 6 eV. Estimate the temperature at which diamond would become intrinsic to an awkward extent, supposing that all other features are the same as for silicon. Is it reasonable to suppose that devices would have useful lifetimes at, or approaching this temperature? Hint: use the data for silicon in $\exp(-E_g/2kT)$ to find '$n_{awkward}$' and take this number over for diamond.

# ▼ Placing the Fermi level in extrinsic silicon ▲

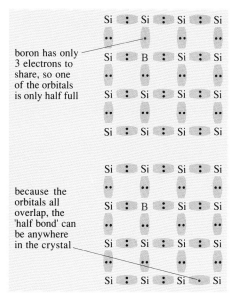

boron has only 3 electrons to share, so one of the orbitals is only half full

because the orbitals all overlap, the 'half bond' can be anywhere in the crystal

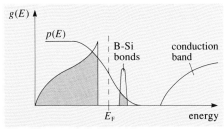

Figure 4.55 (a) A boron atom introduced into a silicon crystal has an incomplete bond. (b) The incomplete bond is mobile

The density of states for p-type silicon is shown in Figure 4.46(d). You'll remember from Chapter 3 that such material has been doped with a trace of a group III

element, such as boron, so not all the silicon atoms can make complete covalent bonds (Figure 4.55). If an electron left a proper Si–Si bond and came to one of these sites to make a Si–B bond it would have a bit more energy. The picture of $g(E)$ in Figure 4.46(d) has a little peak of states in the gap just above the valency band to describe this. Now when the temperature rises from 0 K, these are the first states which electrons leaving Si–Si bonds (that is, valence-band states) can reach. That means that when we put the $p(E)$ curve onto the distribution, $E_F$ is in the narrow gap between the valence band and the little peak, Figure 4.56. The states

Figure 4.56 In p-type semiconductor, electrons from below $E_F$ move into small band of states below conduction band

of the little peak are *localized*, because there are very few boron atoms, so we can't expect them to provide conduction. Instead the **holes** in the valence band, that is electrons jumping from atom to atom trying to fill all the Si–Si bonds, will, given an electric field, tend to go the way the field pushes them. A current flows. Well below room temperature the little peak states are nearly all full, so the strong temperature dependence of intrinsic conductivity has been removed; and by controlling how much boron is put in, the actual conductivity can be chosen. *That is an engineering material*, though as you know the hearts of electronic devices are p–n junctions where boron-doped silicon meets phosphorus-doped material. However, at some much higher temperature, thermally excited electrons reach the conduction band and the exponential rise of intrinsic conductivity dominates the controlled properties. $E_F$ will move to the middle of the gap. Then the p–n junctions in the electronic devices don't behave properly and circuits malfunction. The trick is to make the doping concentration sufficient for this effect to be unimportant at normal operating temperatures.

**SAQ 4.10 (Objective 4.10)**
Alter the argument used for p-type silicon to account for the placement of $E_F$ in n-type silicon at:
(a) room temperature,
(b) high temperatures.
Figure 4.57 shows $g(E)$ for n-type silicon, with all states filled to the Pauli level (that is, at 0 K).

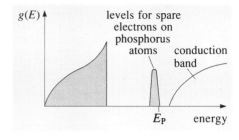

Figure 4.57 Filled states in n-type silicon

**SAQ 4.11 (Objective 4.4)**
The processes of electrical conductivity or resistivity have been covered in this chapter for several materials. All conductivities can be described by

$$\sigma = ne\mu$$

and their temperature variation by

$$\frac{d\sigma}{dT} = ne\frac{d\mu}{dT} + \mu e\frac{dn}{dT}$$

By completing Table 4.7 (with qualitative comments), you should be able to bring the descriptions together.

Table 4.7

| Material | Carrier identity | $n$ large or small? | $\dfrac{dn}{dT}$ | What influences mobility? |
|---|---|---|---|---|
| normal metal | | | | |
| transition metal | | | | |
| ferrous ferrite | | | | |
| indium-doped tin oxide | | | | |
| superconductor | | | | |
| intrinsic silicon | | | | |
| extrinsic silicon | | | | |

## Summary

Electron energy levels in solids have varied distributions. Electrons in solids 'fill' energy levels up to the Pauli level at 0 K. At $T > 0$ K the Fermi–Dirac function describes the probability of states being occupied. Phenomena are explained in terms of the placement of the Fermi level relative to the electron energy level 'ladder' (that is, the density of states function).

# 4.7 Stick or rattle? Where is the balance?

## 4.7.1 An energy description of thermodynamic equilibrium

In this final section of the chapter we come to the events classified as 'sudden' in Table 4.1. By 'sudden' you will remember we meant that these events depend on a critical temperature being reached, not that they necessarily occur quickly. Such events are phase changes, including crystal structure changes, as well as melting and boiling. The key to their modelling at atomic scale is to describe the balance between sticking forces holding atoms into patterns and the chaos generated by thermal rattle. *How* to describe the balance seems to be a problem because our experience in the macroscopic world is of deterministic mechanical systems rather than random ones. We can predict the course of events quite easily. You know that if a clockwork mouse is wound up and let to run it will soon come to rest. A battery goes flat. Water finds its own level. What goes up comes down. And so on. Mechanics tells us that in all these systems the spontaneous trend of events is to the lowest possible energy condition, one of zero kinetic energy and minimum potential energy. This condition we call mechanical equilibrium. As soon as we think about the trends of molecular events this easy view comes apart. Except when they are at absolute zero of temperature, we are invited to suppose that molecules have plenty of kinetic energy and because of their motion they are in general unable to sit in their lowest potential energy condition. The condition such a system reaches of its own accord when left alone is what I mean by thermodynamic equilibrium. For 'rattling' systems, is there an equivalent to the minimum-energy model of equilibrium in mechanical systems?

To answer this we have to invent a way of describing a set of conflicting tendencies. On the one hand, there are forces between particles trying to get them into balance, into an orderly mechanical equilibrium. On the other hand, the persistent kinetic energy associated with temperature is disrupting this well-defined state, insisting on muddle. The balance is expressed by the unfortunately named **free energy equation**. The proposal is to 'invent' an equation, in the dimensions of energy, which will describe the thermodynamic equilibrium of a system by coming to a minimum value. The application of the equation would be similar to using the criterion of minimum potential plus kinetic energy for a mechanical system. To get going I'll imagine:

free energy = random energy + orderly energy − muddle factor.

To test whether a change can happen in the system, the terms on the right hand side are to be worked out and summed for 'before' and 'after'. The initial value is subtracted from the final. If the result is negative the 'free energy' is falling, that is moving closer to a minimum, and the contemplated change is possible. So that this interpretation makes sense I've put in the thing called 'muddle factor' as minus because rattle must cause muddle. If we imagine a rattling but *un*muddled condition and ask how it will reach equilibrium, the answer must be by becoming *more* muddled. The minus sign ensures that the free energy reduces as this happens. The random, or internal, energy, symbol $U$, has been exhaustively dealt with in Section 4.2. What is meant by **ordered energy** will need discussion. Notice too that we have a problem: how to express 'muddle' as an energy-dimensioned quantity. Indeed, that's the tricky bit!

## 4.7.2  Ordered energy

Well-behaved mechanical systems have no random energy and do not generate muddle. The parts do what they are told. Our tentative free energy equation then reduces to

free energy  =  ordered energy

and we know what to do to find equilibrium. So in atomic systems, ordered energy is all that which would be left if the poor things could stop jiggling about. Section 3.3 used two examples of ordered energy as part of the energy of a system: the elastic strain energy of a stretched wire and the surface energy per unit area created by fracture. There are many more examples: magnetic or electrical polarizations in response to applied fields require changes of energy; and the kinetic energies associated with moving charges of an electric current or the moving molecules of a gale are also ordered energies. The potential energy of atomic bonding is probably the most important example of all because it is what has to change in all the phase changes of interest to materials handling. Generally, 'events' are exchanges of energy between the ordered and random constituencies, whereupon the muddle factor changes to bring the free energy down. As examples:

• Relaxation of a stress converts ordered strain energy into thermal energy, which may leak out of the system and give a lower free energy.
• Internal combustion engines reduce chemical energy (ordered). Some leaves as ordered energy of the expansion, doing work $p\Delta V$, but further energy leaves as heat through the cooling system and exhaust.
• Sintering as a way of making powders into continuous solids is driven by surface-energy reduction; that is to say, for sintering this will be the only term in the equation we seek which makes the 'free energy' go down.

To put ordered energies into our equation, notice that they can be made to look like

force $\times$ distance,

that is the product of an intensive parameter describing an applied influence (a stress, a pressure, an electric or magnetic field) and a dependent, extensive parameter, so called because it tells the extent of some feature of the system, such as its length, its volume, its polarization or its magnetization.

Of course several of these ordered energies can belong to one system. A solid responds to stress and it may be polarizable. To make life complicated, but interesting, this solid may be piezoelectric. That is, both stress and electric field produce both mechanical and electric response! So there may be several relevant pairs of extensive quantities $X$ and intensive ones $I$ to take into account in the free energy equation. For now, to be general let's just write them all as a summation,

$$\text{ordered energy} \ = \ \Sigma XI$$

The important chemical ordered energy can be put like this too. Any organization of atoms into molecules (even those giant ones we call crystals) is a particular potential-energy arrangement. For each atom there is some minimum potential energy, which it could achieve if only the rattling would stop, due to its bonding to its neighbours. We can call the bond strength the 'chemical potential' and give it a symbol $\mu$ (measured in joules per mole). If there are $N$ moles in the system then the ordered energy content contribution from them is $\mu N$. When the bonding arrangements change, Figure 4.58, as at a phase change, another value of $\mu$ applies so an ordered energy change of $N\Delta\mu$ will be active in the free energy equation.

If there are several species of atoms or molecules in different amounts, $N_1$, $N_2$, ... the total chemical potential energy is $\Sigma\mu_i N_i$. Chemical reactions, where several $N$ and $\mu$ terms change, are seen primarily as 'driven' by this change of ordered energy. But changing $N$ offers the opportunity for different 'muddle' conditions, and of course a chemical reaction may force changes in the random energy $U$. For example, stiffer bonds have different vibration energies, a gas phase may be formed, and so on. It can all get a bit too complicated for comfort! Let's move on to the last term in our hypothetical equation.

## 4.7.3 The muddle factor and entropy

How can a muddle be described quantitatively? Consider string, the stuff most noted for spontaneous confusion. Its simplest condition is stretched out in a straight line, no part of it crossing another and no variation of its direction. As bought it comes as a neat structure, Figure 4.59(a), with hundreds of equally spaced crossings and a continuous pattern of direction change. A week later, when you come to use it, you find a tangled mess, Figure 4.59(b). This condition probably has the same number of crossings but their regularity has gone and the direction pattern has collapsed. Some numerical trick to do with crossing spacings and directions could measure what has happened. Certainly that number will not be measured in energy units.

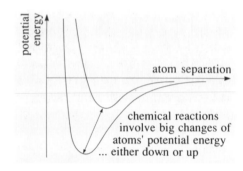

Figure 4.58

(a)

(b)

Figure 4.59 (a) Low entropy. (b) High entropy

For a billion molecules some similar muddle numbers can be imagined. Do the molecules all have the same energy or is it distributed among possible values? Are they all equally spaced or is the spacing a random muddle? All moving in the same direction, or this way, that way and every other? In fact the distributions we have seen in earlier sections have been manifestations of spontaneous muddling with more or less constraint in various circumstances. A crystal, for example, has geometric constraint; electrons have energy constraint. These constraints affect the amount of muddle present at thermodynamic equilibrium in these domains. The extreme case of a perfect gas has no constraints at all, and the Boltzmann distribution describes the condition of *maximum* muddle in the position and momentum domains. In general, the muddle number would be a nasty thing to calculate for a big molecular system, though a simple example will be played with (soon) in 'Statistical games for $W$'. Suppose we know the number, however, and its value is $W$. The problem then is how to make $W$ into an energy. Boltzmann did this with such grace, elegance and downright cheek that he was awarded both ears and the tail. From nowhere he proposed that the thermo-dynamic function of state known as the entropy $S$, which is known spontaneously to increase for the whole universe for all processes (that's the second law of thermodynamics, and you kneel before that), is related to $W$ by

$$S = k \ln W$$

with $k$ being the same constant (Boltzmann's constant), of dimensions joules per kelvin, that we have had in all our previous work. Then, multiplying by temperature we get what we want, an energy dimensioned function of the muddle number:

$$TS = kT \ln W$$

Let me emphasize that this is an axiom of this bit of work. That is to say it cannot be proved from any more self-evident premise. However, it has exactly the right properties to make the model work and you can only play the game if you accept it as a rule. ▼**Statistical games for $W$**▲ calculates $W$ for one simple case.

With the expressions we have derived, the conjectured equation for free energy, denoted $G$, becomes

$$G = U + \Sigma XI - TS \tag{4.2}$$

More often you will see this abbreviated to

$$G = H - TS \tag{4.3}$$

in which all the 'real' energy terms are lumped together as $H$ and called the enthalpy. The 'energified muddle factor' $TS$ is kept separate.

To use this equation as a tool, employ the following dogma. *A spontaneous change in a system will be sustained if it is such that G gets smaller. Equilibrium, a state of no spontaneous change, is therefore a condition at which G has a minimum value.*

# ▼Statistical games for W▲

The argument of this section is for following, not for learning. In most real systems it would be extremely difficult to count up the muddle number. There are so many features which can be randomized by rattle, such as the various energies distributed among the atoms, their positions in space, their patterns of association into different molecules and so on. Among the few relatively simple cases to analyse, and it has some direct applications in materials science, is the entropy of a crystal of identical atoms in which there are some vacant sites. The question to answer is: How many different ways can $N$ atoms be distributed among $N + n$ sites? That means there are $N$ atoms and $n$ vacant sites—or it could be $N$ atoms of one sort and $n$ of another in a mixture; it's the same calculation. Here's how the sum goes.

Suppose to start with that the atoms are labelled from 1 to $N$ to make them distinguishable from one another, and let us imagine loading them one by one into the empty set of $N + n$ crystal lattice points. The number of ways of choosing the site of the first atom is

$$(N + n)$$

For each of these ways, the number of ways of choosing the site of the second atom is

$$(N + n) - 1$$

Similarly, for the third atom

$$(N + n) - 2$$

and so on until we reach the last atom, the $N$th, when we can choose from $(N + n) - (N - 1)$ sites—that's $(n + 1)$.

The total number of arrangements is the product of all these numbers, that is

$$(N + n) \times (N + n - 1) \times$$
$$(N + n - 2) \times \ldots \times (n + 1)$$

which is

$$\frac{(N + n)!}{n!}$$

Note. The symbol ! is read as 'factorial' and means the product of all integers from the specified number down to 1. Thus 6! is $6 \times 5 \times 4 \times 3 \times 2 \times 1$, or 720. It is a compact way of writing long strings. The product string we have is *not* a factorial because it ends at $(n + 1)$, but I have multiplied top and bottom by $n!$ to get the next line.

Now, many of these arrangements will have used the same sites and only appear to be different because of the labels we imagined on the atoms (that is, because of the order we put them in place). Of course that has no real meaning, so the number must be reduced to allow for this. Whichever $N$ sites are occupied by the $N$ atoms, they can be interchanged without affecting the pattern of occupied and unoccupied sites. The number of ways of putting $N$ atoms onto $N$ sites is $N!$ (that's

obvious by putting $n = 0$ in the previous analysis). So the number of different arrangements of the $N$ atoms on $N + n$ sites becomes our muddle number,

$$W = \frac{(N + n)!}{N!\,n!}$$

The corresponding entropy, $k \ln W$, can be expressed more simply thanks to the mathematician Stirling, who showed that for large numbers

$$\ln X! \approx X \ln X - X$$

Using this theorem quickly yields the prettily symmetrical result

$$S = k[N \ln (N + n) - N \ln N + n \ln (N + n) - n \ln n].$$

And, for use in a moment, with $N$ constant,

$$\frac{dS}{dn} = k \left[ \frac{N}{N + n} - 0 + \ln (N + n) \right.$$
$$\left. + \frac{n}{N + n} - \frac{n}{n} - \ln n \right]$$
$$= k \ln \left( \frac{N + n}{n} \right)$$

Does the entropy increase as the number of vacancies $n$ increases?

Yes, it must, because there are more choices of sites for the atoms to occupy. Mathematically it must be that $(N + n)/n$ is greater than 1, so its logarithm is necessarily positive.

## 4.7.4 Critical temperature events

In a system not in equilibrium the descriptors on the right-hand side of the free energy equation can change independently, either up or down, provided the overall result is $\Delta G < 0$, that is $G$ is falling towards a minimum. In principle, even complicated versions of the equation could be examined numerically, but we won't do that here. Rather I want you to feel how changes in different parts of the equation interact to distinguish the possible from the impossible. For a start, the notion of 'sudden' changes, those which happen at a critical temperature, can be interpreted. During such an event there will be a change in $G$:

$$\Delta G = \{\Delta U + X\Delta I + I\Delta X\} - [T\Delta S + S\Delta T].$$

As the temperature stays steady while the change occurs the term in $\Delta T$ is zero. The bit I've lumped into the curly bracket is the enthalpy change ($\Delta H$); that is, the changes in random and ordered energies. In abbreviated form the equation may be written

$$\Delta G = \Delta H - T\Delta S$$

We can consult this equation to determine the temperatures which will give a reduction in free energy. There are four possibilities to consider.

Firstly, $\Delta G$ will be negative at all temperatures if $\Delta H$ is negative and $\Delta S$ is positive. In other words, any change which reduces the 'real' energies and increases the muddle can always happen. Actual examples of this are hard to find because there is always some interatomic force which decreases the muddle at some $T$. A near example is a gas expanding into a vacuum, for which $\Delta H \approx 0$ and $\Delta S > 0$ because the atoms have more space to move around in.

Secondly, if $\Delta H$ is negative and $\Delta S$ is negative, $\Delta G$ can only be negative if the temperature is *low* enough. What this is saying is that with $\Delta S$ negative, $-T\Delta S$ becomes positive so it will overwhelm a negative $\Delta H$ (arising perhaps from tighter bonding) unless the multiplier $T$ is small enough.

Does the phenomenon of freezing fit this description?

Yes. $\Delta H$ is negative because the potential energy of molecular bonding is reduced as the molecules come together to make a crystal. Also the muddle undoubtedly decreases in going from liquid to orderly crystal. And we know that freezing requires a certain critical low temperature to be established.

Here, at last, is the thermodynamic perception of how temperature controls the balance between ordering and muddle. The phenomena we recognized in Section 4.1 as 'sudden' become possible at a temperature where $\Delta G$ switches from positive to negative. Let us finish interrogating our equation.

EXERCISE 4.14   Write the sentence for $\Delta H$ positive, $\Delta S$ positive. Give an example.

A liquid will boil to become a gas if $T$ exceeds a certain critical value. This means that the muddle factor can overwhelm the sticking factor if $T$ is high enough. Notice that here the effect of the work term $p\Delta V$ in the enthalpy is big. A greater pressure means greater $\Delta H$, so $T$ has to go higher to force the change (for example in a pressure cooker). The $p\Delta V$ term is not usually of much importance in freezing because $\Delta V$ is small.

Lastly, if a proposed change has $\Delta H$ positive and $\Delta S$ is negative, it cannot happen spontaneously since $\Delta G$ is not negative at any temperature. If you want a system to do something of this sort it must be coupled into a broader system so that the overall changes come into one of the other categories.

Of course, at precisely the critical temperature $\Delta G = 0$. This means that in the case of a phase change, for example, no change need occur. We can have ice at $0\,°C$, or water at $0\,°C$, or a mixture of the two in any proportions. Changes from one to the other produce no drop in $G$, so no change needs to happen. It is instructive to interpret the free energy of such a system as it freezes or melts in terms of stick and rattle.

Let the system be a fixed quantity of a substance, one mole say, and consider the trends of $G$ as temperature changes for the possibilities that the substance is either a liquid or a solid. We can write

$$G_L = H_L - TS_L$$

and

$$G_S = H_S - TS_S$$

for liquid and solid respectively. Bonding in the solid is tighter than in the liquid, so at any temperature $H_S$ should be lower than $H_L$. The liquid phase is the less organized, so has the higher entropy; $S_S$ is less than $S_L$.

Figure 4.60 sketches the trends for solid and liquid as $T$ changes. To see the meaning of these, neglect the small changes of $H$ and $S$ with temperature so that plots of $G$ against $T$ become straight lines with gradient $-S$. Thus the line for liquid has to be the steeper. (Check this by thinking $y = mx + c$ onto the equation for $G$.)

The intersection is the melting temperature $T_m$. Above it the liquid has the lower free energy; entropy dominates. Below $T_m$ the solid is the stable form; enthalpy takes command. The 'free energy benefit' to the system of changing from liquid to solid, denoted $\Delta G$ on Figure 4.60, gets bigger and bigger as $T$ falls below $T_m$. You can think of this as a 'driving force' for the change.

▼Vacancies in crystals▲, ▼Why the crystal phases of iron?▲, ▼Purification▲ and ▼Ferroelectric effect and how to control it▲ provide further examples of processes where we can understand how $G$ falls by considering what is happening at the atomic scale.

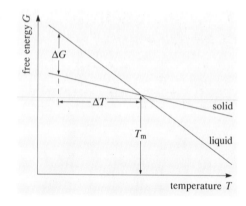

Figure 4.60 Free energy versus temperature for solid and liquid

# ▼Vacancies in crystals▲

In Section 4.5 the mechanisms of diffusion were seen to rely on vacant sites in crystals to allow atoms to jump. Rates of diffusion climb exponentially with temperature partly because more and more atoms get enough energy to jump into a vacancy. But another extreme energy effect is whether there is enough rattle to make sites vacant in a crystal. Since a vacancy is surrounded by broken bonds, energy must be required to create it. Let's say an amount of energy $\varepsilon$ is needed for each. If we can write a free energy equation for a crystal containing $n$ vacancies, it can be used to find how many vacancies there will be in equilibrium at any fixed temperature. The free energy equation is

$$G = U + \Sigma XI - TS$$

At equilibrium $G$ is to be minimum with respect to any increase or decrease in $n$, so

$$\frac{dG}{dn} = 0.$$

We need to identify the bits that change with $n$. The random vibrational energy won't be much affected by a few extra holes, so $dU/dn \approx 0$. Among the $\Sigma XI$ is the chemical ordered energy, $n\varepsilon$. So

$$\frac{dG}{dn} = \frac{d}{dn}(n\varepsilon) - T\frac{dS}{dn} = 0$$

We have an expression for $dS/dn$ from 'Statistical games for $W$', so

$$\varepsilon - kT\ln\left(\frac{N + n}{n}\right) = 0$$

Thus the fraction of crystal sites vacant is

$$\frac{n}{N + n} = \exp(-\varepsilon/kT)$$

which you will recognize as implying that the number of vacancies increases faster and faster as $T$ rises making diffusion faster too.

# ▼Why the crystal phases of iron?▲

Pure iron at room temperature is crystallized into a BCC lattice. As the temperature climbs past 1180 K the crystals all click to the FCC pattern, and on further heating to 1670 K they revert to BCC. The lower temperature change in particular is of the utmost importance in steel metallurgy, and thinking of ways to use it and control it will give you a happy time in the next chapter. But how does this odd behaviour square with the free energy equation and with what we know, or guess, about iron? First note that this change has little to do with iron's magnetism. Only the lower temperature BCC iron is ferromagnetic and it has run out of 'stick' and lost its magnetism at a temperature below the first crystal change.

The changes take place when $T$ is high enough so according to our analysis both $\Delta H$ and $\Delta S$ are positive as we go from BCC to FCC. But that hardly makes sense for a change from the loose BCC structure to the close-packed FCC. The tighter packing implies stronger bonding, that is $\Delta H$ negative, and all the atoms use less space so intuition says there's less muddle not more. What have we forgotten? Remember that iron is a transition metal so an extra source of muddle is the distribution of electrons among the four roles identified in Chapter 3. The change BCC to FCC would make sense if *stronger* bonding came from more electrons being delocalized. They would then have extra spatial entropy—more than enough extra to outweigh the small overall reduction in volume, we trust. And sure enough, FCC needs more delocalized electrons than BCC. So 'electron entropy' is the spur for the change. At the higher temperature the delicate balance of enthalpy and entropy reverses and the crystal reverts to BCC.

# ▼Purification▲

Another point of interest comes from the $dS/dn$ term. The term

$$\ln\left(\frac{N + n}{n}\right)$$

is largest when $n$ is small. Purification processes trying to remove the last few atoms of an impurity are therefore trying to make large reductions of entropy. Even if tighter bonding ($\Delta H$ negative) was promoted by the removal of the foreign atoms, the free energy may only fall if $T$ is low enough. Then process rates become sluggish. The ultra-pure silicon needed for electronics used to be very expensive. Newer vapour phase methods have cut the cost.

# ▼Ferroelectric effect and how to control it▲

Chapter 3 taught you how barium titanate (BaTiO₃) crystallizes in a predictable cubic structure built of octahedral and cuboctahedral cages. Well that's all right at high temperatures when there's plenty of rattle in the crystal, but what makes barium titanate interesting is something less predictable, a crystallographic distortion, that is a phase change, which happens as the crystal cools below 120 °C. The oxygen cages become a little over 1% taller and about 1% narrower, while the positive ions make tiny displacements in the same sense up the 'tall' direction. Figure 4.61 gives you an exaggerated view of this. A 'stretched cube' crystal is called 'tetragonal'. With cations and anions displaced relative to one another the crystal spontaneously polarizes. This is the phenomenon of **ferroelectricity** (so called because it is like ferromagnetism, not because it has anything to do with iron) and is what makes this substance and many like it electronically useful.

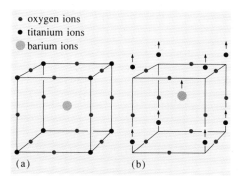

- oxygen ions
- titanium ions
- barium ions

Figure 4.61 Tetragonal distortion of barium titanate (BaTiO₃) crystals. (a) Regular crystal. (b) Distorted, with titanium and barium ions moved relative to oxygen 'cage'

Again let us try to understand this sudden change in the light of the free energy equation. It happens as temperature is falling so the suggestion is that both $\Delta H$ and $\Delta S$ are negative as the change goes in the direction described. The free energy equation for a cell of crystal within a larger volume has several terms:

$$G = U + \Sigma \mu_i N_i + PE + pV - TS$$

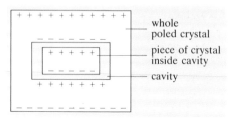

whole poled crystal

piece of crystal inside cavity

cavity

Figure 4.62 Polarized block of crystal in field

The sum term covers all the different ionic bonds of the crystal and the term $PE$ stands for the ordered potential energy of a polarized cell ($P$ is polarization, the extensive factor) lying in the electric field $E$ of the surrounding cells. The $pV$ term is external pressure times volume. We need

$$\Delta G = \Delta U + \Delta\Sigma\mu_i N_i + \Delta PE + p\Delta V - T\Delta S$$

to become negative. For such a small crystallographic change there is not likely to be much change of vibrational energy so let's ignore $\Delta U$. There is indeed a 1% compaction of volume which implies ions holding tighter, so $\Delta\Sigma\mu_i N_i$ is negative. Also, as Figure 4.62 shows, the electrostatic term represents a reduction — you'd have to do work to turn the block of crystal in the diagram the other way up. Because of the volume shrinkage there is also a small mechanical work term $p\Delta V$ which, being work done *on* the crystal by the external pressure, atmospheric say, is an increase in enthalpy. This is the only positive enthalpy change and it's not enough to outweigh the two negative changes. Overall $\Delta H$ is negative, which I shall indicate by $\Delta H^{(-)}$.

Turning to the entropy effects, there should be a bit less muddle in the system because the ions are less scattered in space. But more important is the behaviour of the Ba²⁺ ion. The repulsion between the twelve oxygen ions surrounding it makes the cuboctahedral cage a touch bigger than the ion radii predict. At high temperature the barium ions manage to fill the extra space by rattling all around their cages, that is to say they have high entropy. The distortion restricts them to one side of their cages

reducing their uncertainty of position and giving a substantial negative $\Delta S$ as required. I will denote this by $\Delta S^{(-)}$. For barium titanate the balance between $\Delta H^{(-)}$ and $-T\Delta S^{(-)}$ comes at 120 °C.

If the Ba²⁺ ions are replaced by smaller 2+ ions, then $\Delta S^{(-)}$ will be a bigger negative term. $\Delta H$ is unchanged, so the balance between $\Delta H^{(-)}$ and $-T\Delta S^{(-)}$ happens with a smaller $T$. The effect is for the material to maintain its undistorted structure down to a lower temperature. It turns out that SrTiO₃ flips at about 50 °C. By mixing both barium and strontium in a ceramic, a continuum of phase change temperatures for different grains can be arranged.

Multi-crystalline material, prepared cheaply as a ceramic, is technically versatile. A strong electric field put on the ceramic at $T$ just greater than 120 °C will dictate the polarization direction of each tiny crystal, Figure 4.63, and this will be 'frozen in' if the material is cooled further. Such a poled ceramic is piezoelectric; an alternating voltage put on it will provoke a mechanical vibration and vice versa — just the thing for driving a loudspeaker or converting the wrinkles on gramophone records to electric signals.

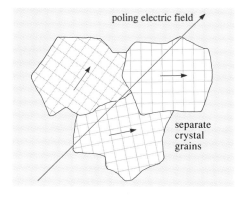

poling electric field

separate crystal grains

Figure 4.63 Poling of ferroelectric ceramic. Above 120 °C the cube axes are randomly oriented. Under the influence of an applied electric field the tetragonal distortion is mostly along the cube axis most nearly parallel to the field

*Unpoled* ceramics should make super dielectrics for capacitors. Just above the phase transition temperature when the crystals are almost ready to click into the polarized configuration, there would be an enormous polarization response to a small electric field. Unfortunately electronic engineers won't buy capacitors which have to be kept at 121 °C in order to work properly! The magic temperature needs to be lower and smeared over a range. The understanding of what happens at the phase transition has told us what to do to keep the high dielectric constant over a wide enough temperature range to be useful (Figure 4.64). (The formulations of the grades used for the diagram are more complicated than the one I have considered.)

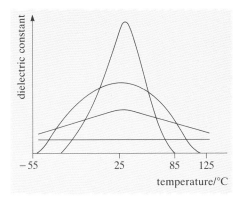

Figure 4.64 Changes of dielectric constant with temperature for various ceramics based on barium titanate

## 4.7.5 How do, or don't, crystals form?

Answering the above question requires us to look at another one: How fast can a system respond to the ΔG driving force? We want to understand crystal formation because many useful materials properties are the direct result of phase changes, either from the molten state or within solids. To control microstructure, and hence properties, it is necessary to understand how crystals develop or, in amorphous solids, do not develop, once the conditions allowing the phase change have been established.

The free energy equation has defined the temperature criterion for any phase change. The equation can be further applied to the atomic and molecular processes occurring close to the true transformation temperature as crystals of a new phase start to form. Atom associations in liquids are ephemeral compared to the persistence of solid structures. We can think of liquids as having small clusters of atoms (perhaps the groups of 4, 6 and 13 making tetrahedra, octahedra and cuboctahedra for starters, though bigger groups too) joining together, breaking apart, swapping atoms, moving past each other. A bigger-than-average cluster might be called an embryonic crystal, and the question of whether it will get bigger still to become a piece of solid or be smashed up and revert to average-cluster-sized liquid bits is all to do with local stick and rattle. ▼**Energetics of crystal nucleation**▲ presents a formal model of how embryonic crystals might emerge as statistical 'flukes', finding at what size the tighter bonding in the solid fragment compensates for the increased surface area that exposes it to disruption. Beyond some critical size of embryo, growth has the greater probability. Up to that size, the embryo is more likely to break up. The model predicts that the rate of formation of growable crystals will increase sharply as $T$ drops some way below $T_m$. This 'monkeys typing Shakespeare' mechanism is not what actually happens. Other easier mechanisms take precedence. But the principles of the analysis, and its predictions that some undercooling is necessary for the formation of crystals and that too much inhibits them, are generally applicable.

Daily experience tells us that a phase does not change instantaneously into another phase. For a start, it takes time because energy, the latent heat of the transformation, has to be transferred. (Think of the time it takes to boil a kettle dry.) Also, for the many crystals of a typical solid to form, the change must have started independently at several places. The 'statistical flukes' model does not recognize any one place as better than another for this. But precisely because either by luck or intent some places *are* better, the numerical predictions of the model are wrong. At some time you will have watched the bubbles forming, growing and breaking away at the bottom of a pan coming to the boil. You will have seen that some points produce a constant stream of bubbles. Something special is making the nucleation easier there. In the model in 'Energetics of crystal nucleation' the surface energy of the new-born crystal appeared as $\gamma^3$. Thus any vagaries which offer reduced surface energy will greatly ease the energetics of formation of nuclei.

217

# ▼Energetics of crystal nucleation▲

At its melting temperature, a substance melts(!) but coming back down to the same temperature, now expected to be the freezing temperature, often nothing happens. Several kelvins of undercooling may be needed to induce solid to precipitate. Again the free energy equation can help us to see why. At the precise freezing temperature, $\Delta G = 0$. Let me write this as

$$0 = (\Delta U - N\Delta\mu) - T_m\Delta S$$
$$= \Delta H - T_m\Delta S$$

where you can see the change of bond energy to put $N$ atoms into the solid and their change in vibration energy bracketed and then put equal to an enthalpy change. This can be measured during the melting transition and is of course the latent heat. So $\Delta S = \Delta H/T_m$ and note that the special thing about $T_m$ is that it is *just* low enough for that amount of energy, $\Delta H$, to carry away the necessary entropy for the transition. At any higher temperature $\Delta H/T$ would be too small to transport the entropy. Crystals begin to form at a lower temperature, with $\Delta H$ and $\Delta S$ essentially unchanged, but with a significant extra amount of ordered work to be done. The embryo must have a surface and that has surface energy by virtue of the mismatch between the solid and liquid atom arrangements. Let $\ell_v$ be the latent heat per unit volume solidified and $\gamma$ be the surface energy per unit area. We want to know the free energy of a small spherical particle of solid, radius $r$, forming from a quantity of liquid initially having free energy $G_0$. The free energy becomes:

$$G = G_0 - \Delta H$$
$$+ \Delta(\text{surface energy}) - (-T\Delta S)$$
$$= G_0 - \left(\frac{4\pi r^3}{3}\right)\ell_v$$
$$+ 4\pi r^2\gamma - T\left(\frac{4\pi r^3}{3}\right)\left(-\frac{\ell_v}{T_m}\right)$$
$$= G_0 - \frac{4\pi r^3\ell_v}{3}\left(1 - \frac{T}{T_m}\right) + 4\pi r^2\gamma$$

Figure 4.65(a) plots this equation. As $r$ starts from zero, $G$ has to *increase*.

Eventually it peaks at a radius $r_c$ when $dG/dr = 0$, that is,

$$0 = -4\pi r_c^2\ell_v\left(1 - \frac{T}{T_m}\right) + 8\pi r_c\gamma$$
$$r_c = \frac{2\gamma}{\ell_v(1 - T/T_m)}$$

From this condition an embryo can reduce its free energy either by growing or by shrinking. Thus $r_c$ is the critical radius beyond which spontaneous and sustained growth is possible. Until this critical radius is reached the reduction of enthalpy as any extra solid deposits on the embryo cannot provide the surface energy it needs *and* dump the entropy necessary to be 'low muddle crystal' instead of 'high muddle liquid'. That means that small embryos are unstable and almost all re-dissolve in the liquid. Beyond radius $r_c$ extra material deposited *reduces* the free energy in the proper manner for spontaneous events. So the difficulty in nucleation is how to 'disobey' the second law of thermodynamics (that is, how to have spontaneously increasing free energy) to get embryo crystals growing to $r_c$. Chance is the answer. Small temporary local increases in free energy are not entirely impossible, just rather improbable.

Putting in sensible figures for metals, $\gamma \approx 0.01\,\mathrm{J\,m^{-2}}$, $\ell_v \approx 10^8\,\mathrm{J\,m^{-3}}$, $T/T_m \approx 0.9$, gives $r_c \approx 2 \times 10^{-9}\,\mathrm{m}$ suggesting that a few hundred atoms have to assemble by chance before stable growth can carry on. The corresponding increase in free energy, $\Delta G_c$ on Figure 4.65(a) is

$$\Delta G_c = \frac{16\pi\gamma^3}{3\ell_v^2(1 - T/T_m)^2}$$

which calculates to about a 1 eV excess. This has some chance of occurring as a statistical fluke, just as a single gas atom may turn up with that much energy. The chance is $\exp(-\Delta G_c/kT)$ so a number proportional to this describes the rate of formation of nuclei able to grow steadily and not at risk of re-melting. Because of the term $(1 - T/T_m)^2$ in the denominator, $\Delta G_c$ gets smaller quite rapidly as $T$ falls so the rate rises, Figure 4.65(b). Nevertheless the rate for spontaneous nucleation is always impossibly slow.

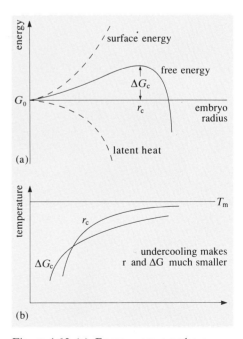

(a)

(b)

Figure 4.65 (a) Energy versus embryo radius. (b) Effect of undercooling

Castings freezing in their moulds find especially easy sites for crystal nucleation on the walls of the mould. In solid-state phase transformations, the favoured sites for the nucleation of a new phase are grain boundaries, dislocations and other strained regions of crystal. Chapter 5 will discuss this in some detail.

## Growth rate

We come now to questions of the growth dynamics of these nuclei. That too depends subtly on the extent of undercooling of the melt below the nominal transition temperature. There are two factors to consider.

1 As a volume of solid is formed some latent heat is liberated. How does this escape from the site of the growth?
2 Molecules moving from liquid to solid, or from the solid back to liquid, have to change their bonding via some intermediate stage of activation. How do the competing rates of freezing and melting vary with temperature?

The first of these issues is easily answered in principle, if not in detail. The highest temperature for the solid to exist is $T_m$. If the surrounding liquid is at the same temperature, there is no temperature gradient to drive a heat flux so no growth can occur. With the liquid undercooled, heat can flow from the growing crystal and details of geometry notwithstanding, a rate of flow proportional to $\Delta T$, the amount of undercooling, can be anticipated. So the rate of formation of new phase will increase with $\Delta T$ if heat flow is the limiting factor. It usually is for easy-to-assemble crystal geometries such as metals have. With more complex structures the activation factor dominates.

Imagine, with the help of Figure 4.66, polymer molecules transforming from liquid to solid state. There are two processes going on all the time. Random, loosely bound liquid folding into neat, tighter bound solid; and the opposite – crystal falling into the liquid state. Each has an activation energy. Intuition suggests that the solid-to-liquid change, involving disruption of several (secondary) bonds, will have a bigger $E_a$ than the liquid-to-solid change, where the initial bonding is weaker. The difference between these activation energies, as Figure 4.66(b) shows, is just the free energy change for the event. Now our theorems from

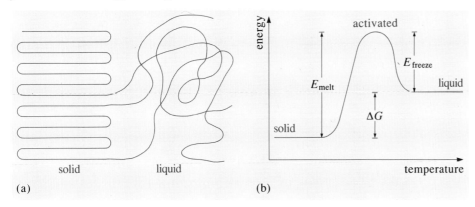

Figure 4.66 (a) Polymer molecule solidifying. (b) Activation energies related to free-energy change

219

Section 4.52 can be applied. If the temperature is lowered, both crystallizing and liquefying processes are slowed down but the one with the greater $E_a$ is slowed more. Consequently the net growth rate of the crystal phase is enhanced. ▼**Activation-energy controlled growth**▲ has a touch of algebra to deduce that the rate of growth will be dependent on the undercooling.

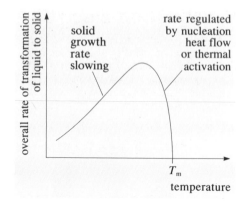

Figure 4.67 Overall rate of solidification just below melting temperature

One way or another, then, through difficulties of nucleation, heat dissipation or molecular organization kinetics, the rate of formation of a new phase after cooling to the thermodynamically defined melting temperature must be zero. Some undercooling is necessary for crystals to form. But too much undercooling will inhibit growth. The exponential slowing of the 'organization' processes, which are thermally activated, must eventually overtake all other influences. All these qualitative ideas are merged to produce Figure 4.67, a general curve of transformation rate versus temperature, which is the key to developments in Chapter 5. Theory, as you will perceive, cannot make this curve quantitative for real systems but can be used to make the following qualitative comments.

To make fine-grained castings, a high rate of nucleation is needed. This is achieved by providing sites which offer low $\Delta G_c$ both at the walls of the mould and in the bulk metal. Innoculating the melt with suitable fine particles on which crystals can grow ensures that the microstructure is not dominated by the few fastest-growing crystals which nucleate at the mould wall. Required properties of an innoculating powder are:

- It must not melt or dissolve in the melt.
- It must be easily wetted by the melt.
- There should be low surface energy for the interface between innoculant particle and the solid phase.

For cast iron a favoured innoculant is cerium oxide. This has a similar crystal structure to uranium oxide, with several vacant oxygen-ion cube cages. Perhaps iron atoms can diffuse into the surface of such a lattice to provide low surface energy in the initial stages of nucleation. Particles of about 1 μm diameter can provide several nucleation points for nuclei about 1000 times smaller, so a few million particles in a casting guarantee fine grains throughout the metal. Another dodge in metal casting is to break up the dendrite crystals growing from the mould wall when they have grown a bit to spread nuclei through the melt.

At the other extreme, to make glasses, that is no crystals at all, nuclei are to be avoided. The melt needs to be cooled rapidly to a temperature below the hump on the nucleation curve and growth processes have to be slow. Growth is slow if complex crystal structures have to be organized or if irregular molecules are involved, as in polymers and silica glasses. To make amorphous metals from the melt, extremely fast cooling rates of the order of $10^6\,\mathrm{K\,s^{-1}}$ are needed to beat the nucleation and growth of the simple crystals that constitute crystalline metals. The earliest amorphous metals were alloys of relatively incompatible metals,

# ▼Activation-energy controlled growth▲

At $T_m$, the rate of formation of solid from liquid is equal to the rate of dissolution of solid into liquid. That is, the decrease of entropy going from liquid to solid ($T\Delta S$) is equal to the enthalpy change ($\Delta H$). Let this common rate be $R_0$.

Both the liquid-to-solid and solid-to-liquid phase changes are thermally activated processes modelled by the Arrhenius rate equation

$$R = A\exp\left(\frac{-E_a}{kT}\right)$$

So the rate of melting is

$$R_{melt} = A_{melt}\exp\left(\frac{-E_{melt}}{kT}\right)$$

and the rate of solidification is

$$R_{freeze} = A_{freeze}\exp\left(\frac{-E_{freeze}}{kT}\right)$$

with $R_{melt} = R_{freeze} = R_0$ at $T_m$.

I want to look at the net rate of phase change, $R_{freeze}$ minus $R_{melt}$ as the temperature falls below $T_m$. At some temperature $T$ below $T_m$ we can say

$$T = T_m - \Delta T$$

$$= T_m\left(1 - \frac{\Delta T}{T_m}\right)$$

so

$$R_{melt} = A_{melt}\exp\left[\frac{-E_{melt}}{kT_m}\left(1 - \frac{\Delta T}{T_m}\right)^{-1}\right]$$

We are certainly considering small $\Delta T$, for which ($\Delta T/T_m$) is much less than 1, so we can use the following binomial expansion and approximation:

$$\left(1 - \frac{\Delta T}{T_m}\right)^{-1} \approx 1 + \frac{\Delta T}{T_m}$$

So

$$R_{melt} \approx A_{melt}\exp\left[\frac{-E_{melt}}{kT_m}\left(1 + \frac{\Delta T}{T_m}\right)\right]$$

$$= A_{melt}\left(\exp\frac{-E_{melt}}{kT_m}\right)$$

$$\times \left(\exp\frac{-E_{melt}\Delta T}{kT_m^2}\right)$$

$$= R_0\exp\frac{-E_{melt}\Delta T}{kT_m^2}$$

You will recall that $R_0$ is the rate of change at $T_m$. It is common to both phase changes at $T_m$.

Precisely the same argument gives

$$R_{freeze} = R_0\exp\frac{-E_{freeze}\Delta T}{kT_m^2}$$

Note that putting $\Delta T = 0$ in both equations gives a rate $R_0$ and that increasing $\Delta T$ (corresponding to dropping below $T_m$) causes both rates to slow exponentially. This agrees with what we know about Arrhenius-type processes. What of the net rate of transformation, $R_{freeze} - R_{melt}$? Clearly this is proportional to the difference

$$\exp\frac{-E_{freeze}\Delta T}{kT_m^2} - \exp\frac{-E_{melt}\Delta T}{kT_m^2}$$

In Figure 4.68(a) I have plotted two decreasing exponential functions. We know that $E_{melt} > E_{freeze}$ so the curve with the faster rate of drop is modelling the melting rate. Figure 4.68(b) plots the difference between these two curves. In the region just below $T_m$, the net rate of growth of solid material, surprisingly, increases with $\Delta T$. Below some optimum amount of undercooling, the rate of growth slows down. You should spend a moment thinking about the physical reasons for this behaviour.

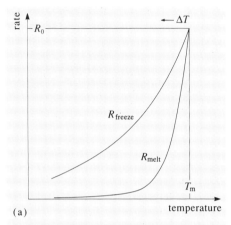

(a)

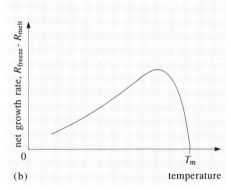

(b)

Figure 4.68 (a) Rates of freezing and melting modelled as exponentially decaying curves. (b) Overall rate of solidification is difference between curves in (a)

such as nickel/niobium, which do not have such simple crystal structures and which have rather low latent heats (small improvement of bonding on becoming crystalline). Some amorphous metals have such advantageous magnetic properties that they are commercially found in a range of applications.

SAQ 4.12 (Objective 4.12)
Use the ideas of this section to define the conditions for growing single crystal silicon for semiconductor applications.

Phase transformations in the solid state, either from one crystal structure to another or in the precipitation of intermediate compounds, also proceed by nucleation and growth. Diffusion rates in solids are much slower than in liquids so the growth of embryos to sustainable nuclei and their subsequent development to new crystals are very slow. Additionally the surrounding solid constrains changes of volume in the growing crystal. This can induce high stresses in the growing embryos. These stresses impose an extra, perhaps dominant, ordered-energy requirement on the phase change. The energy barrier to nucleus formation is raised. Because of these considerations, solid state transformations are generally so slow that thermodynamic equilibrium is the exception rather than the rule. Not surprisingly, grain boundary sites, which offer faster diffusion and less rigid constraints, are favoured for the emergence of new phases.

Later chapters will look at the heat treatments of metals to develop microstructure. All these treatments are designed to control (to encourage or to inhibit) such phase changes and they often involve careful selection of temperature and lengthy process times.

## 4.7.6 Grain growth and sintering

Figure 4.69 shows what happens when a fine-grained crystalline material is held at high temperature for a long time. This spontaneous grain growth has serious consequences for strength, though there are circumstances where it is encouraged, such as reducing the hysteresis loss of a magnetic alloy. Why does it happen? Evidently the change does nothing to the random energy $U$ and the entropy is unchanged in any significant way: a change of macro-organization has virtually no $\Delta S$ compared with changes in atomic muddle because the $W$ numbers involved are so small. All that's left to bring $G$ down is an ordered-energy change. In this case the total surface energy of the assembly of grains is reduced since a few large grains clearly have less surface area than a host of little ones. The mechanism of change is diffusion down the ordered-energy gradient, locally characterized by the curvature of the surface. Atoms tend to jump from sharply curved surfaces to flatter ones. Remember that given a potential gradient, there need be no concentration gradient to support diffusion.

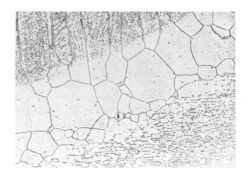

Figure 4.69 Photomicrograph of thermally grown grains

The rate of diffusion and hence of grain growth depends on the potential gradient, in part fixed by the curvature and part by the nature of the interface. The particularly clever thing about the precipitates dispersed in creep-resisting jet-engine alloys is that the crystal spacing of the intermediate compound so closely matches that of the matrix they lie in that the surface energy is very very small. The ultra-fine dispersion is therefore stable against grain growth for a long time.

Sintering, the process by which powders fuse together when held at high $T$, is a special case of grain growth where the contact between the crystals is intermittent rather than continuous. At the contact points diffusion driven by surface energy reduction still occurs. Mostly the technical problem now is to get the rate of diffusion *up* so that the powder quickly fires to a continuum.

Any suggestions?

(a) Maximize contact points – many particles pressed firmly together.
(b) Maximize curvature – use finest powder possible.
(c) Use highest possible $T$ (economics might determine the value to use).
(d) Put liquid between powder particles to enhance diffusion rate and contact area.

If you got all those points you are already a powder technologist.

Nowadays we can control all these process variables, but times were when an advance on one of these fronts heralded a technological breakthrough. For example, when a fusible powdered rock was blended with refractory china clay, porcelain could be made. In contrast, to make ceramics to serve under stress at very high temperature, say as a turbocharger rotor, fluxing aids to sintering are not a good idea. They limit the service temperature range. Hot pressing of ultra fine powders, that is (a), (b) and (c) combined, then gives the best result. Remember also that powder processes are important for metals as well as for ceramics. They allow near net-shape forming, in which you get the shape you want with very little machining cost or waste.

## 4.7.7 Driving force

At the end of Section 4.7.4 I said that you can think of the amount $\Delta G$ by which the free energy of a system decreases during a change as a 'driving force' for the change. You have now met several examples. Table 4.8 lists the $\Delta G$ for various types of phenomena. You will see that the values cover a wide range. I want to consider the implications of these values.

In ordered systems (mechanics), 'force' is regarded as the *cause* of an effect: big forces imply that things will happen fast. That is *not* the implication of this 'driving force', it does not even have the dimensions of 'force'. In random systems (thermodynamics) the 'cause' of change is just the random thermal rattle, which is constantly changing atoms' arrangements. Those arrangements with the higher statistical probability (lower $G$) happen spontaneously.

*Rates* of change depend on activation energies, and we saw in 'Activation-energy controlled growth' that the *net* rate of a change depends on the relative magnitudes of the activation energies for the 'forward' and 'reverse' events. When a 'forward' event has occurred, the activation energy for its reverse is $E_a + \Delta G$. (On Figure 4.66b freezing is the forward change and melting the reverse.) A big $\Delta G$ makes the activation energy of the reverse events much larger than for the forward ones; it will therefore strongly impede the change being undone by random rattle. To this extent, the magnitude of $\Delta G$ gives a measure of the thermodynamic insistence that change will occur in a certain direction — that is, the change is 'forced'.

Table 4.8   Typical values for $\Delta G$

| Phenomenon | $\Delta G/\text{J mol}^{-1}$ |
|---|---|
| chemical reaction with big bond energy change | $10^5$ to $10^6$ |
| 'weaker' chemical reaction | $10^3$ to $10^5$ |
| diffusion down concentration gradients | $10^2$ to $10^4$ |
| solidification with a few K of undercooling | 10 to $10^2$ |
| crystallographic phase changes | 1 to 10 |
| sintering, grain growth, precipitate coarsening | $10^{-2}$ to 1 |

EXERCISE 4.15   A $\Delta G$ of about $1000\,\text{J mol}^{-1}$ is equivalent to about $10^{-2}\,\text{eV}$ per atomic event. But activation energies in the range 0.1 to $1\,\text{eV}$ per atom have been seen as significant in controlling rates of change. Comment on the implications of this comparison on the effectiveness of the $\Delta G$ 'driving force' in directing the course of events.

## Summary

Thermodynamic equilibrium *of a system* is a balance between sticking and rattling tendencies. It is expressed by the free energy $G$ being at a minimum.

Therefore as the system moves from non-equilibrium towards equilibrium the change $\Delta G$ must be negative.

For a change at a fixed temperature $T$,

$$\Delta G = \Delta H - T\Delta S$$

where $\Delta H$ is the change of random plus ordered energies and $\Delta S$ the change in entropy (muddle factor). Change is possible only if $\Delta G < 0$. For different signs of $\Delta H$ and $\Delta S$, Figure 4.70 summarizes how $T$ dictates the possibility of any change.

A phase change becomes possible at the critical temperature at which $\Delta G$ changes sign.

# 4.8 Coda

Chapters 3 and 4 have developed kinetic atomism to bring the separate ideas of chemical bonding and thermodynamics into one frame; and this was done by recognizing the organizing principle (stick) as complementary to the muddling principle (rattle). We have offered a comprehensive view of this complementarity over a wide field of examples which should provide a bedrock and conceptual springboard for further work. Of course, all aspects of these principles are capable of much deeper analysis: no doubt they could be put more simply too. But along that spectrum we have tried to steer a course which will provide images from which more immediately applicable models can be built.

The rest of this book generates and applies some of those models. Much of what has been done here will be taken as a background presumption, rather than the source of a deductive stream. For example, it is self evident that corrosion through a protective oxide layer involves diffusion; equilibrium phase diagrams must be expressions of minimum free energy; and dislocations are 'objects' with describable behaviour and don't have to be seen as particular atom arrangements. So, although engineers often deny the use of the abstract sciences that they have (probably) all studied, the models they use actually contain these principles without necessarily being derived from them. Very often as you use models designed to represent events at a larger scale, you will find it helpful to visualize what is going on in kinetic atomistic terms. Let us now move into the practical world with these insights to guide us.

## Objectives for Chapter 4

You should now be able to do the following.

4.1 Recognize for a thermometer in a practical situation several criteria for its choice. (SAQ 4.1)

4.2 Define temperature in terms of atomic kinetic energy and discuss the criteria which determine whether, or to what extent, an energy form within an atomic assembly is part of the random internal energy. (SAQ 4.2)

4.3 Give an atomic-interaction model of thermal expansion. Deduce the thermal-stress consequences of constraints during temperature change. Discuss several technical examples. (SAQ 4.3, 4.4, 4.5)

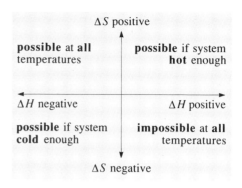

Figure 4.70 Summary of possibilities for change

4.4 Outline, for different classes of substance, appropriate models for electrical conductivity and its temperature variation. (SAQ 4.11)

4.5 Explain and exemplify what is meant by 'thermally activated phenomena', explain why their rates vary exponentially with temperature and explain how their temperature variation depends upon activation energy. (SAQ 4.6, 4.9)

4.6 Extract a value for an activation energy from data presented as a log.-rate versus reciprocal-temperature graph. (SAQ 4.6)

4.7 Describe how structure influences the glass transition temperatures of amorphous solids. (SAQ 4.7)

4.8 State Fick's laws of diffusion (for one dimension only) and explain their practical significance. (SAQ 4.8)

4.9 Describe the mechanisms of diffusion in solids. (SAQ 4.8)

4.10 Describe the energy distribution of electrons in a solid in terms of a density of states function $g(E)$ and the Fermi–Dirac temperature-dependent state occupation probability function $p(E)$ and use this model to understand extrinsic semiconductors, contact potentials and thermoelectric voltage. (SAQ 4.9, 4.10)

4.11 Quote the free-energy equation, outline the argument of its setting up and use it qualitatively to recognize the conditions for 'sudden' thermal effects.

4.12 Combine free energy and activation ideas to describe the development of new crystal phases in either liquid or solid-state transformations and the sintering of powders. (SAQ 4.12)

4.13 In addition to those implicit in the above, define and use the following terms and concepts:

| | |
|---|---|
| absolute zero | hole |
| Arrhenius rate equation | mobility |
| Boltzmann's constant | ordered energy |
| enthalpy | Pauli level |
| entropy | phonon |
| Fermi level | temperature coefficient (of |
| | resistivity, expansion, etc.) |
| ferroelectricity | undercooling |

# Answers to exercises

EXERCISE 4.1 We do it all the time with the weather: the 'sound' of a hot summer day; a halo round the moon tells you it's cold out there. Housemaids used to test flat irons by spitting on them, and I judge oil to be hot enough to fry pancakes when it is smoking. More technically a farrier will judge the temperature of a horse shoe by how it rings as it is hammered. Potters watch little cones which sag when the kiln reaches the right temperature (you can get a series of them for firing different clays).

EXERCISE 4.2 1 E. 2 H. 3 D. 4 A. 5 I. 6 J. 7 B. 8 C. 9 F. 10 G. 11 K. 12 L. If some of these answers surprise you, you should try the exercise again when you have read through the chapter.

EXERCISE 4.3 The text cites Boltzmann's constant as $86 \, \mu eV/K$ so at $300 \, K$ the average kinetic energy is $1.5 \, kT = 1.5 \times 86 \times 10^{-6} \times 300 \approx 0.04 \, eV$. Atomic electron energy levels are much more like $1 \, eV$ apart so this thermal energy is small compared to the energy needed to excite electrons to higher states.

EXERCISE 4.4

• Chemical potential energy due to some bonds being broken.
• Chemical potential due to non-uniform composition.
• Electron potential energy due to excitation or ionization.
• Rotational kinetic energy of parts of molecules.
• Electromagnetic radiation, if the material is red hot, for example.
• Potential energy due to density fluctuations (sound waves).
• Nuclear potential energy (possibility of radio-active decay).

EXERCISE 4.5 There are several properties which change gradually over quite wide temperature ranges in addition to the electrical resistivity of metals and thermal expansion (or its inverse, density) already cited.

Young's modulus depends on atom spacing and that is about proportional to temperature. Materials' yield stresses depend on dislocation mobility which is an average energy effect except when extra slip systems become thermally activated as in some ceramics at high temperature or at phase changes.

Viscous flow of molten materials is another example. The temperature coefficient of this change is larger in polymers than in metals, making temperature control of injection moulding of plastics rather more sensitive than for metal casting processes.

Thermal conduction by which rattle energy is transferred from atom to atom is another readily anticipated instance.

EXERCISE 4.6
Kinetic energy gained $= \frac{1}{2} mv^2$.

Let electron charge be $e$, electric field strength be $E$ and distance travelled be $d$. Loss in potential energy is $eEd$.

$$\frac{1}{2} mv^2 = eEd$$

$$v^2 = \frac{2eEd}{m}$$

$$v^2 = \frac{2 \times 1.6 \times 1.4 \times 2000 \times 10^{-29}}{10^{-30}}$$

$$v \approx \sqrt{(90000)} \, m \, s^{-1}$$

$$= 300 \, m \, s^{-1}.$$

This is the maximum speed reached. The electron reaches it by steady acceleration so the equivalent steady speed is half this, $150 \, m \, s^{-1}$.

EXERCISE 4.7 At $300 \, K$ we have seen $kT \approx \frac{1}{40} \, eV$, so $1 \, eV = 40 \, kT$ and $0.8 \, eV = 32 \, kT$. The relative chances of the two kinds of break are

$$\frac{(32)^{1/2} \exp(-32)}{(40)^{1/2} \exp(-40)} = (0.8)^{1/2} \exp(+8)$$

$$= 0.9 \times 3000$$

$$= 2700$$

The square root term makes only 10% difference. An error of $1 \, kT$ in activation energy would shift the answer by a factor of 2.7.

EXERCISE 4.8 The value of $kT$ at $1000 \, K$ is

$$\frac{1000}{300} \times \frac{1}{40} \, eV = \frac{1}{12} \, eV$$

so the comparison of numbers with energies greater than $1 \, eV$ becomes

$$\frac{N_{1000}}{N_{300}} = \frac{\exp(-12)}{\exp(-40)}$$

$$= \exp(+28) \approx 10^{12}$$

A readily achieved change of temperature produces a massive increase in the number of energetic atoms.

EXERCISE 4.9 Start by merging the two equations of hints (a) and (b) to express the rate of rise of temperature of an element

$$\frac{\Delta T}{\Delta t} = \frac{(\mathcal{Q}_1 - \mathcal{Q}_2)A}{\mathcal{S}_v A \Delta x}$$

Use hint (c) to replace the difference $(\mathcal{Q}_1 - \mathcal{Q}_2)$ of fluxes in and out.

$$\frac{\Delta T}{\Delta t} = \frac{(-d\mathcal{Q}/dx)A\Delta x}{\mathcal{S}_v A \Delta x}$$

The volume $A\Delta x$ of the element now cancels and hint (d) allows for $\mathcal{Q}$ to be re-expressed

$$\frac{\Delta T}{\Delta t} = \frac{-1}{\mathcal{S}_v} \frac{d}{dx} \left( -\kappa \frac{dT}{dx} \right)$$

or, in proper calculus notation

$$\frac{\partial T}{\partial t} = + \frac{\kappa}{\mathcal{S}_v} \frac{\partial^2 T}{\partial x^2}$$

EXERCISE 4.10 The gradient of Figure 4.39

$$\frac{\Delta \ln D}{\Delta (1/T)} = -\frac{E_a}{k}$$

So

$$E_a = -k \frac{\Delta \ln D}{\Delta (1/T)}$$

Using convenient points ($6.8 \times 10^{-4}$, $-36.0$) and $7.8 \times 10^{-4}$, $-39.9$) yields

$$\frac{(-39.9) - (-36)}{(7.8 - 6.8) \times 10^{-4}} = \frac{-E_a}{86 \times 10^{-6}}$$

$$E_a = -86 \times 10^{-6}$$

$$\times \frac{(-39.9) - (-36)}{(7.8 - 6.8) \times 10^{-4}} \, eV$$

$$= 3.35 \, eV.$$

To find $D_0$ we have

$$\ln D = \ln D_0 - \frac{E_a}{kT}$$

227

So

$$\ln D_0 = \ln D + \frac{E_a}{kT}$$

Using the first quoted data point,

$$\ln D_0 = (-36)$$
$$+ \frac{3.35 \times 6.8 \times 10^{-4}}{86 \times 10^{-6}}$$
$$= -9.5$$

So

$$D_0 = 7.5 \times 10^{-5} \, m^2 \, s^{-1}.$$

This answer is very sensitive to your reading of the graph so you may have a different answer within the range suggested.

EXERCISE 4.11 The ratio of the distances diffused in the two cases is

$$\frac{\sqrt{(D_p t)}}{\sqrt{(D_c t)}}$$

and the time for the diffusions is the same, so $t$ cancels. So the distance $L$ diffused by the phosphorus is

$$L = \sqrt{\left(\frac{5.6 \times 10^{-16}}{1.4 \times 10^{-11}}\right)} \times 1 \, mm$$
$$= \sqrt{40 \times 10^{-6}} \, mm$$
$$= 6.3 \times 10^{-3} \, mm.$$

The carbon is moving 160 times faster than the phosphorus.

EXERCISE 4.12 For the lower energy the exponential term becomes $\exp[(E_F - 2.5\,kT - E_F)/kT]$, which is $\exp(-2.5)$ or 0.082. Similarly at $2.5\,kT$ above the Fermi energy we have $\exp(+2.5)$, which is about 12.2. So at $2.5\,kT$ below $E_F$,

$$p(E) = \frac{1}{1 + 0.082} \approx 0.92$$

and an equally short range above $E_F$ brings

$$p(E) = \frac{1}{1 + 12.2} \approx 0.08$$

Here you can see the symmetry of the curves: 8% of the states are occupied at $E_F + 2.5\,kT$; 8% are *un*occupied at $E_F - 2.5\,kT$.

EXERCISE 4.13 Platinum in the 5d transition elements is homologous with Ni but has nine d-electrons, and Rh in the 4d set is under Co but has eight d-electrons. Since rhodium is to be added in *small* amounts, the density of states curve of the alloy will essentially be that of pure platinum. But rhodium atoms bring fewer electrons on their outer surfaces, so not so many states are needed. So the Pauli level

is lowered.

EXERCISE 4.14 You should have written something like this. If $\Delta H$ is positive and $\Delta S$ is positive, then $\Delta G$ can only be negative if $T$ is high enough. The obvious example is the phenomenon of boiling.

EXERCISE 4.15 Rates are very sensitive to the value of $E_a$ especially when $E_a \approx 1\,kT$. (We saw this in Exercise 4.7.) So even a difference of $10^{-2}\,eV$ between doing and undoing a change can be significant and the very weakly 'driven' phenomena will occur, but not with the insistence that might be expected. Thus we saw (Section 4.5.4) that carbon penetrates steel by less than 1 mm even though the individual atoms cover as much as 1 km in diffusion jumps; jumps are nearly equally probable in each direction.

Bigger driving forces of course affect $E_a$ strongly and result in much greater impediment to reversal: oxides do not spontaneously decompose, as you will see in Chapter 7. The chosen level of $1000\,J\,mol^{-1}$ is a reasonable boundary between the thermodynamics being emphatic (big $\Delta G$) or rather more vaguely insistent.

# Answers to self-assessment questions

SAQ 4.1 Your answer should have identified some qualities drawn from the list in the text. Easily found thermometers are domestic oven, refrigerator, central heating and car engine thermometers. Taking this last example we would note that accuracy is a low priority and cost a high one. It needs to be robust and reliable, and electric output is handy. Thermistors (semiconducting resistance thermometer) are widely used.

SAQ 4.2

A   Not coupled.
B   Strong coupling, but modified by other effects.
C   Strong coupling.
D   Slight coupling.
E   Not coupled.
F   Slight coupling, maybe strong.

The thermal energy is any energy which changes 'significantly' with temperature. Since atomic kinetic energy defines temperature, only forms which couple to atomic kinetic energy, that is, respond to collisions, can become random and part of the thermal energy. C is a main component of a solid's energy and F may also be important (polymers above $T_g$). B and D are not main ingredients because only a small proportion of conduction electrons or chemical bonds are affected by atomic collisions at any instant (see Sections 4.5 and 4.6). A and E are definitely excluded until very high temperatures.

SAQ 4.3

Magnesia: Curve C, 2600 °C, $15 \times 10^{-6}\,K^{-1}$.

Aluminium alloy: Curve B, 500 °C, $23 \times 10^{-6}\,K^{-1}$.
Nylon: Curve A, 250 °C, $100 \times 10^{-6}\,K^{-1}$.

Nylon has a low melting temperature because its molecules are held together only by secondary bonds (hydrogen type). That implies a shallow potential well for attractive forces but the repulsion is *always* a steep rise in potential energy. That will emphasize the skewness of the curve, giving big thermal expansion. At the other extreme, the deep well associated with the doubly charged ions in MgO is near symmetrical at the bottom of the well. So the highest melting temperature and the lowest expansion go together. The Al alloy is intermediate.

SAQ 4.4 Make $\alpha_1 \leqslant \alpha_2$ so that the glaze stress is compressive (negative, see equation for $\sigma_1$). This will close up surface cracks and prevent brittle fracture. The balance is delicate though because the $x_2$ in the numerator is big and the glaze can crumple if the compression is overdone.

SAQ 4.5 I have put the quantities we want small on the top of the expression for $M$ and those which should be big underneath, so good thermal shock resistance against cracking fracture will go with $M$ being small.

You should have expected that soda glass would be poor and that metals have no trouble with thermal shock. Nylon rarely finds itself in such service and in any case is so ductile that cracking is not going to be the mode of failure. The ceramics, with their wider range of intermediate values, are where the interesting problems lie. The order of quality I found was steel best, then dural, silica, alumina, porcelain, nylon, and soda glass comfortably the worst. I won't spoil your fun by quoting the values.

SAQ 4.6 Table 4.9 summarizes the data given. Following the method for Ticky Tacky used in text, the rate of cure $R$ is the inverse of the time to cure.

Figure 4.71 plots these and extrapolates both lines to $T = 273\,\text{K}$ (a long stretch, so accuracy is none too good). From the graph, at 273 K (0 °C),

$$\ln R = -9.5 \text{ for Ticky Tacky}$$

$$\ln R = -13.0 \text{ for Bangiton.}$$

The cure time (in minutes) is the reciprocal of $R$, so we get cure times of 9.3 days for Ticky Tacky and 307 days for Bangiton. Only Bangiton is worth the bulk buy discount.

The activation energy for Bangiton can be calculated from the slope of the graph, which can be got from Table 4.9.

$$\text{slope} = \frac{-1.6 - (-3.4)}{(2.5 - 2.68) \times 10^{-3}\,\text{K}^{-1}}$$

$$= -10^4\,\text{K}$$

$$\text{slope} = -\frac{E_a}{k}$$

so

$$E_a = 10^4 \times 86 \times 10^{-6}\,\text{eV}$$

$$= 0.86\,\text{eV}.$$

Notice that Bangiton, with a greater value of $E_a$, has a cure time which is more sensitive to temperature.

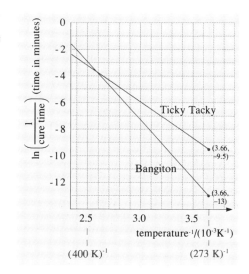

Figure 4.71 Ticky Tacky and Bangiton curing curves

SAQ 4.7 The polymers (a), (b) and (c) have been given in order of increasing $T_g$ as the trade names or applications could have suggested. The answers are as follows.

(a) $T_g = -43\,°\text{C}$. The steric hindrance of the side Cl atom is much less strong than in PVC because it occurs only every fourth atom, not every other atom, along the chain.
(b) $T_g = +105\,°\text{C}$. As in polystyrene, a bulky side group inhibits flexure. This has two side groups on every other atom. Perspex sheet is easily bent if heated with a powerful hair dryer.
(c) $T_g = +230\,°\text{C}$. A very stiff backbone – the rings have only one atom between them. This ICI product is typical of the high-temperature thermoplastics which have become available since about 1970.

SAQ 4.8 After time $t$ the junction is just beyond $3\sqrt{(Dt)}$, while at double the time it has moved further, close to $4\sqrt{(Dt)}$. (The shift is not quite proportional to $\sqrt{t}$ because of the $\sqrt{2}$ in the coefficient of the exponential term.)

Refer to Figure 4.72 overleaf. For boron at uniform concentration $dC/dx = 0$ everywhere so there are no net fluxes of boron atoms. Hence $d^2C/dx^2 = 0$ and $dC/dt = 0$. The uniform concentration remains undisturbed.

For phosphorus the concentration profile shows an inflection. The term $dC/dx$ is everywhere negative but has a minimum value. Fick's first law dictates a positive flux (in the direction of positive $x$). The term $d^2C/dx^2$ has a positive region (accumulation of phosphorus) and a negative region (depletion of phosphorus). The points of matching concentration are in the accumulating region so the junction moves deeper. This is seen also by plotting $C_P$ and $C_B$ for $t$ and $2t$ on the same graph.

SAQ 4.9 For intrinsic conduction

$$n \approx \exp\left(-\frac{E_g}{2kT}\right)$$

and for Si $n$ reaches awkward levels at $T \approx 500\,\text{K}$. So

$$n_{\text{awkward}}$$

$$\approx \exp\left(-\frac{1.1}{2 \times 86 \times 10^{-6} \times 500}\right)$$

$$\approx \exp - 12.8$$

A similar number would obtain for carbon

Table 4.9  Data for Bangiton

| Temperature $T$ | | $T^{-1}$ | Rate $R$ | |
|---|---|---|---|---|
| °C | K | $(10^{-3}\,\text{K}^{-1})$ | (mins.$^{-1}$) | $\ln R$ |
| 127 | 400 | 2.50 | 1/5 | −1.6 |
| 100 | 373 | 2.68 | 1/30 | −3.4 |
| 0 | 273 | 3.66 | ? | |

at

$$12.8 \approx \frac{6}{2 \times 86 \times 10^{-6} \times T}$$

so

$$T \approx 2725\,\mathrm{K}$$

Although this is well below the melting temperature for diamond there are four snags:

1 What happens to the substrate (that is, the material onto which the diamond is deposited)?
2 The diamond would turn to graphite.
3 The graphite would burn.
4 The dopants would diffuse to homogeneity, so junctions would disappear.

The full potential would not be useable, but some higher $T$ than 500 K could be used.

**SAQ 4.10** The little peak for the extra electron on each of the P atoms comes close to the conduction band. These electrons are almost 'free'. As the temperature rises, the first electrons to get to the conduction band are these, so the peak is depopulated as conduction band gets electrons. $E_F$ must be about half way between the peak and the conduction band. At high temperature almost all the little peak levels are empty and some electrons have been promoted from the

valence band. The Fermi level drifts down towards mid-gap as intrinsic behaviour takes hold.

**SAQ 4.11** See Table 4.10.

**SAQ 4.12** The melt must be scrupulously clean of solid particles and should be contained in a vessel not easily wetted. There are not many options; it has to be graphite or silica to avoid dissolving nasties in the silicon. A single seed crystal is provided for all growth to develop from. An adequate growth rate is maintained by arranging for heat flow away from the interface, through the solid.

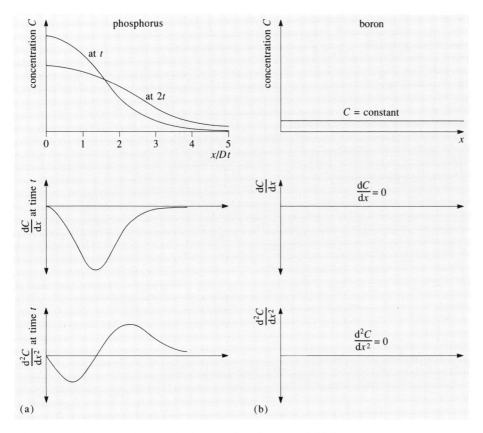

Figure 4.72 Concentration graphs for (a) phosphorus and (b) boron

Table 4.10

| Material | Carrier identity | $n$ large or small? | $dn/dT$ | What influences mobility? |
|---|---|---|---|---|
| normal metal | free electrons | v. large | 0 | phonon interactions only |
| transition metal | free electrons | v. large | can increase | phonon interactions plus traps; when these compensate, $d\sigma/dT \to 0$ |
| ferrous ferrite | hopping electrons | small | 0 | traps, thermal excitation so $\dfrac{d\sigma}{dT}$ positive |
| indium-doped tin | $O^{2-}$ vacancy | small | small | diffusion, so $\mu$ small, but rises with temperature; phonon interactions nil |
| superconductor | electron pairs | small | drops to zero at critical $T$ | $\mu = \infty$ so $\sigma = \infty$ until pairs broken by thermal or magnetic effect |
| intrinsic silicon | equal number of holes and electrons | small | rises by thermal activation | phonon interactions |
| extrinsic silicon | excess of either holes or electrons | controlled, medium | zero at room $T$ | phonon interactions plus traps |

# Chapter 5  Controlling the mix

## 5.1 Introduction

In this chapter we bring together two important ideas developed in earlier chapters — that materials are mixtures of one sort or another and that structural changes depend on the balance between 'stick' and 'rattle'. We use them to explore how the structure of materials, especially microstructure, can be manipulated to give desired properties.

Inevitably, materials are mixtures. All materials contain impurities as vestiges either of their original raw state or of the chemical reactions used in their production. Even materials used in extremely pure form contain traces of impurity: oxygen in high-conductivity copper and device-grade silicon; sulphur in the polymers produced from crude oil; and so on. Incidentally, for thermodynamic reasons, impurities cannot be removed completely. Where high purity is required, the amount of impurity is minimized and remainder arranged to be in places where it does least damage. ▼**Help for Bessemer**▲ shows how this was Bessemer's salvation.

But mixtures are also deliberately made, because they are advantageous. By controlling the ingredients in a mixture, properties and cost can be tailored. That is what this chapter is about.

## ▼**Help for Bessemer**▲

In 1855 Henry Bessemer invented a converter in which air was blown through molten pig iron. This drastically lowered the carbon and impurity content by oxidation and led to the production of large quantities of cheap steel. He made a huge fortune from patent royalties but the eventual success of the technique owed a great deal to Robert Mushet and Sidney Gilchrist Thomas, who didn't do nearly so well out of it. See 'Conventional steelmaking' (Chapter 1) for a reminder of the materials and procedure.

Bessemer steel was originally of inferior quality compared with modern steel, being fairly brittle. Unbeknown at the time, the brittleness was due to small amounts of sulphur and phosphorus in iron ore (very roughly 0.1% in each case). It is now appreciated that such impurities **segregate** to grain boundaries. Figure 5.1, showing the arrangement of atoms at a boundary, indicates why. The structure is more open at boundaries because the boundary is a region of mismatch between neighbouring crystals. Larger atoms are accommodated more readily here than within a crystal.

Mushet empirically solved the sulphur difficulty by adding some spiegeleisen to the molten steel after the converter 'blow'. Spiegeleisen is a cast iron containing manganese. It is now known that manganese reacts with the sulphur to form manganese sulphide, which is insoluble in molten steel and so enters the slag. Nowadays these additions and reactions are carefully controlled.

Bessemer was then able to make steel satisfactorily, but other steelmakers were far less successful. It turned out that Bessemer's pig iron was made from haematite iron ore with a particularly low phosphorus content. Enter Thomas, a chemist. Hitherto, converters had been lined with refractory bricks made from alumina and silica. These were acidic and unable to aid the removal of phosphorus. Thomas reasoned, correctly, that a basic (alkaline) lining, such as lime, would promote reactions in which phosphorus

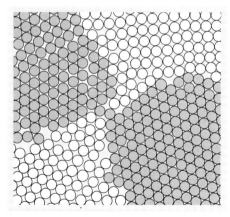

Figure 5.1  A grain boundary on the atomic scale. Note the disorder at the boundary

entered the slag. As a result modern steel converters are lined with basic refractories, which can cope with iron ores containing phosphorus. Because of its phosphorus content, the slag can be used as a fertilizer.

EXERCISE 5.1   Earlier chapters gave examples of mixing to improve properties. Give, in general terms, an example of mixing that alters significantly each of the following properties.
(a) electrical conductivity
(b) viscosity
(c) $BH_{max}$
(d) toughness
(e) Young's modulus

Mixtures can be divided into two classes. The first comprises those produced by the *physical* mixing of components which appear in the product as the same identifiable entities: concrete, GFRP, tungsten carbide particles in cobalt for machine cutting tools, and many more. They are examples of composite materials.

The other class comprises those in which the starting components interact *chemically* to form new constituents or phases (a phase is defined more precisely in Section 5.2.2). For example, in steel, some of the iron and carbon present react to form the compound $Fe_3C$; the phosphorus (say) added to silicon to produce semiconduction substitutes directly in the silicon crystal structure; and so on. Most materials are in this *chemical mixtures* category, including all of those in Figures 2.4 to 2.8. Mixtures like these are the subject of this chapter.

Three variables are used to obtain an optimum mixture:
(a) composition — the number and amounts of the ingredients;
(b) temperature — controlling the amount of rattle that is allowed to occur by means of 'cooking and cooling' processes;
(c) strain, which usually comes into play in the shaping processes, especially the materials-conserving processes such as injection moulding, rolling, and pressing and sintering described briefly in Chapter 1.

In this Chapter we will be concerned with variables (a) and (b); (c) is considered in Chapter 6.

Regulating composition and temperature to produce a desired structure is akin to baking a cake. You know that the basic ingredients for making a cake are sugar, fat, flour and water, each by itself not especially palatable, yet the final cake has a texture and taste quite different from the individual ingredients. Not only is the uncooked mixture different from the ingredients, but the temperature and time of cooking also have a noticeable effect on the finished product. If we were cooking an alloy rather than a cake, the process would be called heat treatment. **Heat treatment** is the term used when property changes are brought about by controlled heating and cooling cycles.

By using different types of sugar and fat, or substituting milk for the water or adding an egg, the texture and flavour of the basic cake can be changed. Clearly mixtures, and particularly 'cooked' mixtures, usually have very different properties from those of their ingredients. This analogy is nothing like as far removed from materials technology as you might think. Until the early part of the present century manufactured materials were produced according to 'recipes' which began with a list of ingredients and went on to describe methods of processing similar to those still used for cake making.

However, as you will see in this chapter, understanding of what happens when materials are mixed chemically has advanced a great deal from the recipe approach. In particular, materials scientists now understand much better how microstructure determines properties and how the required microstructure can be produced by controlling the relevant process variables. Of these variables, temperature is of paramount importance. Figure 5.2 and Table 5.1 illustrate the dramatic changes that can be produced in a steel by heat treatment.

Table 5.1  Data for microstructures in Figure 5.2

| | Typical properties | | Hardness $H_v$ |
| | tensile strength/MPa | % elongation | |
|---|---|---|---|
| (a) slowly cooled from 1030 K (by switching off the furnace) | 930 | 15 | 250 |
| (b) quenched from 1030 K (by plunging into water) | v. low | 0 | 800 |
| (c) quenched from 1030 K and heated at 970 K | 1160 | 25 | 350 |
| (d) quenched from 1030 K and heated at 970 K | 570 | 40 | 170 |

In general, when the ingredients of a mixture interact chemically, the resulting microstructure depends not only on the characteristics of the chemical reactions but crucially on the distribution of the phases produced by the reactions; that is, on the size, shape and spacing of the phases.

Most of the processes used to control microstructure require a 'driving force' (that is, a decrease in the free energy); and their rate is determined by the kinetics of the mechanisms involved — diffusion in the solidification of liquids and in many of the microstructural changes in solids, for instance. In this chapter we shall consider the main ways of controlling driving force and kinetics.

Very crudely, we can consider two types of *solid* materials system. First those in which chemical equilibrium can be reached in practice. (Equilibrium is when the system has a minimum free energy.) Such

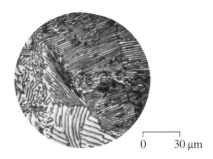

(a) Slowly cooled from 1030 K by swithcing of the furnace

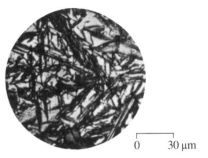

(b) Quenched from 1030 K by plunging into water

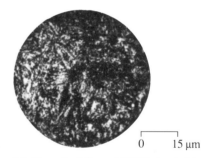

(c) Quenched from 1030 K, heated to 820 K

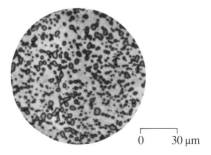

(d) Quenched from 1030 K, heated to 970 K

Figure 5.2 Microstructures and properties of a steel containing 0.80% carbon

systems usually involve single and therefore very mobile atoms or ions. Metals are the prime example. We then consider systems with large and therefore relatively immobile molecules, which tend to exist far removed from equilibrium; inorganic glasses and polymers are examples.

# 5.2 Mixtures that approach equilibrium

Nowadays, the 'recipe' approach can be dispensed with. Instead we use **phase diagrams** to summarize the data on innumerable different mixtures. A phase diagram maps the ranges of composition and temperature over which particular phases are stable in a given system of materials. It enables us to anticipate the alloys that are likely to have useful properties, the treatments we can give to particular alloys to develop the best properties, and the treatments that are likely to be harmful. I shall start with a simple system of great practical value in climates with cold winters.

## 5.2.1 Mixing salt with water

Add a little salt to water and it will dissolve to form a salt solution (brine). Keep adding salt and a point is reached at which it no longer dissolves. The solubility limit of salt in water has been exceeded and the solution is said to be saturated; the excess salt settles at the bottom of the container and co-exists with the saturated brine solution. Warm the brine-and-salt mixture and a little more salt will dissolve — the solubility limit is increased slightly. This behaviour can be plotted in diagram, Figure 5.3.

We can also view Figure 5.3 as a small part of the phase diagram for the $H_2O$–NaCl system. At all temperatures and compositions to the left of the solubility limit a single phase, brine, exists. On the other side, two phases (saturated) brine and solid salt co-exist in equilibrium. So, knowing the composition and temperature of an alloy, the diagram tells us the phase(s) that should be present. ▼**Solubility and free energy**▲ explains the diagram in thermodynamic terms.

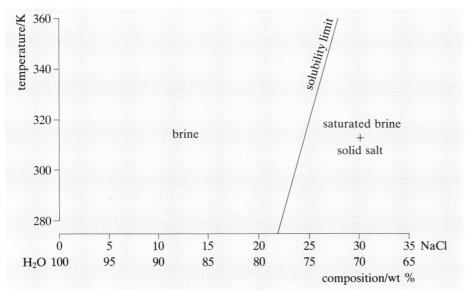

Figure 5.3 The solid solubility of NaCl in $H_2O$

# ▼Solubility and free energy▲

The phase boundary on Figure 5.3 is easily interpreted in terms of the free energy equation you met in Chapter 4. Think of gradually adding solid salt to water at a fixed temperature, say 300 K. The free energy of the two substances separately is:

$$G = G_{salt} + G_{water}$$

As the salt is added, it dissolves in the water: strong ionic $Na^+-Cl^-$ bonds break and are replaced by weaker associations between the ions and the polar water molecules. Apparently $\Delta H_{salt}$ is positive — energy has been added to break the bonds. At the same time the ions have dispersed from their regular arrangement and small volume in the solid into a much more disorderly arrangement, occupying a larger volume, in the solution. The entropy $S$ of the salt has increased. In contrast to these dramatic changes for the salt, not much has happened to the water: $G_{water} \approx$ constant.

So for the system, as salt dissolves

$$\Delta G = \Delta H^+ - T\Delta S^+$$

which will be negative at a given $T$ if the increase in entropy is enough to make $T\Delta S > \Delta H$.

Now we need to think about the changes in the amount of extra entropy generated by adding successive grains of salt. The form of $dS/dn$ derived in 'Statistical games for W' (Chapter 4) tells that the first grain

dissolving gives a bigger entropy change than the next, and the next, etc. It's the same idea as we invoked in 'Purification' in Chapter 4. Eventually a salt grain added to the solution will not produce enough $\Delta S^+$ for $T\Delta S$ to be bigger than the $\Delta H^+$ needed to break up the bonds. The grain does not dissolve and the phase boundary on Figure 5.3 has been reached from the left. If $T$ were raised a few K so that the $\Delta S^+$ which wasn't quite big enough gets a bigger multiplier, the last grain of salt will dissolve. Soon the limit for the higher temperature is reached.

Thus you can see why the solubility line for a solid in a liquid must slope in the direction of higher solubility at higher temperature. Stronger or weaker bonds within the solid or the liquid and the number of particles formed all affect the actual position of the curve.

The significance of this account goes beyond a simple diagram like Figure 5.3. As our phase diagrams develop, several kinds of boundary lines between different mixtures will be drawn. *All of them can be interpreted by the free energy equation.* We shall not follow the pattern of many texts which 'derive' phase diagrams from free energy considerations; but you should be aware that all phase boundaries are expressions of the balance between order and muddle which have minimum free energy.

The H₂O–NaCl system gets more complicated at lower temperatures, see Figure 5.4. We now have a more extended phase diagram. There are a number of important points to note about this diagram:

• It is markedly different from Figure 5.3 because it now includes a third phase — H₂O(s)(that is, ice).
• Adding NaCl decreases the freezing temperature of brine to a minimum at about 250 K (− 21 °C) and a composition containing about 23% NaCl. You can see that in very cold weather a lot of salt would be needed to 'melt' (dissolve) the ice on roads.
• Below about 250 K the two ingredients, solid NaCl and ice, do not interact. They are mutually insoluble and exist as a physical mixture. In practical terms, if you add salt to ice at temperatures below 250 K, once the salt has cooled down nothing more will happen. This is why Scandinavians and Russians don't bother to salt their roads!
• In three of the four regions of the diagrams, two phases co-exist as mixtures. Thus, in the upper left region, saturated brine co-exists with solid ice (a mixture we usually call slush), and in the upper right portion with solid NaCl.
• When brine containing about 23% NaCl is cooled through 250 K it changes directly from a liquid to a solid mixture of salt and ice; it does not go through a slushy range.

EXERCISE 5.2   In winter when snow covered roads are salted, the resulting dirty slush often seems to persist for a long time. Use Figure 5.4 to explain why.

This is the appropriate point to introduce some important definitions.

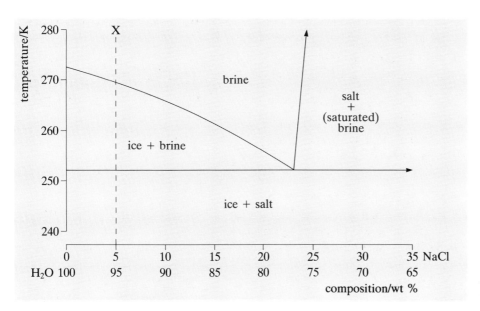

Figure 5.4 The solubility limits of salt and water in brine as a function of temperature

## 5.2.2 Definitions

The **components** of a material are all the **ingredients** that are combined to create it (not the products of the mixing). They are either elements or compounds; $H_2O$ and NaCl are the components in Figure 5.4.

An **alloy** is a particular mixture of components and it has a specific chemical composition. Although, originally, the term alloy was used only to describe mixtures containing a metal, we shall use it in a general way to cover all mixtures of components. Thus alloy X in Figure 5.4 contains 5% NaCl and 95% $H_2O$.

> EXERCISE 5.3  What phases should alloy X in Figure 5.4 consist of at (a) 283 K, (b) 263 K, (c) 248 K.

A **material system** includes all possible alloys made from the given components. Unary systems contain only one component, for example water, iron, silica. Binary systems contain two components, and there can be ternary systems, quaternary systems and so on. For instance, the permanent magnet alloy shown in Figure 2.35(b) belongs to the iron–aluminium–nickel–cobalt quarternary system. Fortunately, in practice binary systems, or approximations of them, are sufficient for our purposes.

A **phase** is a physical entity identifiable by its physical state — solid, liquid or gas — and by its internal structure. It is separated from other phases by a definite boundary.

How many phases are there in the $H_2O$–NaCl system of Figure 5.4?

Three. Ice, solid NaCl and brine (a solution of NaCl in $H_2O$). If we included higher temperatures in the diagram there would be a fourth phase, vapour. In general, vapour will not be relevant to our studies.

On a phase diagram, the **composition** of an alloy is usually expressed as the mass of one of the components in a sample expressed as a percentage of the total mass of the sample. This percentage is conventionally, though incorrectly, called the **weight %**. Sometimes, however, it is useful to give composition in terms of **atomic %**. The difference between weight and atomic percentages can be considerable. See ▼**Converting weight % to atomic % and vice versa**▲. Figure 5.4 follows the conventional choice of axes for plotting composition and temperature on phase diagrams.

You have now met a number of important features of a binary phase diagram. The diagrams for other systems all differ one from another but, fortunately, are all based on a few standard types. All types tell us which phases to expect at particular compositions and temperatures. We deal with the standard types one by one in Sections 5.2.3 to 5.2.6, starting with a diagram rather like that in Figure 5.4. You need to be familiar with all of them.

## ▼Converting weight % to atomic % and vice-versa▲

Consider a particular alloy in a hypothetical materials system A–B with a mass of component A, $m_A$, containing $n_A$ atoms with relative atomic mass $M_A$, and a mass of component B, $m_B$, containing $n_B$ atoms with relative atomic mass $M_B$.

$$m_A = n_A M_A \text{ and } m_B = n_B M_B$$

from which if either the weight % or the atomic % be known, the other can be found.

The weight % of A is then simply

$$W_A = \frac{100\, m_A}{(m_A + m_B)}$$

$$= \frac{100\, n_A M_A}{n_A M_A + n_B M_B}$$

and the atomic % (or molecular %) is

$$A_A = \frac{100\, n_A}{(n_A + n_B)}$$

$$= \frac{100\, m_A/M_A}{(m_A/M_A) + (m_B/M_B)}$$

$$= \frac{100\, m_A M_B}{m_A M_B + m_B M_A}$$

If components have dissimilar relative atomic masses the difference between weight and atomic percentages can be very marked. The important intermediate compound $Fe_3C$ in the iron–carbon (steel) system is a good example.

> EXERCISE 5.4
> (a) What is the weight % C in $Fe_3C$?
> (b) What is the atomic % C in $Fe_3C$?

Remember that thermodynamic equilibrium is assumed under all conditions; in particular, that at all temperatures an alloy achieves the equilibrium composition. Later we shall look at what can happen when non-equilibrium or metastable states exist.

SAQ 5.1  (Objective 5.1)
In the B-C material system represented by the phase diagram of Figure 5.5, which phases will be in equilibrium at the following temperature and compositions?

(a) $T_1$ and $X_1$.
(b) $T_1$ and $X_2$.
(c) $T_1$ and $X_3$.
(d) $T_2$ and $X_2$.

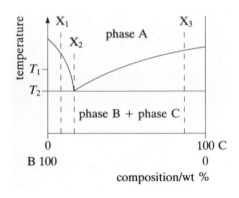

Figure 5.5 Phase diagram of the B–C system

## 5.2.3 Eutectic systems: part one

Here we consider the alloying of two hypothetical components A and B which are completely soluble in one another in the liquid state and completely mutually insoluble in the solid state. (This is unrealistic because there is always some mutual solid solubility in real systems, although it can be very small.) Figure 5.6 is an example of a phase diagram for such a system. It can be viewed in two different ways.

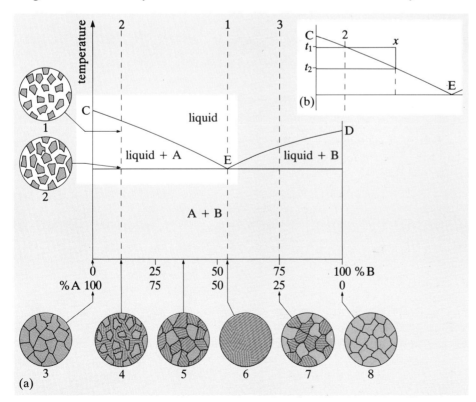

Figure 5.6 A eutectic phase diagram and sketches of typical microstructures in particular alloys. The solid form of component A is red, solid B is grey and the liquid solution of A and B is white

240

Firstly, it is a set of solubility-limit curves. The line ED is the variation with temperature of the solubility of component B in the liquid; in the region below ED the liquid co-existing with solid B is a *saturated solution* of B in A — exactly the same as salt in brine, Figure 5.4. Similarly, the line CE is the solubility limit of component A in the liquid. Below the horizontal line through E all alloys are solid and the components are mutually insoluble.

In the second view, C and D in Figure 5.6 are the melting/freezing temperatures of components A and B respectively. Adding A to B and vice-versa decreases the melting temperature of the component, just as salt does to the melting temperature of ice. In each case a minimum melting point is reached at E.

E is called the **eutectic composition**. The word eutectic comes from the Greek for easy melting. All alloys in the A–B system become solid on cooling to the temperature E. Thus, the horizontal line passing through E is called the **eutectic temperature**. Not surprisingly, the A–B system is called a **eutectic system**. The significance of the eutectic will become clearer by considering what happens when particular alloys are cooled from the liquid.

Alloy 1 in Figure 5.6 has the composition 45% A–55% B, which in this case is the eutectic composition. On cooling, it solidifies at a sharply defined temperature, E, to form a mixture of solid A and solid B. Just as, on cooling, brine containing 23.3% NaCl reaches a eutectic temperature at which it separates into solid ice and solid salt (see Figure 5.4), at E the liquid solution gives rise to a mixture of separated solid A and solid B. This mixture is called a **eutectic mixture** and is produced by a **eutectic reaction**:

liquid → A + B

Not only does the eutectic have a lower melting temperature than its components A and B, but it also retains the fluidity of a single-phase liquid down to this temperature. This is in distinct contrast to all the intermediate alloys, which go through a 'slushy' freezing range.

Now what happens when an intermediate alloy, such as alloy 2 in Figure 5.6, is cooled? See also the enlargement in Figure 5.6(b). Above the line CE, the alloy is totally liquid; in phase diagram jargon, CED is called the **liquidus**. It marks the lower boundary of the liquid phase, that is the limit of solubility for both A and B in the liquid. On cooling to a temperature immediately below the liquidus, the alloy enters a two-phase region in which the (supersaturated) liquid co-exists with solid A. Solidification starts and a small amount of pure solid A is formed as small crystals in the liquid (microstructure 1). Here, I am simplifying the actual solidification process; we consider it in more detail in Section 5.3.1.

Since the remaining liquid has lost some A it is richer in B than the alloy 2 composition line in Figure 5.6; exaggerating, let's say it has a composition $x$ (Figure 5.6b). This new liquid has a lower freezing temperature than the original liquid, and so no more solidification takes place until the temperature reaches $t_2$. Then further pure A solidifies on the existing crystals of A. This process continues with falling temperature, with more pure A being solidified (microstructure 2 for example) and the composition of the liquid *following the line of the liquidus*. When the eutectic temperature is reached, the composition of the remaining liquid is given by E. This liquid then behaves like alloy 1 and solidifies as a eutectic mixture of solid A and solid B. The resulting microstructure (4 in Figure 5.6) consists of particles of pure A in a background (matrix) of eutectic mixture. I have drawn the eutectic as alternate areas of red (component A) and black (component B).

Why does the eutectic take this form?

On reaching the point E, the liquid of eutectic composition is saturated in both A and B and so crystals of both A and B can form. If a crystal of pure A is nucleated, the liquid in the immediate vicinity will be supersaturated in B and therefore encourage the formation of crystals of B. They grow; cause supersaturation in A; hence the growth of crystals of A, and so on. In real alloy systems a eutectic consists of rods, plates or globules of one phase in a matrix of the other. You will meet examples in Section 5.2.5.

▼**Mapping a phase diagram**▲ shows how a phase diagram is constructed from cooling curves.

SAQ 5.2   (Objectives 5.1 and 5.2)
(a)  Explain the changes that occur in alloy 3 during cooling from the liquid to account for microstructure 7 in Figure 5.6.
(b)  On re-heating the alloy, at what temperature would you expect melting to start?

## 5.2.4 Complete solid solubility

Now let's consider two components which are completely soluble in one another in both the liquid *and* the solid state. Here we can relate to real systems. Examples are the metal system Cu–Ni (important in coinage alloys for instance); and the ceramic systems $Al_2O_3$–$Cr_2O_3$ (from which rubies come), MgO–FeO (important for refractory bricks in furnaces) and $Mg_2SiO_4$–$Fe_2SiO_4$ (the olivine system which forms a significant part of the Earth's crust).

We will use the Cu–Ni system as our example, Figure 5.9. Note that here the components are elements rather than compounds.

How many phases are there in the system described?

Two. A liquid solution of copper and nickel, and a solid solution of copper and nickel. The region in between these two phases is a mixture

# ▼Mapping a phase diagram▲

Measuring the temperature of an alloy as it cools from the liquid enables the change of phase to the solid to be detected. When there is no change of phase, the graph of temperature against time (the **cooling curve**) is smooth. The alloy is steadily transferring heat to its surroundings, Figure 5.7(a). The cooling curve of a pure component has a step at the freezing temperature (Figure 5.7b). Here the temperature is constant because the normal rate of heat loss is balanced by the rate of evolution of the latent heat of fusion as the component solidifies. The curve in Figure 5.7(b) is rather idealized.

In practice it would look more like Figure 5.7(c) because of the undercooling required to produce the first stable crystals (nuclei) of solid in liquid. This was discussed in Section 4.7.5.

Figure 5.8 shows cooling curves for some alloys in the A–B system. There is an abrupt change when the liquidus is reached. Except for eutectic alloy and pure components, freezing then takes place over a range of temperatures from liquidus to eutectic temperature $t_e$. At $t_e$ there is a plateau until all the liquid freezes. As Figure 5.8 indicates, the phase diagram can be plotted from the changes of slope

on the cooling curves and from the compositions of the alloys.

'Identifying phases' describes other experimental techniques that can provide information about the shape of phase diagrams, especially for changes in the solid state.

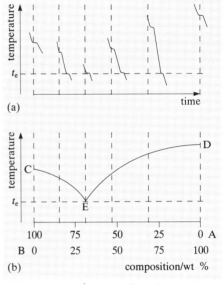

Figure 5.8 (a) A series of cooling curves in the A–B system. (b) Compositions on phases diagram to which they relate

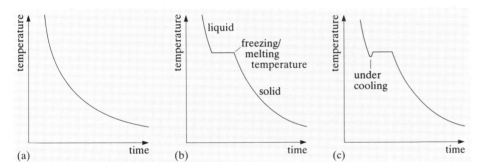

Figure 5.7 Determination of liquid–solid phase changes by cooling curves. (a) Normal cooling, no phase change. (b) Phase change on cooling pure component (idealized). (c) In practice, undercooling initiates phase change

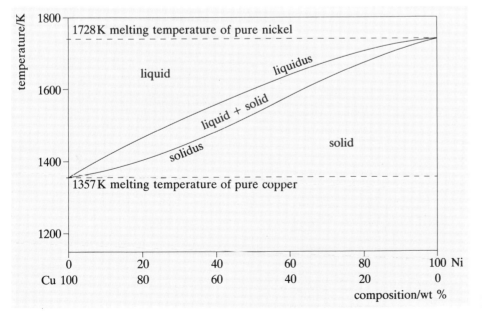

Figure 5.9 The Cu–Ni phase diagram

243

of these two phases, both of which are saturated solutions. Given that crystalline solids are generally close-packed structures, you may wonder what form solid solutions may take. ▼**Arrangements in solid solutions**▲ provides some answers.

EXERCISE 5.5   As you might expect, for extensive solid solubility to occur the two components of the system must have certain characteristics in common. What are these characteristics? Aim for three. Hint. Think of the crystal structure of the pure metals involved and the characteristics of their individual atoms.

# ▼Arrangements in solid solutions▲

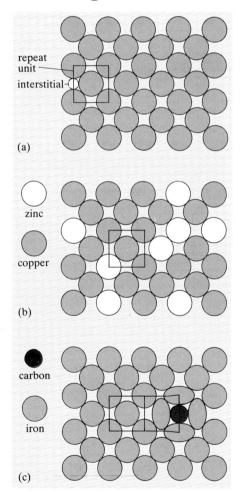

Figure 5.10 (a) A cube face of a face-centred cubic metal structure. (b) Substitutional solid solution. (c) Interstitial solid solution

To form a solid solution, atoms of the second component (the solute), Ni say, must dissolve in the crystal structure of the host component (the solvent), Cu say. There are basically two ways in which solute atoms may be incorporated into the host crystal structure. They are illustrated in Figure 5.10 for a face-centred cubic metal. If the solute and solvent atoms are of similar size, a **substitutional solid** solution is formed — solute atoms simply replace solvent atoms. If the solute atom is much smaller than the solvent atom, it is possible for the solute atom to fit into an interstice between solvent atoms. This is an **interstitial solid** solution.

Substitutional solid solutions are common, although their solubility limits are often quite small as you will see in Section 5.2.5. Interstitial solid solutions are rare because a solute atom has to be very small to fit in an interstice in a close-packed structure.

For instance, you will recall from 'Voids — the spaces in between' in Chapter 3 that even the small carbon atom (diameter 0.154 nm) distorts the octahedral interstitial site in FCC iron (diameter of iron atom = 0.254 nm). This helps to explain why the extent of solubility is quite small, about 7.9 atomic % (2.11 wt %). Nevertheless, it is sufficient to have dramatic consequences for the properties of steel as you will see in Section 5.3.3.

Ionic alloys also form solid solutions and, just as in non-ionic solid solutions, the size of the solute ion determines the way in which the solute is incorporated. Additionally, the need for electrical

neutrality presents another constraint. Figure 5.11 illustrates this for MgO containing iron. When ferrous ($Fe^{2+}$) ions are the solute, they simply substitute for $Mg^{2+}$ ions, Figure 5.11(a). However, if ferric ($Fe^{3+}$) ions are added, in order to maintain the electrical neutrality of the crystal, two $Fe^{3+}$ ions have to replace three $Mg^{2+}$ ions. Thus an extra positive ion vacancy has to be created, Figure 5.11(b).

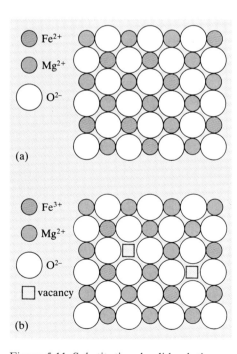

Figure 5.11 Substitutional solid solutions in MgO. (a) $Fe^{2+}$ and (b) $Fe^{3+}$

Returning to Figure 5.9, notice that:

• Adding copper to nickel continuously lowers the melting temperature; adding nickel to copper raises it.

• The liquidus, that is the limit of solubility of both copper and nickel in the liquid solution, is a smooth curve.

• The lower smooth curve is, by analogy, called the **solidus**. It marks the upper boundary of the solid solution phase or, in other words, it is the solubility limit of both copper and nickel in the solid solution.

Let's consider what happens when a solid solution forms from a liquid solution. Figure 5.12 shows the copper-rich end of the Cu–Ni system and includes alloy 1 with the composition 30% nickel. (It is common practice to refer to composition simply as % implying wt %, which is what I shall do here.) Assume that this alloy has been made by melting and mixing the two metals at a high temperature and that it is cooling to ambient temperature.

As with alloys in Figure 5.6, when the liquidus is crossed on cooling (at 1513 K) the first solid appears.

What will be the composition of this solid?

48% Ni–52% Cu. If you look along the 1513 K temperature line in Figure 5.12 you will see that it cuts the solidus at 48% Ni. Now the solidus is the solid solubility limit, so at 1513 K a stable solid solution cannot contain any less nickel than 48%.

Yet the alloy contains only 30% Ni. So what happens? First of all, two points are obvious:

• Above 1513 K, the liquidus temperature for this composition, the alloy is completely liquid with a uniform composition of 30% Ni–70% Cu.

• Below 1438 K, the solidus temperature for this composition, the alloy is complete solid with a uniform composition of 30% Ni–70% Cu.

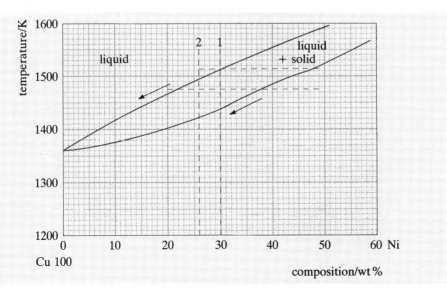

Figure 5.12 Part of the Cu–Ni phase diagram

245

What happens in between, when the alloy is cooling through the two-phase freezing range? In what follows, the important point to remember is that the two-phase region consists of *saturated* liquid solution of nickel in copper and *saturated* solid solution of copper in nickel. When the liquidus temperature of 1513 K is reached, the first solid appears and its composition is 48% Ni. If this is not clear, let me put it another way. The first solid can't have any other composition. It can't contain more Ni because it would have a higher freezing temperature. Equally, it can't contain less Ni otherwise its freezing temperature would be lower. The horizontal line at 1513 K is an example of a **tie-line**. A tie-line crosses a two-phase region and joins together compositions of each phase which are in equilibrium.

Now what about the composition of the liquid? Even though the amount of solid formed at 1513 K will be vanishingly small, since it is richer in Ni than the composition of the alloy, the liquid left must be very slightly deficient in Ni. So it is no longer saturated with Ni. As with alloy 2 in Figure 5.6, no more solid will form until the temperature of this liquid falls to the liquidus. With such a minute change in composition of the liquid, the necessary fall will be infinitesimally small.

As cooling proceeds, crystals of solid solution continue to grow. At a lower temperature, say 1475 K, the solid forming will contain 38% Ni and the residual liquid 22% Ni. If tie-lines are drawn at every temperature as the alloy solidifies, you can see that the composition of the solid moves down the solidus and that of the liquid moves down the liquidus as indicated by the arrows in Figure 5.12. At 1438 K the solidus is reached and the alloy becomes completely solid. Notice that the last drop of liquid to solidify has a composition of only about 14% Ni. Now, the very first solid to form contained 48% Ni. So, to end up with a homogeneous 30% alloy at 1438 K, during solidification the composition of the solid solution has to readjust constantly by the diffusion of Ni out of the previously formed solid (and Cu into it). In this way, the composition of the whole of the solid formed moves down the solidus. This is an important example of the implications of considering changes under equilibrium conditions. The result is a uniform solid solution. In practice, molten solid solutions usually solidify too rapidly for the composition to adjust, so the nickel concentration is higher in the centre of each grain than it is on the outside. Figure 5.13 shows a typical example. The etchant used highlights the variation in composition; the light areas are nickel-rich and the dark areas copper-rich in relation to the overall composition. This phenomenon is called **coring**. The consequences for properties and possible cures are discussed in Section 5.3.1.

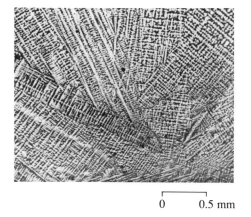

0       0.5 mm

Figure 5.13 Photomicrograph of a 30% Ni–70% Cu alloy produced by casting

SAQ 5.3  (Objective 5.2)
For alloy 2 in Figure 5.12 specify:

(a) the composition and temperature of the first solid to form,
(b) the composition of the liquid at this temperature,
(c) the composition and temperature of the last solid to form,
(d) the composition of the last drop of liquid to solidify.

SAQ 5.4  (Objective 5.1)
What is the essential difference between the first solid formed in alloy 2 in Figure 5.6 and that in alloy 1 in Figure 5.12?

## The lever rule

For a two-phase region, such as that in Figure 5.12, this rule enables us to calculate the relative amounts of each phase (for example solid and liquid) in equilibrium at any temperature. Let's start with alloy 2 in Figure 5.12 as a particular example and think about the relative amounts by mass of solid and liquid present at 1475 K.

Let the fraction of solid be $S$. Since the alloy has been divided into two phases, what is not solid must still be liquid. So the fraction of liquid is $(1 - S)$. The proportion of nickel in each phase is simply the fraction of the phase present times the nickel concentration. From the tie-line, the nickel concentration in the solid is 38% and in the liquid 22%. So, the proportion of nickel in the solid phase is $S \times 38\%$ and in the liquid phase $(1 - S) \times 22\%$. These two together must equal the total proportion of nickel in the sample, that is, 26%.

Hence

$$38S + 22(1 - S) = 26$$
$$(38 - 22)S = 26 - 22$$
$$S = \frac{26 - 22}{38 - 22} = \frac{1}{4}$$

Thus the sample must be exactly one-quarter solid and three-quarters liquid at 1475 K.

Looking at the tie-line at 1475 K you can see that the alloy 2 composition divides the line into these proportions: one quarter to the left and three quarters to the right.

Summarizing then:
• A tie-line is an isotherm across a two-phase field.
• A tie-line links the compositions (solubility limits) of the two phases in equilibrium at that temperature.
• The alloy composition divides the tie-line into two arms and the relative amounts of the two phases is given by the relative lengths of these arms.

A common snag is remembering which arm of the tie-line belongs to which phase. Here is a tip.

Imagine moving the tie-line up or down until one end coincides with the boundary of a single phase. At this temperature the tie-line has only one arm and it must correspond to the single phase now present. Look at the 1513 K tie-line for alloy 1. It has only one arm and it must correspond to all liquid. As the alloy cools imagine the tie-line moving down to, say, 1475 K. The left hand arm (proportion of solid) grows and the right hand arm (proportion of liquid) shrinks.

Example   Consider alloy 1 at 1475 K. Specify:
(a) the compositions of (i) the solid and (ii) the liquid;
(b) the relative amount of solid present.

Answer

(a) (i) The solid composition is given by the point at which the tie-line meets the solidus: 38% Ni. (ii) The liquid composition is 22% Ni.
(b) The left-hand arm of the tie-line is the 'solid arm'. The relative amount of solid is given by the length of the solid arm relative to the whole tie-line:

$$\frac{30 - 22}{38 - 22} = \frac{8}{16} = \frac{1}{2}$$

Glancing at Figure 5.12 you will see that the tie-line splits the alloy composition in half.

You can now see one of the important uses of a phase diagram. For a given alloy composition and temperature, it enables us to calculate the relative proportions by mass of two equilibrium phases.

Complete solid solubility is not common. A few examples are given in ▼**Solid solutions in action**▲. However, some restricted solid solubility occurs between almost any components. It may be as large as a few percent and as small as a few parts of one component in a million parts of the other. This brings us to the next form of phase diagram. You may be relieved to learn that many of its aspects can be viewed simply as a combination of parts of the two we have already considered!

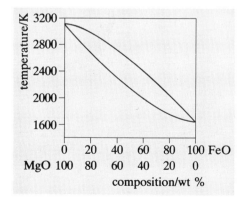

Figure 5.14 MgO–FeO phase diagram

SAQ 5.5   (Objectives 5.2 and 5.4)
Figure 5.14 is the phase diagram for the MgO–FeO system.
(a) How many phases are there in the system?
(b) Label the phase fields and indicate the solidus and liquidus.
(c) Specify the composition of the alloy that completes its solidification at about 2300 K.
(d) Specify the alloy that is roughly a quarter solid at 2300 K.

## 5.2.5 Eutectic systems: part two

Although complete solid solubility is rare, restricted solid solubility is common. The $Al_2O_3$–$ZrO_2$ (alumina–zirconia) system is an example. Figure 5.16 is its phase diagram. 'Restricted solubility' tends to mean solute concentrations of a few % or more for metallic and ceramic alloy systems. But for electronic materials it can mean much smaller concentrations. Nevertheless, these minute concentrations are crucial to the functioning of electronic devices. See ▼**Solid solubilities in silicon**▲

How many phases are there in Figure 5.16?

Three. A liquid solution and two solid solutions. One solid solution, called α, has a small amount of $ZrO_2$ dissolved in $Al_2O_3$, and the second (β) is a solid solution of $Al_2O_3$ in $ZrO_2$. Note that it is conventional to use Greek letters to label the solid phases in a system, usually in sequence from left to right.

Because the components are partially soluble in one another in the solid, the phase diagram is intermediate between one for completely insoluble components (Figure 5.6) and one for completely soluble components (Figure 5.12). Let's note the similarities and then concentrate on the new features.

The liquidus CED is of the same form as that in Figure 5.6 and the two sections CF and DG resemble parts of the solidus in Figure 5.12 (or Figure 5.15). Therefore, alloys containing less $ZrO_2$ than F or more than G will solidify exactly as the solid solutions described in the previous section. The point F denotes the maximum amount of $ZrO_2$ which can

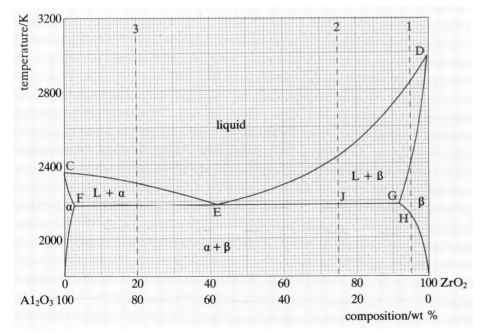

Figure 5.16 $Al_2O_3$–$ZrO_2$ phase diagram

## ▼Solid solutions in action▲

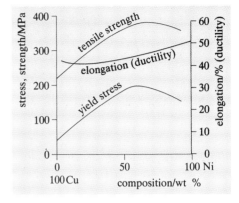

Figure 5.15 The effect of solute concentration in Cu–Ni alloys on (a) yield stress, (b) tensile strength, and (c) ductility (% elongation)

Figure 5.15 illustrates the variation of properties across the Cu–Ni solid-solution system at ambient temperature. Notice in particular that, compared with the pure component, the tensile strength and yield stress increase markedly for a relatively small reduction in ductility. This behaviour is typical of solid solutions, hence their extensive use in all sorts of sheet, tube and pressed components produced by cold working. The solute atoms impede dislocation motion, thus strengthening the alloy (more about this in Chapter 6). Cu–Zn solid solutions (up to about 30% Zn) and Cu–Ni are especially popular because of their excellent corrosion resistance. In the UK, 75% Cu–25% Ni is used for 'silver' coinage.

Copper–gold and copper–silver alloys are interesting examples of solid solutions. Pure gold is extremely soft and malleable (it can be beaten out into very thin sheets) but for most practical purposes is too soft. Its strength, hardness and wear resistance are improved by forming solid solutions with copper and silver; hence the **carat** grades in everyday use. A carat is a twenty-fourth part. Thus 18 carat gold contains 18 parts by mass of gold and 6 parts of alloying elements. The addition of copper produces a reddy gold and silver a pale gold colour. To preserve the colour of gold, copper and silver are added in the approximate ratio three to one.

# ▼Solid solubilities in silicon▲

The concentration of charge carriers (electrons or holes), on which the performance of a silicon p–n junction depends, is controlled by the type and amount of solid solute added during processing. Phosphorus is added to provide electrons, boron to provide holes. The solid solubility in silicon of these dopants, and other elements, is extremely small as Figure 5.17 shows. Dopant concentrations are shown as carrier atoms $m^{-3}$. To put these numbers in perspective, we need to know the number of silicon atoms $m^{-3}$.

> EXERCISE 5.6   The density of silicon is $2.33\,Mg\,m^{-3}$, its relative atomic mass is 28 and Avogadro's number is $6 \times 10^{23}$. How many silicon atoms $m^{-3}$?

There are two important points to note from Figure 5.17 and Exercise 5.6.

1 The useful dopants phosphorus and boron have solubilities of only about 1 atomic % at processing temperatures. As you know from Chapter 3, concentrations much lower than this matter. Other elements have even lower solubilities — for copper it's one part in 5 million at about 900 K for example.

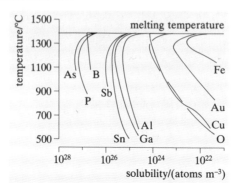

Figure 5.17 Solid solubilities of various elements in silicon

2 The temperature dependence of solid solubility is very marked. This clearly has important consequences for processing of silicon devices, where the solute concentrations at ambient temperature are crucial. Consider oxygen. In the most popular method of growing single crystals, the Czochralski method described in Chapter 1 'Silicon single crystals', oxygen is picked up from the silica crucible used to contain the molten silicon. The solubility just below the freezing temperature is around $10^{24}$ atoms $m^{-3}$ (1 part in 5000), but on lowering the temperature to about 950 K this drops by

two orders of magnitude to $10^{22}$ atoms $m^{-3}$. Just like excess salt in water, the excess oxygen is deposited as particles of oxide. Oxygen is less harmful within a particle than as individual solute atoms, so after the crystals have been grown they are annealed at about 1000 K to encourage them to form oxide. The oxides reside preferentially on faults in the crystal, Figure 5.18. This is a good example of the point I made in the introduction to this chapter: if there have to be impurities, arrange for them to be in harmless locations.

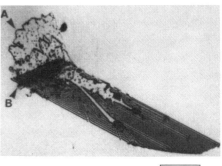

Figure 5.18 A defect (called a stacking fault) in silicon 'decorated' with impurity particles, as seen in a transmission electron microscope. A is a copper-rich particle and B is an iron-rich particle

dissolve in alumina and G the corresponding amounts of $Al_2O_3$ which can dissolve in $ZrO_2$. All the compositions between F and G solidify like those described in Section 5.2.3 with one important difference. Can you spot it? (Hint. Compare $\alpha$ and $\beta$ in Figure 5.16 with A and B in Figure 5.6.)

Because $\alpha$ and $\beta$ are solid solutions, solid phases formed in compositions between F and G will be solid solutions. In Figure 5.6 the solid phases are the pure components A and B.

That covers the solidification of alloys in the system. When the solid is cooled to ambient temperature, further differences appear. Changes occur in the microstructure during further cooling because the solid solubility limits of $\alpha$ and $\beta$ decrease with decreasing temperature. I will consider two examples.

Example 1   Alloy 1 containing 95% $ZrO_2$ will solidify exactly as if it were a solid solution. So, on cooling (under equilibrium conditions) it will consist of homogeneous crystals of $\beta$ phase from the solidus temperature, about 2400 K, until it meets the solid solubility limit at H. This type of solid solubility curve is generally called a **solvus** to distinguish it from the solubility boundary with the liquid state — the solidus.

EXERCISE 5.7

(a) What happens to the microstructure (under equilibrium conditions) when alloy 1 crosses the solvus?
(b) Specify the fraction of α-phase at 2000 K.

The formation of α crystals within the β crystals — the term used to describe it is **precipitation** — occurs at grain boundaries and along particular types of crystal plane. As a result the microstructure has a striking geometric appearance. It is known as a **Widmanstatten** structure after Alois van Widmanstatten who, in 1808, found such a pattern in iron meteorites (Figure 5.19.).

Generally precipitation is not as visible as in Figure 5.19, and some form of microscopy is required to detect it. Identification of a precipitated phase requires X-ray diffraction or X-ray microanalysis. For more details, see ▼**Identifying phases**▲.

Note that precipitation is a change of phase in the solid state, and happens by diffusion. As you know from Section 4.7.5, diffusion processes are very much slower in the solid than in the liquid, and so the achievement of equilibrium is even more unlikely than during solidification. This makes it possible to control the way in which the α phase is precipitated and, as you will see in Section 5.3.2, is put to good use in the development of strong alloys.

Example 2    Alloy 2 in Figure 5.16 contains 75% $ZrO_2$ and will solidify in the same way as alloy 3 in Figure 5.6. After the eutectic reaction has occurred, it would have a microstructure similar to that of microstructure 7 in Figure 5.6. (This alloy was the subject of SAQ 5.2.) The only difference is that in this case the solid phases are solid solutions α and β, with compositions F and G respectively, and not pure components.

EXERCISE 5.8    Use the lever rule to calculate the following from Figure 5.16

(a) the relative proportion of β phase in alloy 2 immediately above the eutectic temperature;
(b) the relative proportion of β phase in alloy 2 immediately below the eutectic temperature.

It is important to remember, as Exercise 5.8 emphasizes, that crystals of β (and of any solid phase in an alloy that undergoes a eutectic reaction) originate in two ways: first from the solidification of the solid solution and then from the formation of the eutectic.

Above the eutectic temperature the mix is β plus liquid. In order to distinguish it, this β is called **primary** β. At the eutectic temperature the remaining liquid is transformed to the eutectic α + β; so below the eutectic temperature the mix is β + (α + β).

Figure 5.19 Polished and etched sample of an iron meteorite showing a Widmanstatten structure. Metallic meteorites consist of iron with, typically, 8 wt % nickel and a little cobalt

# ▼Identifying phases▲

A crystalline phase can be identified by the type and size of its crystal structure and by its chemical composition. X-rays are often used to determine both: **X-ray diffraction** for the former and **X-ray microanalysis** for the latter.

The X-ray diffraction technique was introduced in Chapter 3. A common technique for the analysis of crystals is the **powder method**; the geometry of the set-up is illustrated in Figure 5.20. The material for study, in powder form, is rotated at the centre of a circular camera around the periphery of which is a strip of film. A beam of X-rays is projected at the specimen. Since there is a huge number of crystals in the powder, all possible orientations of particular crystal planes are present. Cones of diffracted X-rays are produced at angles of $2\theta$ for each value of $\theta$ that obeys Bragg's law ($n\lambda = 2d \sin \theta$). Note that in Figure 3.36 the diffracted beam is $2\theta$ from the incident beam. Since it has a particular set of crystal planes, each type of crystal produces a characteristic diffraction pattern (see Figure 5.21).

What can you conclude about the material that produced the pattern in Figure 5.21(b) compared with (a) and (c)?

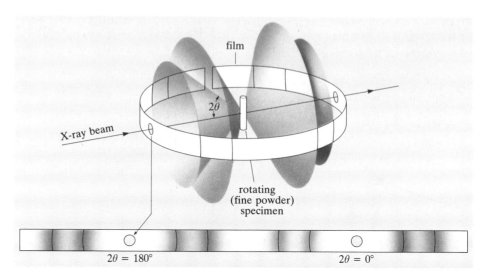

Figure 5.20 The X-ray powder method

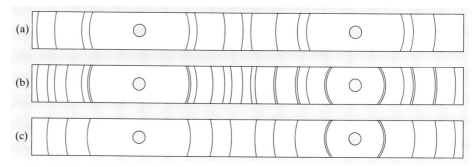

Figure 5.21 The X-ray diffraction pattern produced by (a) a FCC metal and (c) a BCC metal. (b) See text

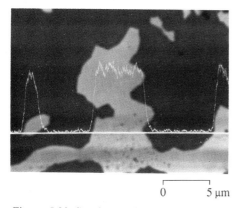

0          5 µm

Figure 5.22 Specimen of Pb–Sn alloy. The trace indicates the Pb composition

The diffraction pattern contained all of the lines on both (a) and (c), so the material must consist of the two phases. From this example you can see two important uses of the diffraction technique:

- identification of individual phases;
- identification of compositions on a phase diagram that contain more than one phase.

X-ray microanalysis is usually operated in conjunction with a scanning electron microscope (see Chapter 2). When a beam of electrons strikes a material, X-rays are emitted. Their energy (or wavelength) depends on the chemical elements present. Each element emits X-rays of characteristic energy. The X-rays are collected in a detector and sorted out according to their energy. The energies 'belonging' to each element can be identified and related to the field of view in the microscope. The trace in Figure 5.22 is of the lead concentration across a Pb–Sn alloy. The light area is lead solution and the dark area is a tin solid solution.

With further cooling the solid solubilities of both α and β decrease. So, again, diffusion is necessary for the compositions and amounts of the two phases to undergo adjustment.

SAQ 5.6  (Objective 5.4)
From Figure 5.16 calculate:
(a) the % increase in the amount of α phase precipitated in alloy 2 between the eutectic temperature and 2000 K;
(b) the fraction of primary α formed in alloy 3.

Eutectics can have advantageous properties, so they are widely used. ▼**Eutectics in action**▲ looks at some of them.

Finally in this section, a eutectic-like reaction can occur completely in the solid state. Such a system is called a **eutectoid**. An extremely important example occurs in the Fe–C system, which is the basis of steels and accounts for much of their versatility. Figure 5.23 shows the relevant part of the Fe–C phase diagram. Note that it covers a limited range of temperature and carbon content. We will meet other parts of the diagram in Section 5.3.3. At this stage I would like you to note the following features of the diagram.

• Iron exists in more than one form; it is polymorphic. At temperatures below 996 K it has a body-centred cubic (BCC) structure (α), and above this temperature it has a face-centred cubic (FCC) structure (γ).
• The solubility of carbon in these two phases is markedly different.

It is γ that undergoes the eutectoid reaction. The eutectoid composition is 0.8% carbon and on cooling from the γ phase field an alloy of this composition will transform to a eutectoid mixture.

The eutectoid reaction is

$$\gamma \; \rightarrow \; \alpha + Fe_3C$$

On slow cooling, an approximation of equilibrium conditions, the eutectoid mix is as illustrated in Figure 5.2(a). The other parts of Figure 5.2 illustrate the changes that can be wrought in this composition; more about this in Section 5.3.3.

SAQ 5.7  (Objective 5.5)
The eutectic system A–B in Figure 5.24 includes four alloys, 1–4. For each alloy:
(a) Sketch a cooling curve and indicate the phase(s) stable in each temperature range.
(b) Comment on why the curve for alloy 4 is different from the others.

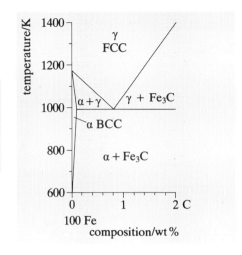

Figure 5.23 Portion of the Fe–C phase diagram

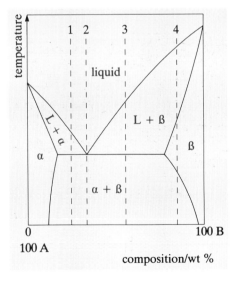

Figure 5.24 The eutectic system for SAQ 5.7

# ▼Eutectics in action▲

Eutectic alloys are common engineering materials. Their melting temperature and mechanical properties make them advantageous both for processing and performance. For processing, their low melting temperature (relative to their components) and high fluidity down to a sharply defined freezing temperature make them suitable for casting. Other alloys in the system solidify through a pasty region.

For service, the properties of eutectics are often superior to those of their components. Their mechanical properties are determined by the properties of both the matrix, or **continuous**, phase and the embedded, or **discontinuous**, phase. Table 5.2 summarizes the possibilities.

The hardening in (i) and (ii) is called **second phase hardening**. You will meet a special example of it, precipitation hardening, in Section 5.3.2.

Figure 5.25 shows three useful eutectics.

Cast irons are suited to casting processes, as the name implies. They are based on the eutectic in Figure 5.25(a). The eutectic temperature is at about 1400 K (iron melts at about 1810 K). Ordinary cast irons have a number of attractive properties for both processing (high fluidity and machinability) and performance (hardness and wear resistance). However their toughness, ductility and tensile strength are low. Cast iron is ubiquitous in drain covers, engine blocks, brake drums and so on. It was very popular in the nineteenth century (see Figure 5.26).

The Al–Si eutectic alloy, Figure 5.25(b), offers similar advantages for casting. It also has good corrosion resistance and thermal conductivity but, above all, its density is much less than iron's. For this reason it is tending to replace cast iron in car engines. It is widely used for other cast products.

Because of its sharply defined freezing temperature, the usual solder for electric circuitry is the Sn–Pb eutectic alloy, Figure 5.25(c). Old domestic cold water systems had lead piping. On these, plumbers used a lead alloy containing 35 wt % Sn to make joints. The wide freezing range gave time to shape ('wipe') the joints. Nowadays copper or PVC piping is used, with different joining mechanisms.

Table 5.2

| Continuous phase | Discontinuous phase | Eutectic mixture |
| --- | --- | --- |
| (i) Ductile | ductile | ductile with improved strength due to (a) obstruction of dislocations at boundaries between particles and (b) benefit gained from 'law of mixtures' (see Chapter 2) |
| (ii) Ductile | hard and brittle | ductile for the reasons above; strengthening even more effective |
| (iii) Brittle | ductile | brittle; alloy behaviour is determined by the continuous phase |
| (iv) Brittle | brittle | brittle |

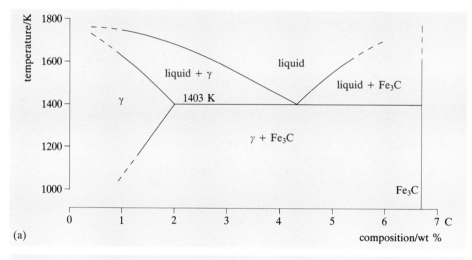

(a)

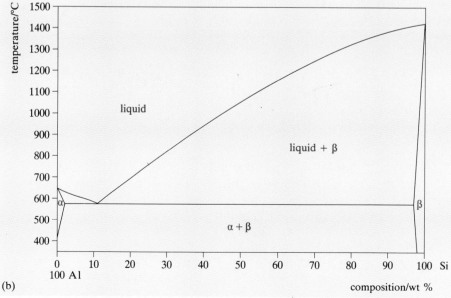

(b)

Figure 5.25 (a) Part of the Fe–C alloy system. (b) The Al–Si system

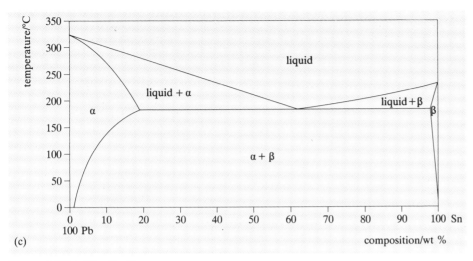

Figure 5.25(c) The Sn–Pb system

The eutectics in some ceramic systems are very useful in the extraction of metals. When mined, a metal ore contains other, unwanted, minerals called gangue. This usually consists of oxides or silicates. These essentially ionic ceramic compounds are generally immiscible with molten metals and float on them. Thus they can be run off separately from the molten metal as a slag. But most gangue minerals have very high melting temperatures and, for reasons of economy, the extraction temperature should be as low as possible. The answer is to add a flux, a substance that combines with the gangue to produce a fusible slag with a low melting temperature. This is why limestone is added to a steel blast furnace. It combines with the silica in iron ore to form calcium silicate (see 'Conventional steelmaking' in Chapter 1). Most fusible slags are at or near the eutectic mix.

Figure 5.26 St. Stephen's church, Istanbul, Turkey, built from cast iron in 1890. Castings 'mass produced' in Vienna were shipped to Istanbul and bolted together

## 5.2.6 Intermediate compounds and phases

In Figure 5.16, the solid solution phases α and β have the crystal structures, respectively, of the components $Al_2O_3$ and $ZrO_2$. The presence of some solute is tolerated and the crystal structure of the phase remains recognizably that of one or other component.

In the Fe–C system, however, we saw that the components iron and carbon could coexist in a solid phase which was quite different from that of either iron or carbon; that is, as iron carbide, $Fe_3C$. This is an **intermediate compound** (or phase). There are many intermediate compounds, but they usually have the following properties.

Intermediate compounds usually have a particular crystal structure corresponding to a fixed chemical composition, $A_xB_y$, say. Thus the intermediate compounds FeO, $Fe_3O_4$ and $Fe_2O_3$ occur in the Fe–O system; NiAl and $Ni_3Al$ in the Ni–Al system; and so on. (Figure 5.27a shows the intermediate compound $A_xB_y$ in the A–B system.)

If either component A or B is soluble to some extent in $A_xB_y$, the intermediate compound's crystal structure will prevail over a range of compositions about $A_xB_y$. We then have an **intermediate phase** or **secondary solid solution**, that is a range of compositions over which the structure differs from that of any other solid phases in the system.

255

The intermediate phase may be wide or narrow, depending on the solubility of the components in the intermediate compound. The important semiconductor compound GaAs can dissolve only about $10^{-4}$ atomic % Ga, for instance. On the other hand, in the $MgO-Al_2O_3$ system, the phase defined by the structure of spinel, $MgAl_2O_4$ (or $MgO.Al_2O_3$) can exist as far as $MgO.3Al_2O_3$.

Intermediate compounds or phases may be formed either directly from the melt or by a reaction between another solid phase already formed and the residual liquid. The latter reaction is called a **peritectic**. I'll take them in turn.

## Directly from the melt

An intermediate compound formed this way has a definite melting temperature and divides the phase diagram in two. In Figure 5.27(a), one part shows the phases that occur between A and $A_xB_y$, and the other the phases that occur between $A_xB_y$ and B. Each part can take the form of the phase diagrams we have already considered. Figure 5.27(b) is an example with the semiconductor PbTe as the intermediate compound. In this particular case each part of the diagram contains a eutectic. The Pb–PbTe eutectic looks rather strange on this scale because the eutectic composition is extremely close to pure Pb.

## The peritectic reaction

In many systems, during solidification a reaction occurs at a definite temperature between the solid (of a particular phase) that has been formed and the remaining liquid. This **peritectic reaction** produces an intermediate compound or phase:

liquid + solid phase 1 $\rightarrow$ solid phase 2
(intermediate compound or phase)

If the bonding in the intermediate compound were stronger than in either the liquid or phase 1, then the enthalpy change $\Delta H$ as the compound is formed would be negative. The entropy change $\Delta S$ would be negative also, because the reaction happens during solidification. Thermodynamically,

$$\Delta G = \Delta H^{(-)} - T\Delta S^{(-)}$$

The reaction is feasible when $\Delta G$ goes negative. This happens when $T$ falls to a critical **peritectic temperature**. ▼**Building a peritectic**▲ develops the phase diagram for a peritectic system.

Figure 5.28 is a phase diagram for a peritectic system with components A and B. The β solid solution reacts with the remaining liquid to form the intermediate compound $A_2B$. The intermediate phase γ occupies a range of compositions around that of $A_2B$. In other words A and B are soluble in $A_2B$, and the γ phase is a secondary solid solution (with the structure of $A_2B$).

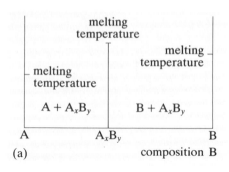

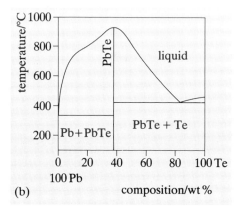

Figure 5.27 (a) $A_xB_y$ is an intermediate compound in the A–B system. (b) Pb–Te phase diagram. PbTe is the intermediate compound

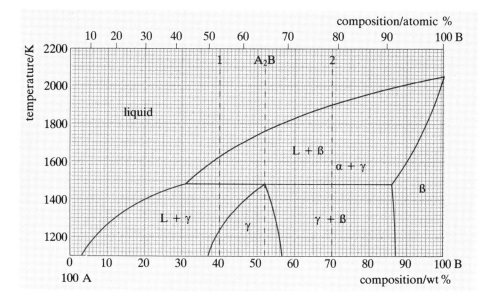

Figure 5.28 The A–B peritectic system. The intermediate phase γ

## ▼Building a peritectic▲

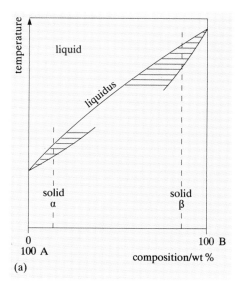

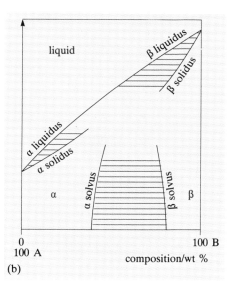

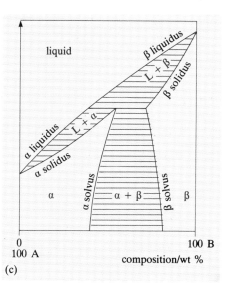

Figure 5.29 Development of a peritectic phase diagram

Figure 5.29(a) represents a material system with a liquidus curve like that of the Cu–Ni system (Figure 5.9). We know that under the liquidus at both ends of the system there must be a solidus that can be linked to the liquidus by tie-lines. Let the solid solution based on component A be α phase, and that on component B β phase. Thus Figure 5.29(a) shows development of the α solidus and liquidus and the β solidus and liquidus.

Now, an intermediate compound has to occur somewhere in the system, so α and β must have limits of solubility. To meet this requirement, in Figure 5.29(b) an α solvus and a β solvus have been added. These two must have tie-lines connecting their phase boundaries at all temperatures. If we extend these solvus boundaries they must meet the solidus boundaries in order to delineate each single-phase region (that

is, α and β regions). This is shown in Figure 5.29(c).

The new feature in this diagram is the horizontal line at which three separate tie-lines meet: α + β, α + liquid and β + liquid. This is the peritectic temperature.

Look at alloy 1. Initially, the alloy will solidify as if it were a solid solution. At 1480 K there is a dramatic change; this is the peritectic reaction. What phase(s), and in what proportion, are present immediately above and immediately below the peritectic temperature?

Above the peritectic, the solid present is $\beta$ in the proportion given by

$$\frac{(40 - 31)}{(86 - 31)} = \frac{9}{55} = 16.5\% \text{ solid } \beta \text{ phase}$$

Most of the alloy (83.5%) is still liquid. Below the peritectic, the solid present is $\gamma$ in the proportion given by

$$\frac{40 - 31}{52 - 31} = \frac{3}{7} = 43\% \text{ solid } \gamma \text{ phase}$$

The remaining 57% of the alloy is still liquid. So at the peritectic temperature, the $\beta$ phase reacts completely to form $\gamma$ and still leaves some residual liquid. The peritectic reaction is

$$\text{liquid} + \beta \rightarrow \gamma$$

Below the peritectic, solidification continues as more of the $\gamma$ solid solution is formed. Eventually the alloy consists entirely of homogeneous $\gamma$ phase at 1240 K.

SAQ 5.8  (Objectives 5.3 and 5.4)
Consider the changes that occur on cooling alloy 2 (70% B) in Figure 5.28. Try to tackle all of the following before looking up the answers.
(a) Does alloy 2 solidify in a similar manner to alloy 1 down to temperatures just above the peritectic temperature?
(b) What is the proportion of $\beta$ phase present just above the peritectic temperature?
(c) What is the peritectic reaction in this case?
(d) What is the proportion of $\beta$ phase present just below the peritectic temperature?
(e) Sketch the microstructure you would expect to see just above and below the peritectic temperature and at about 1200 K. (Note how the solvus boundaries change with decreasing temperature.)

That completes our consideration of the basic types of phase diagram. The alloy systems of engineering materials can be as simple as my examples, but often they appear rather more complex. In fact, they are no more difficult to interpret because they can be broken down into elements similar to the basic types above. Figure 5.30 is an example.

The complete diagram consists of a peritectic which leads into a eutectic. Both $SiO_2$ and $Al_2O_3$ form solid solutions with one another. Between about 50% and 100% $Al_2O_3$, the $Al_2O_3$ reacts with liquid at the peritectic temperature of about 2100 K to form an intermediate compound $3Al_2O_3.2SiO_2$ (or $Al_6Si_2O_{13}$) called mullite. Notice that it is a

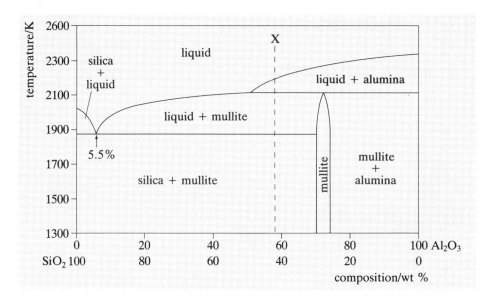

Figure 5.30 Alumina (Al$_2$O$_3$)–silica (SiO$_2$) phase diagram. A number of compositions in this system are used for refractory bricks for furnace linings

solid solution; it can dissolve limited quantities of SiO$_2$ and Al$_2$O$_3$ and thus exists over a range of composition. The liquid-plus-mullite phase field then leads into a eutectic just below 1900 K.

We can see that Figure 5.30 is a combination of more basic diagrams by simply considering alloy X, for instance. It starts to solidify in exactly the same way as alloy 1 in the A–B system of Figure 5.28. Below the peritectic temperature it consists of mullite plus liquid. On cooling further it undergoes a eutectic reaction, just like alloy 2 in the Al$_2$O$_3$–ZrO$_2$ system (Figure 5.16), and should end up as mullite crystals in a eutectic of mullite and SiO$_2$.

EXERCISE 5.9   Figure 5.31 shows two important metal alloy systems: Fe–C, the basis of steels and cast irons; and Cu–Zn, the basis of brasses.
(a) Complete the labelling of the phase fields for the Fe–C diagram.
(b) Describe each one in terms of the basic types of phase diagram of which it is composed.

So, given the phase diagram for a materials system, for any composition and temperature we can say which phases are present and in what proportions, *provided that equilibrium has been maintained throughout*. In reality, solid engineering materials are seldom in equilibrium. This is either inevitable or by design; inevitable because it is impracticable to allow castings, injection mouldings, sintered compacts or whatever to cool infinitely slowly — even assuming they reached equilibrium at the (elevated) processing temperature; by design because some properties of a material can often be improved vastly by deliberately creating a particular non-equilibrium microstructure. We explore some important examples of such structures in the next section.

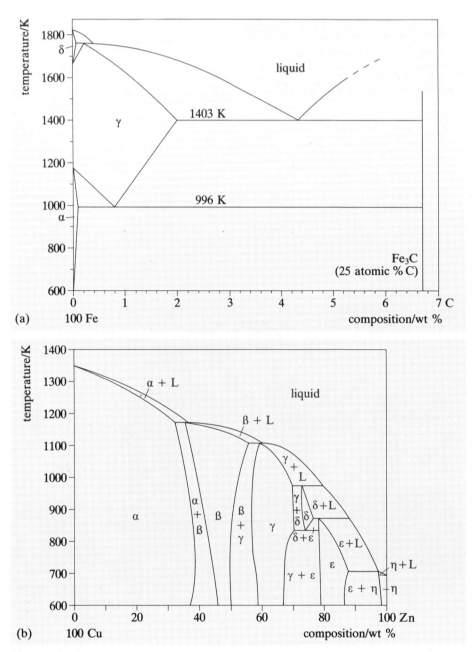

Figure 5.31  (a) The important part of the Fe–C phase diagram. (b) The Cu–Zn phase diagram

# 5.3  Juggling with equilibrium

If non-equilibrium structures are so prevalent, why do you think equilibrium phase diagrams are so important?

Because they give a clear indication of the direction in which microstructural changes are likely to occur when an alloy is heated and cooled, that is, as it moves *towards* equilibrium.

In this section I am going to illustrate the versatility and importance of the idea of producing non-equilibrium structures by exploring three aspects: solidification, precipitation hardening, and quenching and tempering. But before doing this it is useful to consider briefly some of the important structural features of materials in terms of equilibrium.

EXERCISE 5.10 Imagine an alloy which has the following microstructural features at ambient temperature:
(a) a cored solid solution (as in Figure 5.13 for instance);
(b) many small particles of second phase (on a finer scale than those in Figure 5.2(c) for example);
(c) many small elongated grains;
(d) many dislocations.

What would you expect to happen if the alloy were heated to provide sufficient 'rattle' for changes towards equilibrium to occur?

Hint. Remember that, because some atoms are away from their equilibrium sites, structural features which disrupt the regular order of a crystal have a higher energy associated with them.

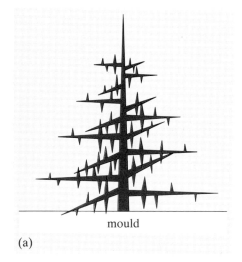

mould

(a)

## 5.3.1 Solidification

In casting, and other processes, material solidifies from the melt; and the microstructure so produced depends on conditions during solidification.

What happens during solidification?

Look back to Figure 5.7(c) to remind yourself of the characteristics of a cooling curve for a pure metal.

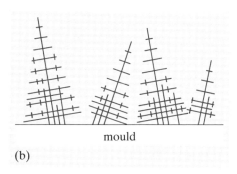

mould

(b)

SAQ 5.9  (Revision)
Explain the origin of (a) undercooling and (b) the horizontal plateau in a typical cooling curve of a pure metal.

So, when crystals grow in a melt they give out latent heat. If this is not removed from the interface between the crystal and the liquid, the temperature rises locally and solidification stops. In practice the latent heat flows away by convection in the liquid and by conduction through the growing solid and the mould material. The rate of heat flow will be very different in a metal mould compared with, say, a sand mould.

The first liquid to touch the walls of a conducting mould is cooled very rapidly — the latent heat is readily conducted away. The resulting undercooling leads to the **heterogeneous nucleation** of many nuclei on the mould walls. Usually, each crystal nucleus grows out as a **dendrite** (from the Greek *dendron*, tree). Figure 5.32(a) is a schematic illustration of a dendrite and clearly indicates that it grows preferentially on certain types of crystal plane. In face-centred and body-centred cubic structures,

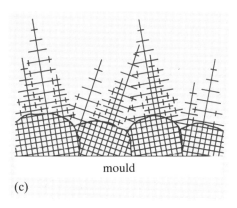

mould

(c)

Figure 5.32 Stages in the solidification of a casting. (a) and (b) Dendritic growth. (c) Chill crystals between dendrites; growth of columnar crystals into melt

261

for instance, the cube faces are the preferred planes: the dendrites grow in the direction of the cube edges. In Figure 5.13 the highlighting of the coring also reveals clearly the dendritic form of the cast alloy.

The dendrites grow until they meet other dendrites from neighbouring nuclei, Figure 5.32(b). The space between the dendrites then solidifies thus producing a layer of small equi-axed crystals, called **chill crystals**.

As the layer of chill crystals thickens, and the mould warms up, the adjacent liquid cools more slowly because the heat is conducted away less quickly. So it undercools less and crystal growth rather than nucleation dominates. Then, Figure 5.32(c), those crystals growing perpendicularly to the mould wall 'edge out' the others and grow into the melt as larger elongated (columnar) crystals. In effect, the grains are elongated in the direction of the easiest heat loss. This is a common feature in castings. Figure 5.33 illustrates the resulting cast structure. But is it good enough to meet all requirements? See ▼**Requirements of a good casting**▲

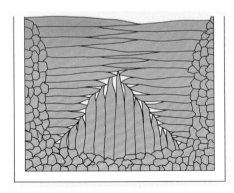

Figure 5.33 Sketch of cast structure showing chill crystals around edge and interior columnar crystals

## Coring and Castings

The problem of coring in cast solid solutions is an excellent example of a non-equilibrium structure that is usually undesirable. The variation of properties it engenders across the section of a casting can be considerable. Mechanical and corrosion properties, for instance, may be affected and in extreme cases it can considerably reduce melting temperatures in the solute-rich regions. Where high production rates are needed, it may not be practical to use slow cooling to prevent coring. Can annealing be used to remove coring?

Annealing is a diffusion process. Consider the cast structure in 70% Cu–30% Ni alloy shown in Figure 5.13. I will take the required diffusion distance as that from the centre of a dendrite (the light regions rich in nickel) to the centre of the interdendritic region (the dark ones rich in copper). On average, this spacing is about 0.25 mm on the photograph and hence, taking into account the magnification, about $10^{-5}$ m in the specimen. Let's assume the alloy is annealed near its melting temperature, at which the diffusion coefficient $D$ for the diffusion of nickel in copper is about $10^{-13}$ m$^2$ s$^{-1}$. For how long must it be annealed to be fully homogenized?

Using the relationship $l = \sqrt{(Dt)}$ you met in Section 4.5.4, we get

$$t = \frac{l^2}{D}$$

$$= \frac{(1 \times 10^{-5})^2}{10^{-13}} \text{ s}$$

$$= 10^3 \text{ s or about 20 min.}$$

This is a practical proposition, and is called **homogenizing**. Notice the importance of the rate of solidification. Slower feezing would cause larger dendrites. Since the time taken for homogenization is proportional to the square of the distance the atoms must travel, twice the dendrite arm spacing would require four times as long. A two-stage process of quick freezing and annealing is, in practice, the favoured way of producing homogeneous alloys. But *extremely* fast cooling rates can have a dramatic effect and produce a quite different microstructure.

# ▼ Requirements of a good casting ▲

The cast structure illustrated in Figure 5.33 has serious shortcomings because of its grain structure. For instance it will be weaker along its centre line where the columnar crystals meet and along the planes extending at 45° from the bottom corners. The structure can be manipulated by changing the casting conditions. For instance, if the liquid is **superheated** (raised to a temperature well above its melting temperature) and poured in, the initial layer of chill crystals may remelt and disappear. This happens because the latent heat cannot be removed fast enough, so the temperature rises. When freezing recommences, cooling is slower and a coarse columnar structure grows inwards from the mould surface.

At the other extreme is a thin section of metal cast in a cold metal mould. The fast cooling rate permeates the liquid so that the entire section solidifies as small equi-axed chill crystals. Such a structure is usually desirable for two reasons. First, it is uniform across the section, and second, the finer the grain size the harder and stronger is the material. This is because, as we shall see in Chapter 6, grain boundaries act as obstacles to moving dislocations. Thus the yield stress of low-carbon steel is doubled by a ten-fold decrease in the grain size.

That's fine for thin sections, but how could you induce a fine grain structure in a large casting?

By **inoculating** the melt with appropriate solid particles, often called **grain refining agents**, that encourage mass heterogeneous nucleation (see Section 4.7.5). The greater the number of nucleation particles that grow into crystals, the more crystals there will be and hence the finer the grain size. For this reason TiC is added to aluminium alloys. Inoculation is also used to produce a fine distribution of the two phases in a eutectic alloy. Small additions of Na to the Al–Si eutectic and of Mg to cast irons are important examples.

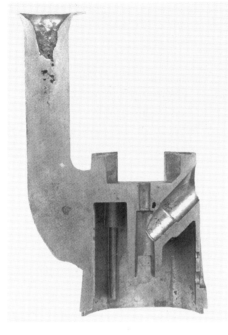

Figure 5.34 Shrinkage porosity in a casting

Besides grain structure, the quality of a cast microstructure is strongly affected by shrinkage and segregation. **Shrinkage** occurs because, with odd exceptions like bismuth and water, liquids contract on freezing. This is why, as Figure 1.21 showed, a runner and a riser are attached to a mould. They allow additional liquid to top up the casting as it solidifies and shrinks. Figure 5.34 shows a section through the body of a small microscope produced by gravity die-casting. The runner is still attached. The hollow and the holes in the runner are created by liquid flowing into the mould as the casting contracted. Notice that there is also a hole within the microscope body. Such holes are called shrinkage porosity.

**Segregation** refers to all non-equilibrium compositional variations in a casting. Coring is an example. Another is gravity segregation, which can arise when a casting cools slowly through its liquid-plus-solid region. If the liquid and solid have different densities they will separate. Figure 5.35 shows a 90% tin–10% antimony alloy in which particles of the intermediate compound SbSn floated to the surface before solidification was complete.

Segregation can occur on a very fine scale too. For instance, alloying elements may segregate to grain boundaries, and you are already aware of the coring of dendrites (Figure 5.13). Microscopic segregation is almost unavoidable but can be removed by subsequent annealing of the casting.

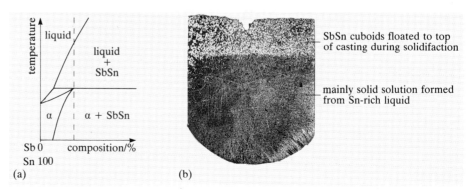

Figure 5.35 Gravity segregation in a tin–antimony alloy

## Rapid solidification

Solidification, like other diffusion-controlled phase transformations, can be mapped by a TTT (time–temperature–transformation) diagram such as Figure 5.36. Above the melting temperature $T_m$ the liquid is clearly stable. When it is cooled quickly to a temperature below $T_m$ there is an incubation period before crystallization starts. Eventually crystallization is complete. The TTT curve shows that the incubation and completion periods for the transformation change markedly with decreasing temperature below $T_m$. This is what you should expect from our consideration in Chapter 4 of the way the rate of a phase transformation varies with temperature (see Figure 4.67). ▼**Rate of transformation and TTT diagrams**▲ considers the relationship between Figure 4.67 and TTT diagrams. As you will see in later sections, TTT diagrams have a practical use in the design of effective heat treatments for metal alloys.

For many materials, especially glasses and polymers, the 'transformation start' curve is well to the right compared with metals. In other words, the incubation time for the transformation is long. Essentially, this is because it is difficult for long and complex molecules to group together as crystals (more about this in Section 5.4). Hence such materials may easily be cooled past the 'nose' of the TTT diagram to give an amorphous structure at room temperature. Theoretically, such a 'supercooled liquid' is an unstable structure at ambient temperature, but usually the incubation period for the start of crystallization is so long that it is of no practical consequence. However, a very ancient glass ornament or window may begin to crystallize (or **devitrify**) and the accompanying contraction in volume can cause cracking.

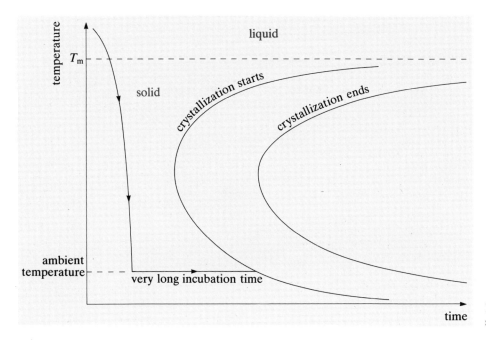

Figure 5.36 Schematic TTT diagram for solidification

As Chapter 4 showed, it is possible to make amorphous metals; but because an individual metal atom is much more mobile than a large molecule, the required cooling rate is extremely high, usually in excess of $10^{10}$ K s$^{-1}$ for pure metals. However, certain alloys can be made amorphous at rather slower cooling rates — a mere $10^5$ to $10^6$ K s$^{-1}$! See ▼Amorphous metals▲

# ▼Rate of transformation and TTT diagrams▲

Figure 5.37 shows general curves for (a) the variation of the rate of a phase transformation with temperature below the equilibrium transformation temperature, and (b) the TTT (time–temperature–transformation) curve. In (a) the axes of Figure 4.67 have been transposed to facilitate comparison with (b). You can see that the TTT curve is related to the rate of transformation curve.

The 'transformation starts' curve of (b) is a plot of the time taken for crystallization to be detected at various temperatures. Since the rate of transformation is zero just below $T_t$ and around $T_0$, the start curve approaches infinity in these ranges. Around $T_t$, the 'driving force' is low and so few crystal nuclei are produced. Around $T_0$, the 'driving force' and hence the nucleation rate is high, but growth of the crystal nuclei is restricted by the very low rate of diffusion. The time before any

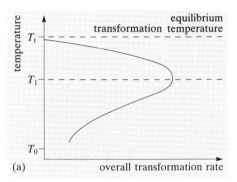

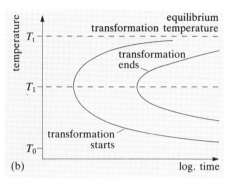

Figure 5.37 (a) Rate of transformation curve, for comparison with (b) TTT curve

transformation is observed is shortest when the transformation rate is a maximum, that is at temperature $T_1$, where both the 'driving force' and the rate of diffusion are fairly high.

Notice that the existence of a period at $T_1$ before the transition gets under way

means that, in principle, if a material is cooled sufficiently fast through the temperature range $T_t$ to $T_0$ the 'nose' of the 'transformation starts' curve is missed — the transformation is suppressed completely. You will meet important practical examples of this shortly.

# ▼Amorphous metals▲

Amorphous metals are produced by various **rapid solidification processes**, which entail the rapid removal of heat. This requires that the material be small in at least one dimension, as in wire, powders, and strip up to about 100 μm thick. **Planar flow casting**, illustrated in Figure 5.38, is one practical technique. Here molten metal is delivered onto a rapidly rotating cooled drum. The speed of casting is at least 10 m s$^{-1}$ and sheets up to 0.5 m wide can be produced.

Amorphous metals are finding a number of novel commercial applications, mainly because some of their properties are markedly different from their crystalline equivalents. A particularly important example is the behaviour of amorphous magnetic alloys. Because they can be produced, in principle, in a completely

homogeneous and structureless form, they should present no obstacles to domain wall movement — no particles of second phase, grain boundaries or dislocations for instance. So they have extremely low hysteresis loss, and high permeability. They are used for tape recorder heads, relays, small transformer cores and so on. This is an excellent example of PPP: a new process provides new microstructures which lead to different products.

EXERCISE 5.11 Eutectic compositions are particularly suitable for quenching to the amorphous state. Can you explain why? Hint. Think about $T_m$ (the melting, or crystal temperature) and $T_g$ (the glass transition temperature).

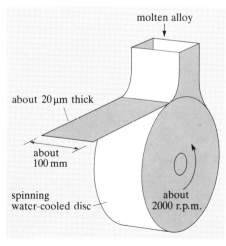

Figure 5.38 The melt spinning process

So, control of solidification allows non-equilibrium structures to be designed, either as advantageous alternatives to equilibrium structures or as novel structures in their own right.

SAQ 5.10  (Objective 5.6)
Figure 5.39 shows the configuration used to produce an unusual type of casting. The whole mould is filled with molten alloy and the heaters keep the upper section at a controlled temperature. Describe, giving your reasons, the sort of grain structure you would expect to form in (a) the lower section and (b) the upper section. Hint. Think about the direction of heat flow and the point of having a narrow constriction between the two sections.

SAQ 5.11  (Objective 5.7)
A molten solid solution alloy is poured into a narrow metal mould. After solidification it is annealed at 0.6 $T_m$ for a substantial time. What grain structure, porosity and homogeneity would you expect in the product? Explain your answers.

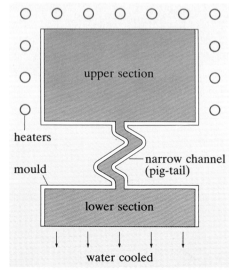

Figure 5.39  A special method of casting

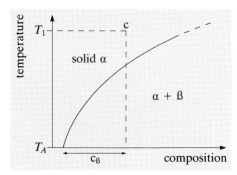

Figure 5.40  A temperature-dependent solvus

## 5.3.2 Precipitation hardening

We have seen the importance of rate of cooling to the type of structure formed during solidification. In this and the next section I want to show that this rate is also crucial in the manipulation of microstructure in the solid state. We start with the development of alloys based on precipitation or, as it is often called, **age hardening**. Figure 5.40 shows a portion of a general phase diagram in which the solubility of one phase (α) in a second phase (β) increases with increasing temperature.

Consider the alloy C. Remember, we are concerned only with its behaviour as a solid. Let's assume that it is held at a temperature $T_1$ until it has been homogenized, that is coring has been removed. $T_1$ is called the **solution treatment temperature**. Now, what happens when it is cooled at different rates? Figures 5.41(b), (c) and (d) show in simplified form three distinct possibilities compared with the homogeneous solid solution α at $T_1$, Figure 5.41(a).

Under equilibrium cooling conditions, at ambient temperature $T_A$ an amount of β phase, related to the 'lever arm' $C_β$, is precipitated in the microstructure. If the alloy is cooled very slowly to $T_A$, the resulting structure will resemble that of Figure 5.41(b): large particles of β widely distributed. If the cooling is fairly quick, the particles will be smaller and more finely distributed, Figure 5.41(c). Cooling the alloy very rapidly, in practice by quenching, may completely suppress precipitation. The microstructure at $T_1$ is retained at $T_A$, Figure 5.41(d), and the alloy is said to be **supersaturated**.

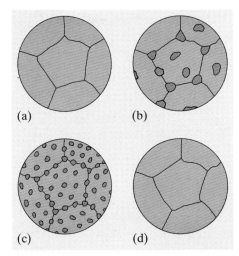

Figure 5.41  Effects of cooling rate on the microstructure of a solid solution.
(a) Homogeneous solid solution.
(b) Slowly cooled. (c) Fairly quickly cooled. (d) Quenched

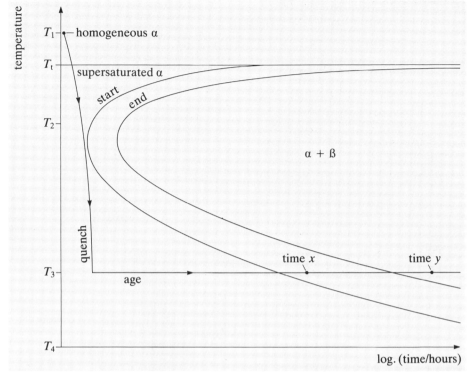

Figure 5.42 A schematic TTT curve for alloy C in Figure 5.40

SAQ 5.12   (Objective 5.8)
(a) Interpret the microstructure (b) to (d) of Figure 5.41 in terms of the rate of transformation curve shown in Figure 5.37(a).
(b) Figure 5.42 illustrates the general features of a TTT curve for the precipitation of β phase in alloy C of Figure 5.40. Describe and explain, using a sketch of the microstructure, what you would expect to happen to the quenched alloy, Figure 5.41(d), if it is subsequently heated (this is called **ageing**) to $T_3$ for time $X$ and time $Y$ as indicated.

Notice that in Figure 5.41(b) most of the second phase precipitates on the grain boundaries. This implies that they are preferred sites for nucleation or that growth rates there are much faster. ▼**Preferred nucleation sites**▲ looks at why certain sites are favoured for precipitation.

Now, as you know from 'Solid solutions in action', the supersaturated solute strengthens the alloy and, of course, different distributions of the second phase lead to further changes in properties. So, the solute concentration and the ageing conditions provide direct means of manipulating the structure to give the properties required. Basically, the more particles there are and the closer they are together, the stronger is the alloy. Such a distribution provides very effective barriers to dislocation movement. We shall consider the particular mechanisms in Chapter 6. ▼**Precipitation hardening in action**▲ gives a brief introduction to the properties and uses of some commercial alloys.

# ▼Preferred nucleation sites▲

In practice the homogeneous nucleation of a solid in a liquid is extremely rare. Heterogeneous nucleation on suitable particles or sites on the mould surface is the norm. The nucleation of a second phase in a solid is analogous. Here the heterogeneities that foster nucleation are regions of disorder such as grain boundaries and dislocations. This is because some of the atoms or molecules around them are out of their equilibrium sites, and so these regions have a higher internal energy. Put another way, these regions are physically more open and thus provide less resistance to the growth of a particle. Their relative openness also means that they provide easier paths for diffusion of the atoms of the new phase. In general, grain boundaries are regions of greater disorder than dislocations and are more likely to act as nucleation sites. In the absence of sufficient disorders, particular types of crystal plane can be the preferred nucleation sites. The possibilities are illustrated schematically in Figure 5.43.

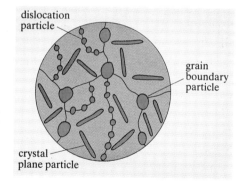

Figure 5.43 Some preferred nucleation sites

# ▼Precipitation hardening in action▲

A wide variety of commercial alloys rely on precipitation hardening for their strength. Alloys based on aluminium and nickel are important examples. Many of them, especially nickel-based alloys, are complex and contain a range of alloying elements each added as a means of optimizing the materials property profile. Here I shall simplify the picture.

Aluminium–magnesium–silicon precipitation-hardening alloys are fairly straightforward and illustrate well the main effects of ageing treatment. Figure 5.44 shows the variation of tensile strength with both ageing time and temperature of a typical alloy in this system. Hardness values would show similar trends. The composition is about 1.5% Mg, 0.5% Si, with traces of other elements. The precipitating phase is $Mg_2Si$. The following points are worth noting.

- Hardening occurs with time at ambient temperature and continues to increase slowly at very long ageing times. To facilitate the shaping of a component, such as a rivet or an aeroplane panel for example, the material needs to be relatively soft, so the solution-treated alloys would have to be refrigerated.
- As the temperature of ageing is raised, the time to maximum strength decreases and, up to a certain temperature, the maximum strength increases.
- Above about 136 °C a longer period of ageing results in softening. The alloy is said to overage. This is due to coalescence of the particles. Coalescence reduces the total interfacial energy of the material, bringing the structure nearer to equilibrium. (Interfacial energy is the energy associated with the region of disorder between the particles and the matrix. See Exercise 5.10.)

- At higher temperatures, for example 240 °C, large, widely spaced particles are created. The rate of softening is great and the alloy becomes softer than the supersaturated solid solution.

The family of aluminium alloys to which this belongs is in common use because of its reasonable strength, formability by rolling and extrusion, toughness and corrosion resistance. Its applications vary from architectural structures to container bodies.

Other aluminium precipitation-hardening alloys based on copper (about 2% to 5%) and on zinc (about 5% to 6%) plus magnesium (2% to 3%), both with crucial traces of other elements, can attain significantly higher strengths. They are used extensively in aircraft structures. In these alloys the precipitation process is usually more complicated. The maximum strength is achieved by a dispersion of precipitates so fine that it can only be observed directly using electron microscopy (see Figure 5.45). Furthermore, the precipitates are not actually the equilibrium intermediate compound, basically $CuAl_2$ and $MgZn_2$ respectively, but an embryonic form of it. In fact, depending on the ageing temperature, the type of precipitate may go through various incarnations with time.

The most complex precipitation hardening alloys are the so-called superalloys based on nickel. Essentially, they are strengthened by precipitates of $Ni_3Al$ and contain a wide range of alloying elements each with a role to play, including the control of the amount, size and shape of the precipitates. Figure 2.40(b) in Chapter 2 is an electron micrograph of such an alloy. Notice the extreme proximity and regularity of the precipitates. These alloys

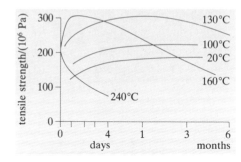

Figure 5.44 Ageing curves for an alloy in the Al–$Mg_2$Si system

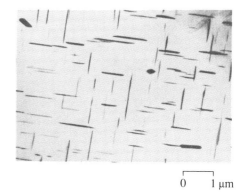

Figure 5.45 Electron micrograph of minute precipitates in an Al–4% Cu alloy

maintain their strength at high temperatures and are used extensively in turbine blades for jet engines. They are shaped by investment casting.

A disadvantage of precipitation-hardened alloys is that they will overage and soften during subsequent processing or in service if they are heated above the ageing temperature. So, for instance, the aluminium alloys cannot be welded unless the quenching and ageing treatments are repeated.

---

SAQ 5.13  (Objective 5.9)
In the quest for a good precipitation hardening alloy, comment on

(a) the solute concentration required,
(b) the solution and ageing temperatures in terms of practicable times, temperatures and service conditions.

## 5.3.3 Heat treatment of plain carbon steels

The main difference between steels and other precipitation hardening systems is that, in steel, quenching not only produces a solid solution, but also provokes a different type of solid-state phase transformation. But before looking at these non-equilibrium structures, let's have a closer look at the iron–carbon phase diagram and consider the properties of slowly cooled plain carbon steels.

### The iron–carbon diagram

The labelled Fe–C diagram of Exercise 5.9 is reproduced as Figure 5.46.

For historical reasons concerned with the elucidation of the microstructure of steels, the different phases have been given names — just as geologists gave names to rocks. Here are a few to start with.

- $\alpha$ is $\alpha$ **ferrite** or just plain **ferrite**; BCC crystal structure.
- $\gamma$ is **austenite**; FCC crystal structure.
- $\delta$ is $\delta$ **ferrite**; BCC structure.
- $Fe_3C$ is **cementite**.

In addition, the particular phase mixture consisting of alternating platelets of ferrite and cementite ($\alpha + Fe_3C$) is called **pearlite**, because under the microscope it can have an iridescent mother-of-pearl appearance. Cooled sufficiently slowly, a steel of eutectoid composition will consist solely of pearlite. Figure 5.2(a) is an example. Mixtures of $\alpha$ and $Fe_3C$ can arise in other ways, as you will see shortly. These do not have the same appearance and they are not pearlite.

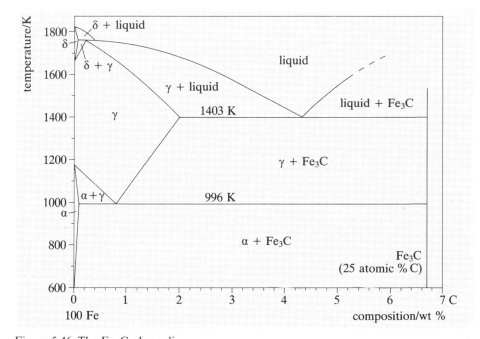

Figure 5.46 The Fe–C phase diagram

**SAQ 5.14** (Revision)
Refer to Figure 5.46.

(a) What is the reaction that occurs at 996 K called?
(b) How does it differ from a eutectic?
(c) Describe the reaction in terms of the names of the phases and combinations thereof.

Solid–solid reactions, such as in a eutectoid, occur much more slowly than liquid–solid reactions. There are two reasons for this, one concerned with nucleation and the other with growth. First, the unit volume of a nucleus of a new phase is different from that of the old phase. Thus, the old solid resists the growth of a new nucleus to the critical radius.

Where would you expect new phases to nucleate in a solid–solid reaction?

In regions of disorder, such as grain boundaries and dislocations.

Secondly, once nucleation has started, the diffusion rate, and hence growth, is much slower than in a liquid.

As you know from Exercise 5.9, the iron–carbon diagram consists mainly of a eutectic, which is to do with cast irons, and a eutectoid, which gives rise to steels. This implies that steels have a carbon content up to 2 wt %, but in practice the normal limit is about 1.5 wt %.

## Some properties of slowly cooled steels

In Figure 5.47 I have summarized some of the basic mechanical properties of plain carbon steels as a function of carbon content up to 1.2 wt % carbon. The data is for steels in a hot-worked and annealed condition. (The annealing is carried out by slow cooling in a furnace which has been switched off. The resulting microstructure is nearly that produced by true equilibrium cooling.) As the carbon content rises, the amount of cementite in steel rises, which accounts for the variations in properties shown in Figure 5.47. Cementite, in common with many intermediate compounds, for example $Mg_2Si$ and $CuAl_2$ (see Section 5.3.2), is both hard and brittle for the reasons that were explained in Section 3.9.4.

EXERCISE 5.12 The white regions surrounding the regions of pearlite in Figure 5.48(a) and (b) are different phases. Which phase is which? Hint. Consider the cooling of the two alloys; Figure 5.46 will be useful.

In annealed steels containing less carbon than the eutectoid composition (0.8 wt %), the cementite is present only as a constituent in the lamellar pearlite. But its properties, and the intimate lamellar microstructure of pearlite, confer on pearlite quite different properties from those of the

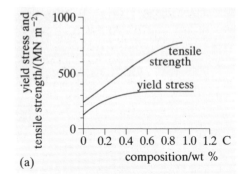

(a)

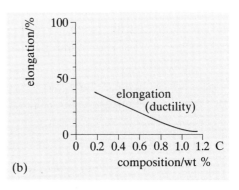

(b)

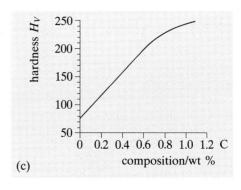

(c)

Figure 5.47 Typical properties of annealed plain carbon steels. In (b) percentage elongation at fracture is used as a measure of ductility. Note. All alloys have the same grain size

soft and ductile ferrite. It is much harder, stronger and less ductile, as the values for the eutectoid composition in Figure 5.47 illustrate. So, in such steels a pearlite region may be regarded as behaving like a hard particle of second phase. Applying a law of mixtures argument (see Chapter 2), as the amount of ferrite decreases (and the amount of pearlite increases) there should be an increase in hardness and strength but a fall in ductility. Figure 5.47 illustrates this.

In the alloys containing more carbon than the eutectoid composition, the pearlite regions are surrounded by brittle cementite. Ductility is impaired because plastic deformation is now only possible in the ferrite lamellae. A further increase in hardness and strength is also expected, and observed.

This loss of ductility is important because, for most uses of steel, ductility is essential. ▼Plain carbon steels in action▲ provides some explanation.

With this background we can now consider, briefly, how inhibiting the eutectoid reaction by rapid cooling helps to confer even greater versatility on steel.

## Quenching

The first blacksmith to drop red-hot steel (at about 1100 K) into a bucket of water must have been very surprised on subsequently trying to

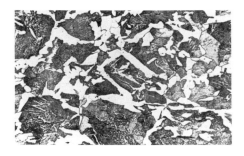

(a)

(b)

Figure 5.48  (a) 0.4 wt % C,  (b) 1.1 wt % C

# ▼Plain carbon steels in action▲

Steels are used for a wide variety of artefacts. Table 5.3 gives you a flavour.

The ductility of low-carbon steels is especially important because it implies good formability. As a result such steels are extensively used for beams in the construction industry, and for sheet in the motor industry and in canning.

Imagine what steps must be involved in converting a ten-tonne ingot into thin sheet for car body panels or beer cans. The early shaping has to be done by hot working so that manageably low forces can be used. But hot working produces oxide scale and makes it impossible to maintain a good surface and accurate dimensions. Hence cold working will be involved sooner or later. Cold working results in work hardening and, if carried too far without some intermediate softening process, causes cracking.

Softening can be achieved by heating the cold-worked steel to a temperature at which it **recrystallizes**. This phenomenon, which is common to all metals that have

Table 5.3   Some uses of different plain carbon steels

| Steel | Carbon content/% | Applications |
|---|---|---|
| low carbon (mild) | 0.1–0.25 | beams, angles, car bodies, cans |
| medium carbon | 0.2–0.5 | general forgings, shafts, rotors, tools such as spades |
| high carbon | 0.5–0.65 | railway lines, car springs, hammers, chisels |
|  | 0.65–0.95 | saws, drills, dies, cutting tools |

been sufficiently cold-worked, results in a set of completely new, soft grains. (It is covered in Chapter 6.) Now, as it happens, the temperature range for recrystallization spans the eutectoid temperature, so three possible treatments arise:

1 **Subcritical annealing**. This means heating at a temperature just below the eutectoid. It causes recrystallization of the ferrite, without changing the distribution of the pearlite.

2 **Full annealing**. This means heating the steel to a temperature at which it enters the austenite region, holding it there for

long enough to allow new austenite grains to form, and then cooling slowly to obtain near-equilibrium products. The prolonged cooling allows time for the growth of austenite crystals. The resulting *coarse* pearlite improves ductility. It is used on the sheet steel used in can making.

3 **Normalizing**. This is similar to full annealing, but the steel is heated to a slightly higher temperature, and cooled in air (that is, more rapidly). It produces fine pearlite and a small grain size in the ferrite, hence strengthening. Normalizing is quicker and cheaper, and lends itself to mass production.

work it. Our blacksmith would have found it very hard compared with the same steel left to cool in air, but also very brittle. In fact the high hardness values found in steels quenched this way extend the properties of steel dramatically. Compare Figure 5.49 with Figure 5.47(c). How are they achieved?

The crucial point is the very different solid solubility of carbon in $\gamma$ and $\alpha$ iron: up to 2 wt % in $\gamma$ at 1400 K, and about a hundred times less in $\alpha$ at the eutectoid temperature (even less at ambient temperature). When any Fe–C alloy is cooled slowly from the austenite region, virtually all the carbon initially in solid solution in the FCC austenite is incorporated in the cementite ($Fe_3C$) created by the eutectoid transformation. However, when the alloy is quench-cooled, there is insufficient time for this to happen. With the carbon trapped the result is quite different. The austenite crystals undergo **martensitic transformation**, that is structural change involving the shearing of planes of atoms over one another. As you know, shear processes involve a change of shape. Figure 5.50 is evidence of this. The new phase, called **martensite**, does not appear on the phase diagram because it is not an equilibrium structure.

Even in pure iron the FCC to BCC transformation will occur by the martensitic route if the cooling is sufficiently rapid. The resulting martensite is soft and ductile. However, when alloys containing carbon are quenched, the carbon trapped in the martensite distorts it into a non-cubic structure which is hard and brittle.

The martensite structure is body-centred but has one axis longer than the other two. It is called body-centred tetragonal (BCT) and is illustrated in Figure 5.51(b). $BaTiO_3$, which you have met in Chapters 3 and 4, also has a BCT structure. The possible interstitial positions for a carbon atom in this new BCT structure are the octahedral ones inherited from the FCC austenite (Figure 5.51a), and the degree of elongation of the cell will depend upon the amount of carbon in solution.

Because of the difference in shape of the old and new structures, the strain energy involved in the martensitic reaction is enormous and a large undercooling is necessary to produce stable nuclei. This is why the mechanism is only invoked by rapid cooling; the normal diffusion transformation to pearlite is by-passed.

As the material cools there will be a temperature where the largest embryos will become stable nuclei. Once this critical temperature is reached the nucleus can grow. The martensite–austenite interface can only move in one direction easily and a plate, or lath, of martensite is formed within an individual grain of austenite. As I said earlier, martensite is produced by shear transformation, and just like slip — the other crystalline shear process we have considered — it is achieved by the movement of dislocations. Because diffusion is not involved, the transformation is extremely rapid once started.

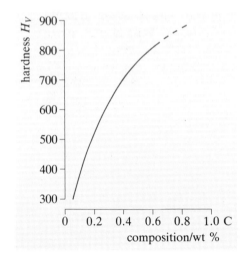

Figure 5.49 Hardness values for quenched steel

The surface of the steel was scratched with emery paper. These scratches were straight when the sample was fully austenitic.

A lath has cut the surface, displacing the scratched lines. | martensite

Figure 5.50 Rumpling, produced by quenching, on the surface of a sample of steel

The structure of martensite is quite difficult to interpret on an optical micrograph. Within each prior austenite grain a number of martensite laths will have formed. Any section through these will cut them at random and the structure looks something like a mass of needles. Figure 5.2(b) is an example.

Martensitic transformations are not exclusive to steels, as ▼Other martensites▲ shows.

## Tempering of martensite

Martensite is very hard but brittle. The carbon in supersaturated solid solution distorts the structure considerably. This is the source of the hardening and also, since it prevents dislocation movement, of the lack of toughness.

How do you think some ductility can be restored to the alloy without sacrificing too much of the hardness?

Heat the alloy to a temperature at which the carbon atoms diffuse. Then precipitates of iron carbide will form and the regions around them, now depleted of carbon, will relax to BCC ($\alpha$) phase — thus providing a ductile matrix. This is a similar process to that used for precipitation-hardened alloys, except that in this case it causes softening! The process is called **tempering** and the product is called **tempered**

# ▼Other martensites▲

All transformations which occur by a shear rather than a diffusion mechanism are called martensitic. Another example in metals is in titanium. On rapid cooling it transforms from BCC to hexagonal close-packed (HCP) by a martensitic reaction. In steels, the high distortion produced by the supersaturated interstitial solid-solution of carbon makes the martensite hard. There is not an equivalent in titanium alloys because the alloy elements form *substitutional* solid solutions and so the martensite is quite soft.

A martensite which is exploited in a ceramic system is zirconia ($ZrO_2$). It is used as a means of toughening other ceramics and has been particularly successful with alumina ($Al_2O_3$). Alumina is a very useful engineering ceramic because of its high melting temperature, hardness and chemical stability but, as with many other ceramics, its fracture toughness is low. Alloying with about 15 wt % $ZrO_2$ doubles the toughness. The

reason for this is the precipitation of some $ZrO_2$ solid solution as fine particles on the $Al_2O_3$ grain boundaries.

Why should this precipitation occur? Hint. Look at Figure 5.16.

Below the eutectic temperature, the solubility of $ZrO_2$ in $Al_2O_3$ decreases with decreasing temperature. So, on cooling, the proportion of $\beta$ phase increases. As you know from 'Preferred nucleation sites', grain boundaries are favoured.

On cooling, the $ZrO_2$ would normally undergo a martensitic transformation, but in the alloy the elastic constraint provided by the $Al_2O_3$ matrix prevents the required change of crystal structure from taking place. Now for the fascinating bit. When a crack passes close to a $ZrO_2$ particle it releases the elastic constraint and the material opens up. This allows the $ZrO_2$ martensite to be created suddenly, which puts the crack into compression and closes it. Hence the toughening.

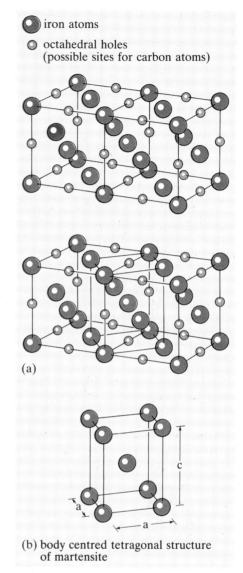

○ iron atoms
◎ octahedral holes
(possible sites for carbon atoms)

(a)

(b) body centred tetragonal structure of martensite

Figure 5.51 A unit cell of martensite (b) in relationship to FCC cells of austenite (a)

**martensite**. Figure 5.2(c) and (d) are examples. Tempering at higher temperatures leads to coarsening of the precipitates and to the equivalent of overageing, as Figure 5.2(d) illustrates. The data in Figure 5.2 and Table 5.1 give a good indication of the variation in properties that can be achieved by the manipulation of steel microstructures.

Note that the two-phase microstructures produced by the tempering of martensite are not the same as pearlite formed from austenite.

### Intermediate cooling

What happens when cooling is quicker than equilibrium cooling but slower than quenching? The TTT diagram for the eutectoid transformation (Figure 5.52) gives a good indication. (Note that a TTT diagram is not a precise graph, because of the way it is determined. See ▼TTT **diagrams for steels**▲.)

---

EXERCISE 5.13   Sketch the microstructures you would expect in a eutectoid steel:

(a)  after annealing it at 1100 K,
(b)  immediately after quenching it to 850 K,
(c)  ten seconds after quenching it to 850 K.

---

If the steel is cooled at about $200\,\mathrm{K\,s^{-1}}$ the 'nose' is missed and martensite is produced — starting at about 550 K (the $M_S$ temperature). $M_F$ is the temperature at which transformation to martensite is complete. So at ambient temperature the alloy still contains some unstable austenite.

## ▼TTT diagrams for steels▲

To obtain a TTT diagram, samples of the specified alloy are quenched from the austenite phase into baths of molten salt or tin at predetermined temperatures. Each sample is held at a particular temperature for a particular time and then quenched into water. The sequence is shown in Figure 5.53. The final structure of the alloy is then determined by optical microscopy, quantitative X-ray diffraction or dilatometry. The TTT diagram is built up using sets of many results. (Dilatometry is the measurement of the change in length (dilation) of a specimen. Because atoms or molecules move into different relative positions during phase transformation, accompanying volume changes can be monitored by dilatometry.)

It is important to appreciate that the accuracy of the 'start' and 'end' curves in a TTT diagram depends on the sensitivity of the analytical techniques used to detect the onset or cessation of crystallization. 'Resolution limit of a microscope' (Chapter 2) showed that a transmission electron microscope could detect a particle of a new phase some 1000 times smaller than that observable in an optical microscope.

A TTT diagram, being based on discontinuous cooling, does not strictly reflect what happens in continuous slow cooling. Provided such cooling does not extend too far below $T_t$ (Figure 5.42), the diagram is a good guide.

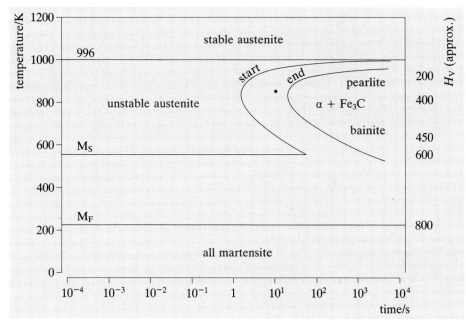

Figure 5.52  The eutectoid TTT diagram

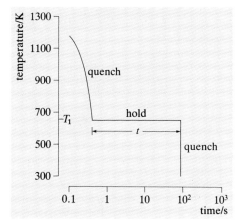

Figure 5.53 Typical cooling curve used to determine a TTT diagram

The form of the $\alpha + Fe_3C$ produced depends on the cooling rate. With little undercooling, the pearlite is coarse — wide spacing between the cementite lamellae. At lower temperatures (more undercooling) the pearlite is finer. At still lower temperatures the carbide appears as smaller particles and the $\alpha$ phase may be more like martensite than ferrite. This structure is generally known as **bainite**. Note on Figure 5.52 the range of values of hardness for from pearlite to bainite.

SAQ 5.15 (Objective 5.10)
Summarize briefly the main microstructure and hardness changes that can be produced in a plain carbon steel of eutectoid composition by heat treatment from the austenite phase. Hint. Think in terms of the rate of cooling.

SAQ 5.16 (Objective 5.11)
Summarize and comment on the major mechanical property characteristics of (a) solid solution alloys, (b) eutectic alloys and (c) other two-phase alloys.

# 5.4 Mixtures far from equilibrium

So far we have looked at materials systems in which, except for martensitic reactions, changes are brought about by the movement of single atoms. Not surprisingly, large molecules are very much more sluggish than single atoms. Whereas, generally, metals always crystallize, even when cooled quickly, some polymers never crystallize, even when cooled slowly. Other polymers may crystallize partially to give a microstructure of mixed crystalline and amorphous regions. So, although the driving force for complete crystallization may exist, the required mobility may not.

For all practical purposes, no commercial polymer can be made 100% crystalline, even though some are often (loosely) referred to as crystalline. Polymers are either completely amorphous or partially crystalline.

The degree of crystallinity of a polymer (usually expressed as a percentage) is a major determinant of the properties of a given polymer. For instance, amorphous polypropylene (PP) is a semi-tacky rubbery wax of little value except as road-paint base. But partially crystalline PP is an important polymer, similar to HDPE but superior in certain respects — its softening temperature is above 373 K, for example, hence its widespread use for articles that have to be sterilized.

Table 5.4 shows the variation of properties with degree of crystallinity for the two main types of polyethylene (PE) and for polymethylene. Polymethylene is a special, non-commercial form of PE that, compared with other varieties, has particularly long and unbranched molecules.

Table 5.4   Some polyethylene and properties

| Polyethylene type | $CH_3$ groups per 1000 carbon atoms | Density/kg m$^{-3}$ | % crystallinity | Melting temp/K | Young's modulus/GPa | Tensile strength/MPa | Permeability to $H_2O$/ $(10^{-15}$ mol mN$^{-1}$ s$^{-1})$ |
|---|---|---|---|---|---|---|---|
| Polymethylene | 0 | 0.98+ | 86+ | 414 | 2+ | ~35 | low |
| HDPE | 5–7 | 0.955–0.970 | 70–80 | 403 | 0.55–1.00 | 20–37 | 4 |
| LDPE | 20–35 | 0.910–0.935 | 40–55 | 381 | 0.15–0.24 | 7–17 | 30 |

Note. The number of $CH_3$ groups per 1000 carbon atoms is an indication of the number of branching points along a molecule; more about this later.

Why would you expect polymethylene to have a higher degree of crystallinity than HDPE?

Unbranched molecules are more readily able to pack together in ordered arrays.

It is important to appreciate that the values in Table 5.4 are very approximate. There are two reasons for this.

Firstly, two of the properties included in the table, Young's modulus and the melting temperature, are difficult to assess. Most polymers don't obey Hooke's law and so an alternative method of measuring the elastic modulus is used. Also, many polymers melt over a range of temperature, and others do not exhibit a melting temperature at all.

▼Volume changes on cooling▲ provides an insight into the differences between polymers and fully crystalline materials.

Secondly, the degree of crystallinity of a particular type of polymer depends on the chemical make-up of its molecules and on how the polymer is processed. In this section I shall explain the implications of these factors for a variety of polymers.

Full crystallinity, as we have seen, is a state of microstructural equilibrium. For polymers, as for metals, the balance between stick and rattle determines how closely a microstructure approaches equilibrium. However, because of the size and shape of polymer molecules, and the variety of their bonding mechanisms, polymer crystallization is much more complex than metallic crystallization.

SAQ 5.17   (Revision)
In the style of Section 3.7 comment on how the following molecular attributes affect the ability of polymer chains to form crystals: (a) short, (b) linear, (c) smooth, (d) regular, (e) slippery, (f) stiff.

The degree of crystallinity of a polymer is determined both by its chemical composition and by its processing history. This includes such things as:

• polymerization (which determines the structure of the molecule);

# ▼Volume changes on cooling▲

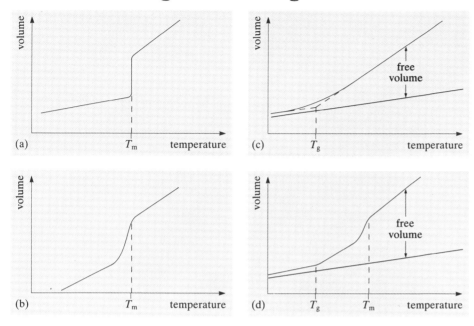

Figure 5.54 Volume changes with temperature for (a) a pure metal, (b) 'crystalline' polymer, (c) amorphous polymer, and (d) partially crystalline polymer

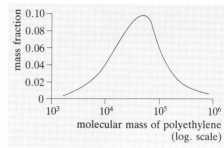

Figure 5.55 Molecular mass distribution of a polyethylene

When atoms are the mobile species, as opposed to large molecules, there is usually a sharp decrease in volume as the material crystallizes. This is shown schematically in Figure 5.54(a).

Even if a polymer crystallized completely, its freezing would spread over a range of temperature and its volume contraction would not be sharply defined. See Figure 5.54(b). This spread arises from the variation in the molecular mass (therefore length) of the molecules in a polymer. The **molecular mass distribution** is thus an important determinant of polymer properties, and Figure 5.55 shows an example. (This will be a major topic in Chapter 7.)

In some polymers, for example polycarbonate (PC), crystallization does not occur during normal cooling. Such polymers have no melting temperature. In effect, the solid polymer is a supercooled liquid. In the molten state, the polymer molecules are in constant motion because of thermal agitation (rattle). The volume of the 'liquid' is greater than that of the solid polymer, in which molecular movement is restricted. This 'extra' volume is called the **free volume**, see Figure 5.54(c). As the polymer cools, there is less thermal energy available and the free volume decreases. Now, molecular bond rotation requires significantly more free volume than, say, bond bending or stretching. Below $T_g$, the glass transition temperature, bond rotation ceases and the free volume becomes much less.

Partially crystalline polymers will exhibit a blurred melting temperature for the crystalline fraction and a glass transition temperature for the amorphous fraction, Figure 5.54(d). Hence the degree of crystallinity has a marked influence on properties.

• in a thermoplastic, the conditions under which the polymer melt was cooled;
• mechanical forces during processing (during injection moulding or drawing into a fibre, for instance).

Cooling conditions are relevant to this chapter; the mechanical forces involved in polymer processing will be considered in Chapter 6.

Before looking at crystallization in more detail, we should review the pertinent characteristics of polymers. As you know from earlier chapters, a thermoplastic is a polymer that flows on heating, and will, in principle, continue to do so when it is cooled and reheated. On the other hand, a thermoset will not melt because it contains a rigid three-dimensional network of atoms, or, put another way, a network of strong, covalent cross-links between previously independent molecules. If we look at the ways in which the covalently bonded carbon atoms may link up we can understand the reasons for the differences in behaviour. Figure 5.56 shows some molecules at a scale where the individual atoms are not seen.

For clarity, the shape has been simplified. Molecule (a) is a completely linear polymer with no branch points. The branched polymer in (b) has one or more branch points (shown as ●) at which three or more chain segments meet. However there are no closed circuits, as there are in (c). The latter is a small section of a **network polymer**, which is, in effect, an infinite molecule in which a vast number of closed circuits exist.

Figure 5.56 is a static picture. It does not show any secondary bonds (such as Van der Waals or hydrogen bonds) between molecules. Nor does it indicate the shape of the molecules between branch points, or whether the molecules are changing shape. Thermoplastics consist of a collection of either linear or branched polymer molecules. When they are heated sufficiently the molecules become mobile and can flow past one another. There are two classes of network polymer:

- thermosetting plastics, where the degree of crosslinking is such that the network is a rigid unchanging one;
- elastomers (rubbers), where the crosslinking is much lower in frequency and where above $T_g$ the material can change its shape.

Here for simplicity I shall be concerned mainly with thermoplastic polymers.

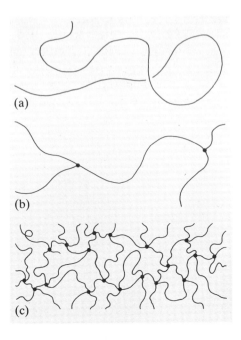

Figure 5.56 Schematic illustration of (a) a linear polymer molecule, (b) a branched polymer and (c) part of an infinite network

## 5.4.1 Molecule conformations and configurations

Figure 5.57(a) shows a 'ball and stick' model of the shape of part of a polymer chain. The balls represent the centres of carbon atoms and the sticks represent the strong (directional) bonds between carbon atoms. Other bonds and side groups have been left off for clarity (this is the case for many of the figures from now on). In this arrangement, or **conformation** as it is called, the atoms form a zig-zag and lie on the same plane. Figure 5.57(b) is a view of the molecule looking along the bond between atoms 2 and 3.

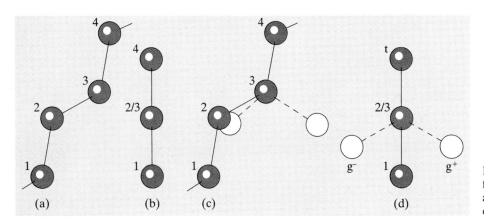

Figure 5.57 Bond rotation in a four-carbon-atom polymer. Atoms 1, 2 and 3 are fixed. Atom 4 can take three different positions.

This isn't the only possible conformation, though. Above $T_g$ every C–C bond is free to rotate about its axis, and will settle preferentially into sites which are energetically favourable. When a carbon atom forms four single covalent bonds, as in diamond, the bonds ($sp^3$ hybrid orbitals) are disposed towards the corners of a tetrahedron (see Figure 3.21). The angle between any two of the bonds is about 109.5°. Imagine that atom 3 is at the centre of a tetrahedron and that atoms 2 and 4 in Figure 5.57(a) are on two of the corners. Now rotate the bond between atoms 2 and 3 to put atom 4 into one of the other two tetrahedral sites shown in Figure 5.57(c); Figure 5.57(d) is the view along atoms 2 and 3. The original position of atom 4 (Figure 5.33a) is called the **trans** (t) conformation; the two alternatives are called **gauche plus** ($g^+$) and **gauche minus** ($g^-$).

In general these three options are possible for each carbon atom along the polymer chain. So a chain of four carbon atoms has 3 possible arrangements, a chain of five atoms has $3^2$ possibilities and one of $n$ atoms has $3^{n-3}$ possibilities. Clearly, if all choices have equal probability, the number of possible conformations is vast. As an example, Figure 5.58 shows a computer-generated model of a randomly constructed molecule of 400 carbon atoms.

However, the conformation of some molecules is more straightforward because not all t, $g^+$ and $g^-$ are equally probable.

Can you see why this should be so?

The answer lies in the influence of the side-groups on a molecule. In some conformations, side groups may be forced into closer proximity than in others. Such conformations are energetically less favourable. For instance you can see from Figure 5.57 that side groups attached to atoms 1 and 4 would be closer together in gauche positions than in the trans position. For this reason, in crystalline regions polyethylene molecules opt for the trans position and the conformation of the molecule is then the planar zig-zag of Figure 5.57(a). On the other hand, polypropylene (PP) molecules can adopt the helical configuration illustrated in Figure 5.59. The size of the $CH_3$ side-groups and their regular distribution along and around the chain prevent free rotation about the C–C bonds; alternate bonds take the t and $g^+$ conformations and this leads to a helical configuration.

However, these relatively simple (!) molecular structures are only possible for the particular cases where the side-groups are regularly arranged along the molecules. As you know from Chapter 3, there are other possible configurations and we will now consider them in more detail.

## Chain regularity

There are three main kinds of irregularity that modify 'ideal configurations' such as that of Figure 5.59. These are listed overleaf:

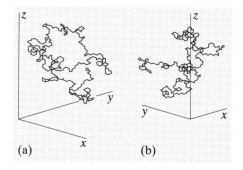

Figure 5.58 Two views of a random polymer chain

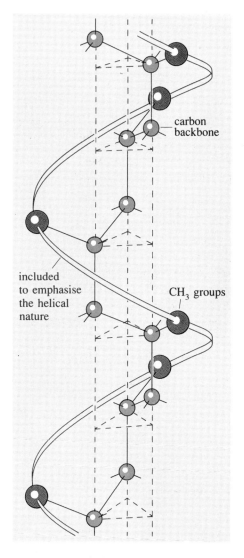

Figure 5.59 A helical polypropylene molecule

carbon backbone

$CH_3$ groups

included to emphasise the helical nature

279

- side branching,
- the disposition of the side groups,
- the presence of copolymer units.

The number and length of the side branches contributes to the configuration of a molecule; and configuration can have a very marked effect on microstructure and properties. It is particularly significant in polyethylene as Table 5.4 demonstrates. A polyethylene molecule with a number of short side branches is illustrated schematically in Figure 1.15. You can see readily why complete alignment with similarly configured molecules is unlikely. The amount of side-branching depends on the type and operating conditions of the polymerization process, as 'Processing polyethylene' in Chapter 1 indicates.

The different configurations of a repeat unit give rise to what are called **stereoisomers**. Imagine a repeat unit similar to polyethylene but with an additional side group X (Cl for PVC; $CH_3$ for PP; $C_6H_5$ for PS and so on). There are three basic kinds of stereoisomer as Figure 5.60 shows. For simplicity, the carbon chain is shown in the planar zig-zag (trans) form and the plane is included.

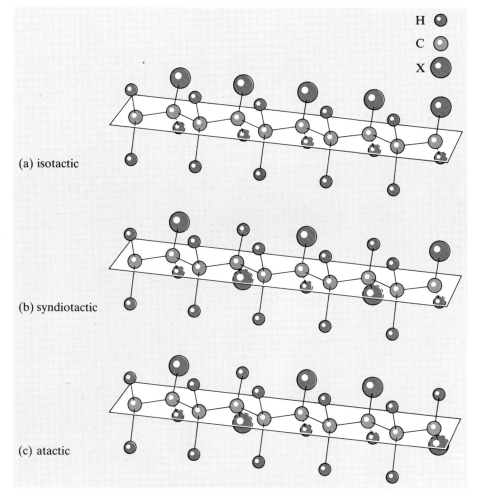

(a) isotactic

(b) syndiotactic

(c) atactic

Figure 5.60 The three kinds of stereoisomer

In the **isotactic** arrangement all the side groups are on the same side of the chain; in the **syndiotactic** they alternate about the chain; in the **atactic** they are randomly disposed. The former two are regular, and therefore likely to facilitate crystal formation; the latter is irregular. The differences can be extremely important. For instance, the PP I referred to earlier is less useful because it is atactic; the valuable version is isotactic, as is the one in Figure 5.59. In practice no commercial plastic is completely stereoisomeric. So PP, for example, always has a small proportion of atactic units. Clearly this influences the degree of crystallinity.

A **copolymer** is produced by polymerizing two or more monomers together so that segments of the monomers appear within each molecule. Normally the monomers form into a **random copolymer**, in which the two monomer units have a random sequence, Figure 5.61(a). With special polymerization techniques the units can be produced in alternating sequences. These are **block copolymers**, Figure 5.61(b).

As you would expect, copolymerization can be used to manipulate properties as ▼**Some copolymers in action**▲ indicates.

(a)

(b)

○ monomer $M_1$     ● monomer $M_2$

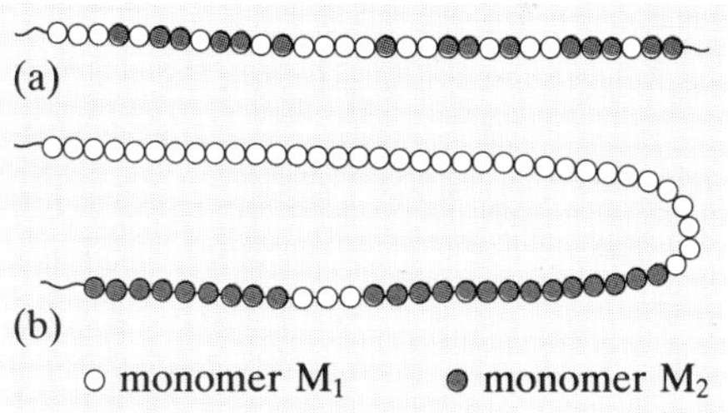

Figure 5.61 (a) Part of a random copolymer. (b) Part of a block copolymer

# ▼Some copolymers in action▲

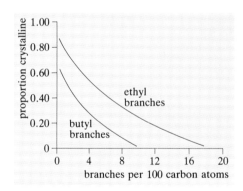

Figure 5.62 The effect of copolymerization on the degree of crystallinity of polyethylene

First, a polyethylene random copolymer. Polyethylene is by far the most commonly used polymer (followed by PVC, PP and PS). Since its first commercial production a variety of types of polyethylene have been developed, but the dominant varieties are HDPE and LDPE. LDPE molecules are extensively branched (Figure 5.56b) but by copolymerizing ethylene with butene or hexene, ethyl ($—C_2H_5$) and butyl ($—C_4H_9$) short-chain branches are produced. The copolymer is called medium density polyethylene (MDPE) if the density is about 940 kg m$^{-3}$, and linear low density polyethylene (LLDPE) if it is between 910 and 930 kg m$^{-3}$. The short two- and three-carbon atom branches militate against crystal formation and so the degree of crystallinity decreases with the amount of ethyl and butyl branching, see Figure 5.62. Here copolymerization is being used to control the number and length of the side branches on a polyethylene molecule. Compare these with the numbers of $CH_3$ side-groups given in Table 5.4 for HDPE and LDPE, which are the result of the type of polymerization method.

Poly(vinyl chloride) (PVC) is also a ubiquitous polymer. One of its many variants is as a copolymer with poly(vinyl acetate) (PVA). On its own, PVC is an unstable, rigid polymer. PVA is very soft. It is no use as a moulding material, but is used as an adhesive and a substitute for starch. Randomly copolymerized, they make a material used extensively for gramophone records and floor coverings.

The polymer styrene-butadiene-styrene rubber (SBS) is an intriguing example of a block copolymer. The centre of each molecule consists of about 1000 butadiene units

$$—CH_2—CH=CH—CH_2—$$

and both ends consist of about 100 styrene units (see Table 4.5).

Polybutadiene (BR) and polystyrene (PS) are mutually insoluble, so the microstructure has a two-phase structure of PS blocks in a matrix of BR (Figure 5.63). The BR ends of the molecules are trapped in the PS phase. Below $T_g$ the material acts as a rubber and, as in a composite, the PS blocks provide strengthening and stiffening. Above the $T_g$ of the PS, the whole microstructure flows, allowing the melt to be moulded as a thermoplastic. Thus SBS is a thermoplastic elastomer. It is increasingly used to replace conventional cross-linked rubbers in shoe soles for example.

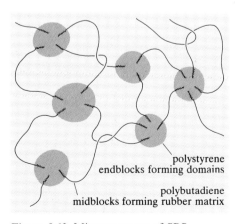

polystyrene endblocks forming domains

polybutadiene midblocks forming rubber matrix

Figure 5.63 Microstructure of SBS

SAQ 5.18  (Objective 5.12)
What sort of molecular structure would you expect in an amorphous polymer? Think of:

(a) conformation,
(b) configuration.

## 5.4.2 The microstructure of polymers

### Amorphous polymers

In a polymer melt, a typical molecule (at rest) will have the sort of random shape shown in Figure 5.58. Because of the relatively high temperature, any bonding between neighbouring side groups or molecules is insufficiently strong to prevent the C–C bonds rotating. In consequence, the molecular coil is constantly changing shape and Figure 5.58 can only be regarded as a snapshot at an instant of time.

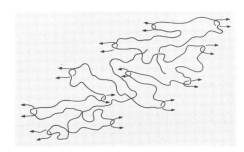

Figure 5.64 A typical molecule in a flowing melt showing how entanglements with neighbouring molecules affect its shape

One consequence of the open coil shape of the molecules is that, to achieve the molecular packing required to account for the observed densities, the outline 'spheres of influence' of neighbouring molecules must overlap. So, a molecule containing about 600 carbon atoms or more will be entangled with many neighbours, Figure 5.64. These entanglements are either physical constraints (knots) or the net effect of the secondary forces between neighbouring pieces of polymer chain at particular points (think of the spaghetti model). When the melt is made to flow, the changes of molecular shape that allow disentanglement take a certain time. The longer the molecule, the greater the number of entanglements. So we expect the viscosity of a polymer melt to increase with increasing molecular length.

The shape of a polymer molecule does not change markedly on cooling through $T_g$, so we may assume that Figures 5.58 and 5.64 are typical of an amorphous polymer below $T_g$. Since fabrication processes involve flow, they require the polymer to be fluid. The shape is fixed by cooling in an amorphous polymer, and by crosslinking in a rubber or a thermoset.

### Partially crystalline polymers

Partially crystalline polymers such as polyethylene and polypropylene generally have far more complex microstructures than either amorphous polymers or metals, and they are still not completely understood. Thus our understanding of how to manipulate the structure to improve properties such as toughness is incomplete. However, the models I use here are useful and generally accepted.

In a polymer crystal the molecular chains are in regular forms such as the planar zig-zag of polyethylene and the helix of polypropylene. Seen at a lower magnification the chains appear as straight 'rods' with the side groups distributed in a regular pattern on the outside of the rod. These rods pack together in a parallel fashion like rows of matches in a box, although with slight axial shifts and rotations to achieve the maximum packing density. The consequence of this is that the crystal bonding is highly anisotropic. It is conventional to denote the axis of the rods as the $c$ (chain) direction, so there is only continuous covalent bonding along the $c$ axis. In the other two $a$ and $b$ directions the chains are only held together by secondary forces. The anisotropy of bonding is shown for polyethylene crystal by the elastic moduli: $E_a = 8\,\text{GPa}$, $E_b = 5\,\text{GPa}$ and $E_c = 250\,\text{GPa}$.

The polymer crystals are in the form of lamellae (little plates) of thickness of the order of 10 nm, each equivalent to a row of matches in a box. In each lamella the polymer chain folds back and forth in a regular way, rather like continuous computer paper. Figure 5.65 illustrates, highly schematically, an edge-on view of a number of lamellae.

In a cooling polymer melt, at a temperature at which nucleation and growth occur, a number of such linked lamellae grow into the melt with layers of amorphous material between them. There is a similarity between this arrangement and the lamellae of $Fe_3C$ in pearlite. The rate of cooling of the polymer melt is usually too fast for a randomly coiled molecule, like Figure 5.58, to completely adopt the folded form of a single lamella. So a molecule is usually part of a number of lamellae. The misfitting parts in between the lamellae and in the folds make up the amorphous regions. In this way, the lamellae are physically interlinked.

As with other liquids that crystallize on cooling, nucleation is usually heterogeneous. It begins on foreign particles, which may be additives (such as pigments) or innoculants (such as sodium benzoate in polypropylene). The lamellae continue to grow outwards from the nucleus, branching as they go and forming a sheaf-like sphere called a **spherulite**, Figure 5.66.

The growth of a lamella is rather like dendritic growth in metals. However, dendrites grow in preferred crystal directions; the branching of lamellae is rather different. New lamellae are nucleated in the widening amorphous regions between existing lamellae. The spherulites grow, in the fashion of grains, until they contact one another on all surfaces. Figure 2.29(a) is a view, using polarized light, of the spherulite structure in a thin sheet of polyethylene. As you can see, spherulites occupy the whole of the polymer, even though it may be only, say 70%, crystalline. The crystalline and amorphous phases are intimately mixed within the spherulites.

Most methods of shaping polymers, especially thermoplastics, involve heating in order to reduce viscosity. How does the rate of cooling affect the microstructure?

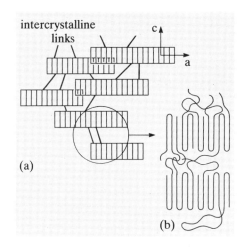

Figure 5.65 Lamellae and the amorphous fraction between

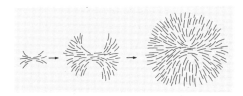

Figure 5.66 The formation of a spherulite

283

## 5.4.3 Crystallization in practice

SAQ 5.19 (Objective 5.13)
Have a go at sketching a schematic TTT diagram for the crystallization of a thermoplastic polymer. Include $T_m$ and $T_g$. Comment on the consequences of $T_g$ being above and below ambient.

The TTT diagram provides a useful context in which to point out some important characteristics of practical polymers.

With only (small) hydrogen side groups and no stereoisomers, crystallization occurs very rapidly in polyethylene. In fact, it cannot be quenched to give an amorphous state (the rate of cooling is limited by its low thermal conductivity). In comparison, polycarbonate, which has larger —$CH_3$ side groups and a stiff backbone containing benzene rings, is for practical purposes amorphous. For these two extremes the rate of cooling is not important.

The position is different for polymers which come between these extremes, such as polypropylene and some nylons. The conditions under which they are processed and subsequently cooled will affect the degree of crystallinity, and thereby properties. For instance, a particular variety of nylon can be 50–60% crystalline when slowly cooled and then annealed below $T_m$, but only 10% crystalline in rapidly cooled, thin-walled mouldings. A particular application of poly(ethylene terephthalate)(PET) is an excellent example of the importance of controlling the rate of crystallization. See ▼The PET bottle▲.

For polymers which partially crystallize, the size of the spherulites can be very important. It is sometimes necessary to have high crystallinity, to improve stiffness for example, and also a small spherulite size. The spherulite size is particularly significant for transparency. If it is larger than the wavelength of light (about 500 nm) it scatters light and causes the polymer to look opaque. This is why, for instance, polyethylene and polypropylene products are usually opaque. To achieve transparency, as in thin film for example, it is necessary to cool the material rapidly (in an air stream) immediately after shaping. In this way the spherulites are kept small because the film is cooled rapidly to below the nose of the TTT curve.

The size of spherulites can also change across a moulding. For example, a nylon injection moulding can have small spherulites at the surface and larger ones in the relative more slowly cooled centre.

By comparing with metal casting, can you think of a solution to this problem?

Use a nucleating agent. For nylon, small additions of silica or various phosphorus compounds can be used.

# ▼The PET bottle▲

This is another case of PPP *part excellence*. The development of a new product stemmed from a surplus of a material, and processing methods were used which enabled the material to meet the requirements of the product. The material is poly(ethylene terephthalate) (PET) and the product is the transparent plastics drink container, Figure 5.67. Its introduction virtually doubled the sales of some popular drinks.

Plastics can offer advantages over glass in terms of weight and toughness, but other

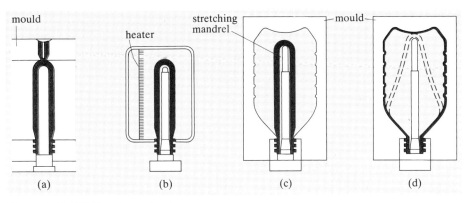

Figure 5.68 PET bottle production

Figure 5.67 PET bottles

aspects of the product design specification present considerable problems. For example, the material should have:

* high transparency (for marketing),
* impermeability to gases (fizzy drinks must not lose $CO_2$ pressure too quickly),
* creep resistance (the container must not lose its shape on the shelf).

If cooled quickly after moulding, PET is amorphous. In this condition, that is with relatively loosely packed molecules, it is clear but too permeable and has very poor creep resistance. If cooled slowly, crystallization occurs and the resulting material has adequate creep resistance and impermeability, but is opaque. The solution lies in the processing. The procedure is shown in Figure 5.68.

First, a parison is made by injection moulding (a), which is cooled rapidly in the mould to below $T_g$ (340 K). At this stage the PET is amorphous. The parison is then heated to about 400 K (b). This is sufficiently above $T_g$ to facilitate plastic flow but not so high that extensive crystallization occurs. It is then stretched in length (c) and blow-moulded (d). (Injection and blow moulding were described briefly in Chapter 1.)

The stretching and blow moulding encourage some crystallization, about 15–20%, because these processes tend to pull the molecules into line. The crystals which form are not spherulitic and are too small to affect the clarity, but they give sufficient creep resistance and impermeability to $CO_2$.

If $T_g$ is below ambient temperature, crystallization can continue when an artefact is in service. For this reason some types of nylon can exhibit shape changes for several years after manufacture unless they are annealed above $T_g$. This produces the crystallization that will 'never' be achieved in service below $T_g$.

Finally, a reminder about phases and mixtures. The crystalline and amorphous regions in a polymer are different phases. Although they have the same chemical composition, they have different physical structures, as do austenite and martensite for instance. So a partially crystalline polymer is a physical mixture. Of course, polymers can also contain phases of different chemical composition, so they may be chemical mixtures also. Many commercial plastics are. ▼Plastics in action▲ introduces a few examples.

# ▼ Plastics in action▲

It is often said that a plastic is a polymer plus all its additives. Additives are used for various purposes, including cost reduction. Table 5.5 gives the main ones added during processing to modify bulk and surface properties. I have concentrated on the big four commodity plastics, PE, PP, PVC and PS. One of the advantages of manufacturing with plastics rather than metals is that finishing processes, such as polishing and painting, are eliminated. A pigment will colour the artefact all the way through.

Notice that PVC can be manipulated in a number of ways. In fact, it couldn't be processed at all without being stabilized because during production it can degrade by losing HCl.

About 50% of PVC sold incorporates a plasticizer. Essentially plasticizers are polymer solvents. They work by insinuating their molecules between the polymer molecules, so reducing the bonding between neighbouring polymer molecules. This generates more 'free' volume and therefore leads to a reduction in $T_g$. Generally, plasticizers are organic, high-boiling-temperature liquids of low molecular mass (about 100–1000). Dibutyl pthalate is a plasticizer for PVC. However, other substances may be suitable. Water plasticizes nylon, and camphor was used to plasticize cellulose nitrate in cellulose, one of the earliest plastics.

Toughening of, say, PS is achieved by copolymerization with a rubber-forming monomer in such a way that a dispersion of rubber particles is created on cooling, rather as $CuAl_2$ and $Fe_3C$ are precipitated in aluminium and steel, respectively. The rubber particles act as crack stoppers. High impact polystyrene (HIPS) is an example. Its microstructure is illustrated in Figure 2.39(b).

Table 5.5    Property modifying additives

|  | Property requirement | Examples of additive | Typical amount by mass | Polymer |
|---|---|---|---|---|
| Bulk properties | stiffness | glass fibre | 10–30% | PP |
|  | hardness | chalk, clays | 20–50% | PVC |
|  | toughness | rubber | 20–50% | PVC, PS |
|  | softness | plasticisers | 10–50% | PVC |
| Surface properties | colour | pigments | 0.1–1% | many plastics |
|  | friction (low) | PTFE | 1% |  |
|  | anti-static | conducting waxes | 1% |  |
| Chemical properties | stability in UV light | carbon black | 0.5–2% | all plastics |

# Objectives for Chapter 5

You should now be able to do the following:

5.1  From a phase diagram, specify the equilibrium phases to be expected in particular alloys at particular temperatures and compositions. (SAQ 5.1, 5.2, 5.4)

5.2  Explain in terms of a phase diagram the changes in the phases and their compositions that occur in specified alloys cooled from the melt under equilibrium conditions. (SAQ 5.2, 5.3, 5.4, 5.5)

5.3  Account qualitatively for the equilibrium microstructures produced in particular alloys by solidification and phase transformations in the solid state. (SAQ 5.1, 5.8)

5.4  From a given phase diagram, calculate the relative amounts of the phases present at specified temperatures. (SAQ 5.5, 5.6)

5.5  Explain how cooling curves and X-ray microanalysis can be used to determine a phase diagram. (SAQ 5.7)

5.6 Describe and explain (a) the grain structure to be expected in a typical large casting, (b) the effect on grain structure of the rate of heat flow from a cooling liquid, (c) the influence of grain structure on properties. (SAQ 5.10)

5.7 Describe and explain a general approach to producing a fine-grained, homogeneous casting with low porosity. (SAQ 5.11)

5.8 Use a TTT curve and/or a transformation rate versus temperature curve to explain the microstructure produced by various heat treatments. (SAQ 5.12)

5.9 Specify and explain the main requirements for a good precipitation hardening alloy. (SAQ 5.13)

5.10 Describe and explain the types of microstructural and property changes that can be achieved in steels by heat treatment. (SAQ 5.15)

5.11 Describe and explain the basic advantages and disadvantages of solid solution alloys, eutectic alloys and other two-phase alloys. (SAQ 5.16)

5.12 Specify and explain the characteristics of a polymer molecule that facilitates crystallization. (SAQ 5.18)

5.13 Summarize the crystallization of a polymer in terms of a TTT diagram. (SAQ 5.19)

5.14 Define and use the following terms and concepts:

| | |
|---|---|
| alloy system | molecule conformation |
| atactic | martensite |
| austenite | overageing |
| bainite | pearlite |
| block copolymer | peritectic reaction |
| cementite | phase |
| chill crystals | phase diagram |
| columnar crystals | polymer lamellae |
| component | precipitation (age) hardening |
| cooling curve | random copolymer |
| coring | solid solution |
| dendrite | solidus |
| entanglements | solution and ageing temperature |
| eutectic reaction | spherulites |
| eutectoid | stereoisomer |
| ferrite | syndiotactic |
| intermediate compound | tempering |
| isotactic | tie-line |
| liquidus | TTT diagram |
| molecule configuration | X-ray microanalysis |

# Answers to exercises

**EXERCISE 5.1** Important examples are:

(a) Adding boron and phosphorus to silicon to make an electrical insulator into a semiconductor (Chapter 3).
(b) Adding oxides of calcium, lead and so on to silica to lower the viscosity of glass for processing purposes (Chapter 4).
(c) Alloying iron to produce a very fine distribution of small particles, for example Alnico (Chapter 2).
(d) Adding glass fibres to a polymer matrix to produce GFRP (Chapter 2).
(e) Fibre reinforced plastics again.

**EXERCISE 5.2** As Figure 5.4 shows, for a given amount of salt the mixture of ice and brine is stable over a wide range of temperature. For instance, if the salt content is 5%, the slush is stable from about 270 K (above which the ice phase melts) to about 250 K (below which it will freeze). In addition, of course, for ice to melt the latent heat of melting has to be provided from the surrounding (cold) atmosphere. More about this later.

**EXERCISE 5.3** From Figure 5.4:

(a) at 283 K (10 °C) alloy X will consist of brine only;
(b) at 263 K (− 10 °C) it will be a slush of solid ice and liquid brine;
(c) at 248 K (− 30 °C) it will be a mixture of solid ice and solid salt.

**EXERCISE 5.4**

(a) The weight % of carbon in $Fe_3C$ is

$$\frac{100\, n_C M_C}{n_C M_C + n_{Fe} M_{Fe}}$$

$$= \frac{100 \times 1 \times 12}{(1 \times 12) + (3 \times 56)} = 6.67\%$$

(b) The atomic % of carbon in $Fe_3C$ is given by

$$\frac{100\, m_C M_{Fe}}{m_C M_{Fe} + m_{Fe} M_C}$$

$$= \frac{100 \times 6.67 \times 56}{(6.67 \times 56) + (93.33 \times 12)}$$

$$= 25\%$$

**EXERCISE 5.5** There are three important characteristics:

1 Both components must have the same crystal structure. Copper and nickel are both face-centred cubic.
2 The size of the atoms of the components must be similar. In this way the crystal structure of the solvent is not severely distorted by the atoms of the other component. In practice, an empirical 'rule of thumb' is that the atomic diameters should not differ by more than about 15%.
3 The components must be chemically similar so that there is no tendency to react to form a chemical compound within the system. A good guide is that the components have similar valencies.

**EXERCISE 5.6** The relative atomic mass of a material, expressed in grams, contains $6 \times 10^{23}$ atoms. That is 28 g of silicon contains $6 \times 10^{23}$ atoms. A cubic metre of silicon has a mass of $2.33 \times 10^6$ g, which is

$$\left(\frac{2.33 \times 10^6}{28}\right) \times 6 \times 10^{23} \text{ atoms}$$

$$= 5 \times 10^{28} \text{ atoms m}^{-3}$$

**EXERCISE 5.7**

(a) The alloy enters a two-phase region of α and β. In other words, the solvus is the limit of solid solubility of $Al_2O_3$ in $ZrO_2$. Once the temperature falls below H, the $ZrO_2$ becomes supersaturated with $Al_2O_3$. So crystals of α must begin to form in the β microstructure. Because the solubility of $Al_2O_3$ in $ZrO_2$ falls with decreasing temperature, more α must be formed as cooling proceeds.
(b) At 2000 K the alloy will consist of β phase containing about 98% $ZrO_2$ and α phase containing about 1% $ZrO_2$. The predominant phase is β. The proportion of α phase will be:

$$\frac{98 - 95}{98 - 1} = \frac{3}{97} = 3.1\%$$

Note that this, and similar calculations, give a proportion by mass. What is visible in a microstructure is a volume fraction.

**EXERCISE 5.8**

(a) The relative proportion of β phase in alloy 2 just before the eutectic reaction is given by

$$\frac{JE}{GE} = \frac{75 - 42}{92 - 42} = \frac{33}{50} = 66\%$$

(b) Immediately after the eutectic reaction, the relative proportion of β phase in alloy 2 is given by

$$\frac{JF}{GF} = \frac{75 - 3}{92 - 3} = \frac{72}{89} = 81\%$$

Remember that this is shared between primary β and β in the eutectic.

**EXERCISE 5.9**

(a) The labelled Fe–C diagram is given in Figure 5.69. A useful tip here is that single-phase regions are separated by two-phase regions. So if you draw an isotherm across a whole system (avoiding reaction temperatures) it passes through alternate single and two phase fields. Try this in Figure 5.31(a).

(b) The Fe–C system shown consists of a peritectic at 1760 K (which produces the γ solid solution), a eutectic at 1403 K and a eutectoid at 996 K. You can now see how the eutectic portion in Figure 5.25(a) and the eutectoid part in Figure 5.23 fit into the whole. Cast irons are based on the eutectic and steels on the eutectoid.

The Cu–Zn diagram looks complicated but is essentially a solid solution α and a sequence of five peritectics. The peritectics produce the five intermediate compounds β, γ, δ, ε, η. The δ phase undergoes a eutectoid reaction at about 830 K.

**EXERCISE 5.10**

(a) Gradually the cored solid solution would become more homogeneous as solute atoms diffuse down the

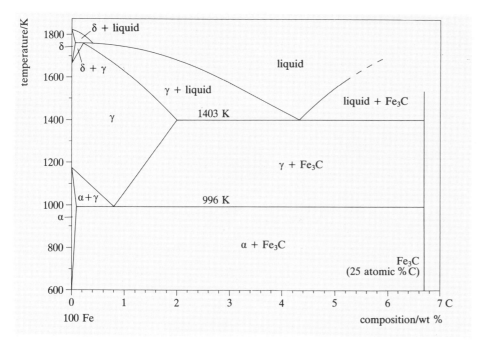

Figure 5.69 Fe–C phase diagram

concentration gradient from the solute-rich to the solute-poor regions. More about this in Section 5.3.1.

(b) The interface between a particle and the matrix is a region of disorder. The resulting higher internal energy of the alloy will be lowered if some of the particles grow at the expense of others, thus reducing the total interfacial area. More about this in Section 5.3.3.

(c) As with (b), the energy of the alloy is reduced if the total grain boundary area is reduced. To achieve this reduction, grains attempt to form three-dimensional arrangements analogous to those found in stable soap bubble foam: the boundaries try to form flat planes which meet at 120°. More about this in Chapter 6.

(d) Again, energy is reduced if the dislocations join to form networks that reduce the total length of disolocation. (Dislocations can also annihilate one another, as you will see in Chapter 6).

EXERCISE 5.11  As you know from Chapter 4, $T_g$ is the temperature below which the atoms or molecules in amorphous solid effectively become immobile. So the idea of choosing a eutectic composition is to minimize the gap between $T_m$ and $T_g$. Since $T_m$ is low, such an alloy has only to be cooled rapidly through a narrow temperature range.

EXERCISE 5.12  The white regions are ferrite ($\alpha$) in Figure 5.48(a) and cementite ($Fe_3C$) in Figure 5.48(b). If you don't follow this answer, study again Section 5.2.3.

EXERCISE 5.13  The microstructure expected for austenite annealed at 1100 K is illustrated, schematically, in Figure 5.70(a). It is a single phase polycrystalline material. Immediately after quenching to

850 K the microstructure will still be like that of Figure 5.70(a).

Ten seconds after quenching to 850 K about 75% of the austenite will have transformed to pearlite, as indicated by ● on the TTT diagram of Figure 5.52. So the microstructure would resemble that of Figure 5.70(b).

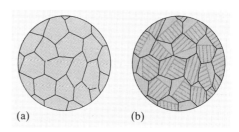

Figure 5.70 Schematic microstructure of (a) austenite and (b) 25% austenite–75% pearlite

# Answers to self-assessment questions

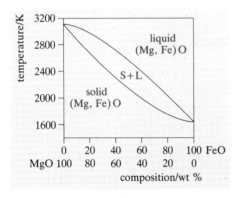

Figure 5.71 MgO–FeO phase diagram

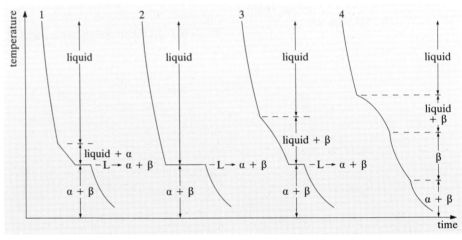

Figure 5.72 Cooling curves for alloys 1–4

**SAQ 5.1** The equilibrium phases present are:
(a) A and B.
(b) A.
(c) A and C.
(d) A, B and C.

The point $T_2X_2$ is rather special. It is where a region of one phase meets three regions each of which has two phases in equilibrium. At this point phase A changes into phases B and C. Whilst these changes are occurring, three phases co-exist.

**SAQ 5.2** The solidification of alloy 3 is analogous to that of alloy 2 except that component B is involved rather than A. On cooling to the liquidus ED (Figure 5.6), some crystals of solid B form. As cooling proceeds through the liquid + B region, the crystals of B grow. At the eutectic temperature the remaining liquid has a composition E. As with alloy 1, it solidifies as a mixture of alternating regions of solid B and A (microstructure 7).

On reheating, the alloy will start to melt when it reaches the eutectic temperature. The eutectic reaction occurs:

crystals of A + crystals of B

→ liquid of composition E

**SAQ 5.3**
(a) The first solid forms at 1495 K and contains 42% Ni.

(b) At 1495 K the liquid contains 26% Ni.
(c) The last solid forms at 1420 K and contains 26% Ni.
(d) The last drop of liquid to solidify contains 11% Ni.

**SAQ 5.4** In the eutectic system of Figure 5.6 there is no solid solubility, so the first solid to form in alloy 2 has to be *pure* A. In alloy 1 in Figure 5.12 there is solid solubility and so the first solid to form is a *solid solution*.

**SAQ 5.5**
(a) There are two phases in the system: a liquid solution and a solid solution of MgO and FeO. The phase field in-between is a mixture of saturated liquid and solid solutions.
(b) The labelled diagram is given in Figure 5.71.
(c) The alloy contains 40 wt % FeO. This composition crosses the solidus at 2300 K.
(d) The alloy contains 60 wt % FeO. This composition divides the tie-line at 2300 K roughly into $\frac{1}{4}$ and $\frac{3}{4}$. The liquid is in the majority and so the alloy is about a quarter solid.

**SAQ 5.6**
(a) The percentage of $\alpha$ phase immediately below the eutectic temperature is 19%. This follows from the conclusion that the % of $\beta$ phase is 81% (Exercise 5.8). At 2000 K, the percentage of $\alpha$ present in

equilibrium (in terms of wt % $ZrO_2$) is given by

$$\frac{98 - 75}{98 - 1} = \frac{23}{97} = 23.7\%$$

So, the increase in the amount of $\alpha$ phase precipitated is 23.7% − 19% = 4.7%.

(b) The primary $\alpha$ is that formed as $\alpha$ solid solution during solidification. The maximum amount of primary $\alpha$ is that in equilibrium with liquid immediately prior to the eutectic temperature. At this temperature the fraction of $\alpha$ phase is given by

$$\frac{(42 - 20)}{(42 - 3)} = \frac{22}{39} = 56.4\%$$

**SAQ 5.7**
(a) The cooling curves for alloys 1–4 are shown in Figure 5.72 together with the temperature ranges over which particular phases are stable.
(b) Two features of the cooling curve for alloy 4 are distinctive. First, the composition of alloy 4 means that it does not undergo the eutectic reaction. Hence, unlike the other alloys, it does not exhibit an arrest at the eutectic temperature. Second, it crosses the $\beta$ solvus and so has an arrest as some $\beta$ phase transforms to $\alpha$.

**SAQ 5.8**
(a) Yes. The $\beta$ phase appears when the temperature falls to the liquidus at 1900 K.

The compositions of the liquid and β follow the liquidus and solidus respectively.

(b) The proportion of β phase just above the peritectic temperature is

$$\frac{70 - 31}{86 - 31} = \frac{39}{55} = 70.7\% \ \beta$$

$$\approx 71\% \ \beta$$

(c) The peritectic reaction is liquid + β → γ. All alloys between 31% and 86% B will undergo the same reaction. In alloy 2, just above the peritectic there is much more β present than liquid. As a result, during the peritectic reaction all the liquid is consumed in the production of γ and some (unreacted) β is left.

(d) Just below the peritectic temperature γ is in equilibrium with β. The proportion of β phase present is

$$\frac{70 - 52}{86 - 52} = \frac{18}{34} = 52.9\% \ \beta$$

$$\approx 53\% \ \beta$$

This is less than above the peritectic and confirms that at the peritectic some β combined with liquid to form γ.

(e) The expected microstructures are illustrated in Figure 5.73.

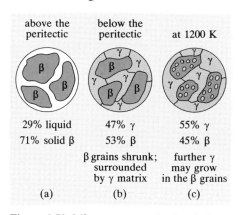

| above the peritectic | below the peritectic | at 1200 K |
|---|---|---|
| 29% liquid 71% solid β | 47% γ 53% β | 55% γ 45% β |
| | β grains shrunk; surrounded by γ matrix | further γ may grow in the β grains |
| (a) | (b) | (c) |

Figure 5.73 Microstructures in the A–B peritectic system

SAQ 5.9
(a) Undercooling is concerned with the nucleation of the first solid in the melt. A decrease in the temperature of the melt below the equilibrium melting temperature is necessary in order to provide the change in free energy ($\Delta G_c$) needed to establish a stable solid nucleus.

(b) The plateau is related to the growth of solid nuclei. In this region the heat lost by the molten metal to its cooler surroundings is offset by the latent heat released as the liquid solidifies.

If you are unsure of these points, study again Section 4.7.5.

SAQ 5.10 The configuration of Figure 5.39 is designed to produce a strongly directional flow of heat through the bottom of the lower section of the mould. So solidification in the lower section is dominated by the growth of columnar crystals, possibly preceded by the formation of some chill crystals on the cooled surface. Now, when solidification in this section is complete, it proceeds along the 'pigtail'. It is clearly possible that only one of the columnar crystals will continue to grow into the pigtail. In this event, a single crystal will emerge from the pigtail and gradually grow throughout the entire upper section of the mould. This way of producing 'single crystal castings' is a modification of investment casting (see Section 1.5.1). It is used in the manufacture of a relatively new generation of nickel-alloy turbine blades for jet engines.

SAQ 5.11 The casting is likely to be fine grained. Casting into a metal mould means that a thin casting will solidify very quickly. Chill crystals should be formed throughout.

The casting will be cored. Annealing 0.6 $T_m$ should allow diffusion to occur sufficiently rapidly for coring to be removed. The alloy should then be homogeneous.

With liquid poured directly into the mould (not via a runner) the casting will contain obvious shrinkage porosity in the form of a hollow at the top and probably other holes down the centre-line (the last part to solidify).

SAQ 5.12
(a) Figure 5.41(b). When the alloy is slowly cooled at temperatures just below $T_t$ the undercooling is low and the rate of nucleation is low. After a long delay, β phase is nucleated at a few sites and grows rapidly into large particles.

Figure 5.41(c). If cooling is quicker, the undercooling rapidly increases and so

therefore, does nucleation rate. This leads to a fine dispersion of small particles.

Figure 5.41(d). Here the alloy is cooled to low temperatures so quickly that growth of any nuclei is prohibited. In fact, on cooling, the alloy misses the nose of the TTT diagram of Figure 5.42.

(b) Having missed the nose of the TTT curve, the alloy remains a solid solution at ambient temperature. It is clearly a non-equilibrium structure because it contains far more than the equilibrium concentration of solute. The driving force to precipitate this as β phase is large, but diffusion is negligible. To encourage precipitation, some rattle is required. At $T_3$, well below the nose of the TTT curve, nucleation rate is very high and sufficient diffusion can occur albeit relatively slowly. Consequently, β phase is precipitated as an extremely fine distribution of particles. The typical microstructure to be expected at time $X$ is illustrated in Figure 5.74(a).

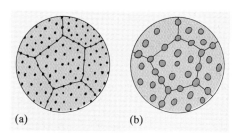

(a)          (b)

Figure 5.74

At much longer times such as time $Y$ in Figure 5.42, the alloy moves closer to equilibrium by agglomeration of the fine β particles, Figure 5.74(b). Eventually there are relatively few, but relatively large, precipitates and the structure is not dissimilar from that for the slowly-cooled alloy, Figure 5.41(b).

SAQ 5.13
(a) For a good alloy, a solid solution is required in which there will be a high degree of supersaturation on quenching to ambient temperature. A large amount of second phase is then precipitated during ageing, which enables a distribution of precipitates to be obtained in which the distance between precipitates is small. For high supersaturation, the solid solubility at the solution treatment temperature should be high and decrease to a low value at ambient temperature.

(b) The solution treatment temperature must be sufficiently high for the alloy to be homogenized in a reasonable time.

If the alloy is for use at ambient temperatures, the required ageing temperature must be sufficiently above ambient to preclude significant ageing (and therefore softening). For use at a high temperature, the alloy must have an ageing temperature much higher than the service temperature, otherwise it could overage rapidly in service.

### SAQ 5.14

(a) The reaction is a eutectoid.
(b) A eutectoid only differs from a eutectic in that it occurs entirely in the solid state.
(c) On cooling, austenite transforms to a mixture of ferrite and cementite, that is pearlite.

### SAQ 5.15

Slow cooling in a furnace produces relatively coarse pearlite; the lamellae of cementite ($Fe_3C$) and ferrite ($\alpha$) are quite thick and widely spaced. It has a moderate hardness, about $220\,H_V$, see Figure 5.47(c).

Quenching to ambient temperature fast enough to miss the nose of the TTT diagram (Figure 5.52) means that the martensitic reaction will not go to completion. The alloy will consist of martensite together with some supersaturated austenite. The martensite is very hard (roughly $800\,H_V$). The austenite will be substantially harder than pure iron because it is supersaturated with carbon.

Quenching and tempering leads to the precipitation of iron carbides. The resulting hardening offsets to some extent the drop in hardness of the martensite. The actual hardness will depend on the temperature and time of tempering. The sample of Figure 5.2(c) has a hardness of $350\,H_V$ for instance. When the precipitates coarsen the alloy overages and softens, see Figure 5.2(d).

Cooling at different intermediate rates provides a structure varying from coarse pearlite to lower bainite with an associated increase in hardness from about $250\,H_V$ to $600\,H_V$. See Figure 5.52.

### SAQ 5.16

(a) Compared with the pure host metal, solid solutions have increased strength, depending on the concentration of solute, whilst still retaining useful ductility. The amount of strengthening is limited.
(b) In a particular materials system, the eutectic has the advantage as a casting alloy over all the other *alloys* because of its high fluidity, and over the pure components because of its low melting temperature. Strength can be relatively high, especially with rapid cooling to encourage a fine distribution of the two phases, but ductility is fairly low.
(c) Two-phase alloys (other than eutectics) give far more scope than solid solutions for manipulating the microstructure to alter properties, notably strength, ductility and toughness. In general, for a given amount of second phase, finer distribution gives higher strength, lower the ductility and, hence, lower toughness. Figure 5.2 gives an idea of the range of property values that can be achieved.

### SAQ 5.17

(a) Short molecules. An analogy with cooked, buttered spaghetti is useful. A pile of long spaghetti is a random arrangement. Cut it into short lengths and they can be aligned readily, as is appreciated by many people learning how to eat it! For the same reason, candle wax is more highly crystalline than polyethylene, even though both have the same chemical nature.
(b) Linear molecules. Molecules that are linear and without side branches can pack in an orderly fashion more readily than those with branches.
(c) Smooth molecules. The bigger the side-groups (knobs) on the molecule, the less easily they can align.
(d) Regular molecules. Crystallinity is aided by any side-groups being regularly distributed along and around the molecule.
(e) Slippery molecules. The stronger the bonding *between* molecules, the greater the tendency for them to adjust their relative positions and align.
(f) Stiff molecules. Spaghetti again. Uncooked pieces align easily and won't get tangled up in doing so.

### SAQ 5.18

(a) Conformation. The more irregular the conformation of a molecule (that is, the more irregular the shape of shape of the molecular backbone), the less likely it is to be able to pack closely with other molecules to form a crystal. Irregularity is produced by random switching between trans (t), gauche plus ($g^+$) and gauche minus ($g^-$) positions at each carbon atom in the polymer chain.
(b) Configuration. The factors that encourage the production of an amorphous polymer are the antitheses of those that facilitate crystallinity. So the molecule should have a highly irregular configuration; that is, lots of side branches, atactic repeat units and randomly copolymerized repeat units. Such configurations are a cause of the irregular conformations in (a).

### SAQ 5.19

Since the times to reach equilibrium are extremely long at any temperature, the TTT diagram can only be very schematic. See Figure 5.75. Compared with the TTT diagrams you have met so far, the curves are moved a long way to the right because crystallization in polymers is sluggish. The irregular structure of polymer molecules, due to their branching, atacticity, large side groups and so on, means that arrangement into ordered arrays needs time. For some polymers, the 'nose' of the TTT curve can be missed comfortably. This is why polymers such as polycarbonate are commonly amorphous. The action ceases at $T_g$. Below this temperature an amorphous polymer is to all intents and purposes stable. Below $T_g$, bond rotation ceases and so rearrangement becomes impossible.

Quenching to produce an amorphous polymer is possible if crystallization is sluggish and $T_g$ is above ambient.

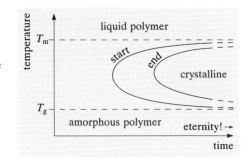

Figure 5.75 A possible TTT diagram for the solidification of a polymer

# Chapter 6 Mechanical properties for processing and use

## 6.1 Introduction

In the last three chapters the emphasis has been on the structure of materials from the atomic scale upwards, how structure determines properties and how temperature can be used as a means of manipulating structure to give desired properties. The emphasis now moves to properties — mechanical properties in this chapter and chemical properties in the next. Both chapters reinforce the links between product, process and properties because the properties with which we are concerned are important not only in the performance of a product in service, but also in the process route used for the manufacture of the product. For instance:

• In thermoplastics, high rigidity is usually sought after in the product, but it is not at all desirable in forming processes.

• In ceramics, the useful service properties of hardness and temperature resistance are the root cause of processing problems.

• In metals, a high yield stress is usually needed in service, but metals are more readily worked into shape when they are ductile with a low yield stress.

Notice that in these examples the mechanical property requirements for service actually conflict with those for processing. Our PPP relationships are becoming more complex! Let's unravel them a bit with the aid of Figure 6.1.

In solely physical terms a finished product (or its components) comprises a shape and a microstructure. You will well appreciate by now that a microstructure is concomitant with a profile of properties. These materials properties in combination with the shape of the product provide the required product properties. For instance, the stiffness of a plastic knife depends on the knife's shape and on the elastic modulus of the chosen plastic.

Both the shape and the microstructure of a product are determined by how the material is processed. In very simple terms, we can view the whole transformation of raw material into products as a sequence of materials processing which at each stage results in a particular shape and a particular microstructure (hence property profile). The primary goal is to provide the required shape and microstructure for the product. The secondary goal is a processing route which facilitates these ends whilst ensuring ease of production.

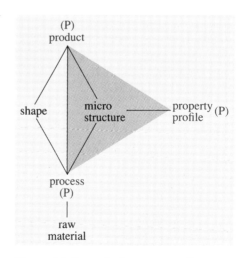

Figure 6.1 Some further aspects of PPP

The conflict between mechanical properties for processing and for service stems from the role of stress. On the one hand, stress is experienced by materials in service and, on the other, stress is the agent in many shaping processes — rolling, blow moulding, machining and so on. Often the same mechanical property is involved in both stages; ductility, elastic modulus, fracture toughness for instance. It is not surprising, therefore that the same mechanisms of flow and fracture are often fundamental to creating a shape (processing) and also to destroying that shape (failure in service). In this chapter we explore some of these conflicts and the possibilities for resolving them. But first I want to return to the idea of a product being the combination of a shape and a property profile, because it provides some useful insights into the relative importance of processing and properties in different types of product.

# 6.2 Products as shapes and microstructures

Thinking specifically about the materials aspects of a product, all products will be one of the following:

- microstructure dominated
- shape dominated
- dominated neither by shape nor by microstructure.

Let's consider each in turn.

## 6.2.1 Microstructure dominated products

Some products are only possible because of the very special properties of certain materials. One example is the tungsten filament in an incandescent light bulb, which you met in Chapter 1. Other examples are the platinum meshes used as catalysts in many chemical process plants and the recently discovered superconducting oxide ceramics. In order to capitalize on this particular part of a property profile, ways have to be found of coping with the rest of the profile and, in particular, with producing the material in an acceptable shape. Often these 'imposed' properties and limited shape possibilities do not lend themselves to straightforward production.

Taking the incandescent light bulb as an example, the useful part of the property profile of tungsten includes its electrical resistivity and high melting temperature, but some other properties present problems. The high melting temperature precludes casting processes and shaping of the solid is made difficult by the lack of ductility. Rapid oxidation in air at its operating temperature means that the familiar glass envelope has to contain an inert gas — another complication in the manufacturing process.

## 6.2.2 Shape dominated products

Sometimes the shape of the product is paramount. The shape may be controlled by aerodynamics or fluid dynamics, or by mechanical function as in gears, screw threads and scissors. The basic shape of scissors is fixed by their mechanical function, but the materials they are made of can be surprisingly varied. Those in Figure 6.2 have cutting blades of very fine grain alumina bonded to a glass reinforced nylon frame. In such cases the strategy is to fix the important shape details and then select the material profile and processing method to suit the technical and financial constraints.

Figure 6.2 Scissors made from alumina and glass reinforced nylon

This is illustrated by comparing another two products — the control flaps of the wing of a Boeing 757 airliner and a particular mass produced squash racket. Both are made from carbon fibre reinforced plastic (CFRP) and both are shape dominated; the flaps by the required aerodynamic performance and the racket by the rules of the game.

CFRP is used in the wing because it is light and if the fibres are long and aligned parallel to the directions of maximum stress, it has excellent strength, stiffness and resistance to fatigue (fracture under cyclic loading). At present, it is made by carefully arranging woven carbon fibre cloth and strands in moulds, and introducing thermosetting or thermoplastic polymer around the fibres. Another way of making it is to pre-impregnate the fibres with a thermosetting 'dough', which is then compressed in heated dies to trigger polymerization. Both of these techniques are laborious, expensive and without any cost savings as batch size increases; these costs are obviously acceptable in this 'strategic' product. (Figure 6.3 shows the product.)

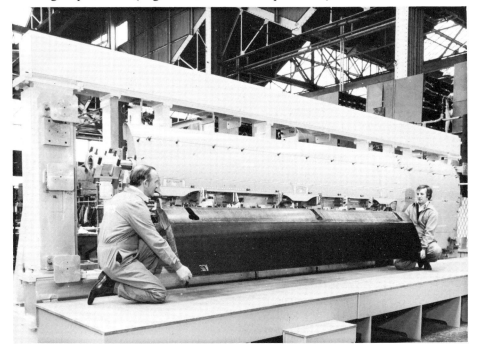

Figure 6.3 CFRP control flaps being removed from a mould (courtesy of Shorts, Belfast)

At the other end of the spectrum is the mass produced squash racket (Figure 6.4). In this case the shape is achieved by injection moulding a mixture of thermoplastic (nylon 6,6) and short reinforcing carbon fibres. With long fibres, the polymer becomes too viscous for injection moulding to be used. The resulting more random arrangement of fibres gives a less efficient structure than fully aligned carbon fibres running parallel to the rim of the racket. However, after the initial capital, tooling and development costs have been recovered, the process becomes relatively inexpensive to operate.

If the control flap were made by injection moulding like the tennis racket, many thousands would be needed before the tooling costs would be recovered. And the flap would have to be much heavier to counteract the reduction in structural strength due to the random fibres. Since aircraft control flaps are only needed in small numbers, injection moulding is unlikely to be adopted.

## 6.2.3 Products without dominant microstructure or shape

With many engineering components the exact shape is not critical. Although some dimensions or surface textures may be crucial, other aspects of a design specification can usually be varied to allow a variety of materials (properties) to be considered, making it easier to optimize material choice and manufacturing cost. Examples are car bodies, bridges and knives. There are also many products for which the properties are not crucial, and for which a wide variety of materials (microstructures) are acceptable as long as the combination of shape and properties make the product fit for purpose. Cups, gates, window frames and containers of all sorts spring to mind.

Figure 6.4 Mass produced squash racket (courtesy of Dunlop Slazenger International Limited)

With such products, which do not operate close to any geometric or microstructural design limits, the quantities needed and therefore the type of production system become important in determining the microstructures, detailed shapes and processes used. Figure 6.5 shows various differently shaped components all designed for use as refrigerator compressor connecting rods, but each optimized for different production systems, and made with different materials. In such situations the final decision on the material and manufacturing route may be strongly influenced by other factors. For instance:

(a) The use of expensive equipment and materials may give high production rates which offset high overhead and labour costs.

(b) The use of bought-in standard parts may be cheaper.

(c) Conservative practices may be adopted because experience and previous investment in particular materials and processes win out over a new and better material or process.

Figure 6.5 Refrigerator connecting rods (1) 1-piece cast iron (2) 2-piece die cast Al alloy (3) 1-piece die cast Al alloy (4) furnace brazed sheet steel with cast iron slider (5) ball-ended steel rod from resistance welded stock items (6) steel plate (7) bulge-formed and resistance welded steel tube

SAQ 6.1   (Objective 6.1)
Is each of the following products shape dominated, microstructure dominated or neither?

(a)  gas-turbine blade
(b)  domestic waste bin
(c)  spectacle lenses
(d)  elastic band
(e)  nuclear fuel element
(f)  key

Now, let's return to mechanical properties.

# 6.3 The mechanical behaviour of solids

This is about how a material responds to stress. From earlier chapters you will know that an important means of assessing mechanical characteristics is the stress–strain curve obtained from a tensile test (Figure 6.6).

## SAQ 6.2 (Revision)

Consider Figure 6.6 and assume that the sample fractures at B.

(a) At A, is the elastic strain in the sample OE or OA?
(b) Is the elastic strain in the sample at fracture DB, OD, or DC?
(c) Is the plastic strain in the sample at fracture DC, EC, OD or OC?
(d) Is Young's modulus given by (EA/OE), (EA/OA) or (CB/DC)?

But by no means does a stress–strain curve tell all, especially for materials tested under conditions where their response to stress is time dependent. For instance, the material portrayed in Figure 6.6 obeys Hooke's Law: on removing the stress in the elastic region, the material returns instantly to its original shape. For such materials Young's modulus can be unambiguously defined and determined. Now the design of a variety of products is determined by the Young's modulus of the material used, and for Hookean solids like this selecting the appropriate material is fairly straightforward; ▼The material for a beam▲ provides a simple illustration.

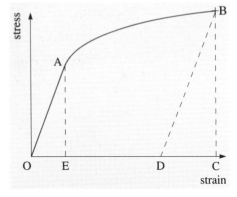

Figure 6.6 A typical stress–strain curve for a metal

# ▼The material for a beam▲

Beams form components in all sorts of products. They are usually required to be stiff, that is not deflect by more than a certain amount; and in many applications (aeroplanes for example) weight saving is also an important element in the design specification.

How do we choose a material of minimum weight for a given stiffness? Consider the simple example, depicted in Figure 6.7, of a solid beam supported at its ends and carrying a force $F$ at its midpoint.

From your previous studies you will know that for the beam configuration in Figure 6.7 the deflection $\Delta$ at the midpoint of the beam, length $l$, produced by a force $F$ is:

$$\Delta = \frac{Fl^3}{48EI} \qquad (6.1)$$

where $E$ is Young's modulus and $I$ is the second moment of area. Since for a square sectioned beam of thickness $t$:

$$I = \frac{t^4}{12}$$

$$\Delta = \frac{Fl^3}{4Et^4} \qquad (6.2)$$

Now the mass of the beam is

$$m = lt^2\rho \quad \text{or} \quad t = \left(\frac{m}{l\rho}\right)^{1/2}$$

where $\rho$ is the density of the material. Substituting for $t$ in equation (6.2):

$$\Delta = \frac{Fl^3 l^2 \rho^2}{4Em^2}$$

from which the mass is

$$m = \left(\frac{F}{\Delta}\frac{l^5}{4}\right)^{1/2}\left(\frac{\rho^2}{E}\right)^{1/2}$$

So, for a given stiffness $F/\Delta$, the lightest beam of fixed dimensions is that with the smallest value of $(\rho^2/E)^{1/2}$, or $(\rho/E^{1/2})$. This is another example of a merit index; you met others in Chapter 1.

EXERCISE 6.1 Using the data in Table 6.1, which material offers the best solution?

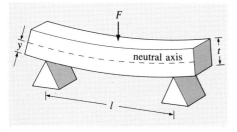

Figure 6.7 A beam of square section

Table 6.1

| Material | $\dfrac{E}{\text{GPa}}$ | $\dfrac{\rho}{\text{kg m}^{-3}}$ |
|---|---|---|
| wood (spruce along the grain) | 10 | 600 |
| CFRP | 120 | 1500 |
| aluminium | 70 | 2700 |

Figure 6.8 Creep in old lead plumbing

On the other hand, in some materials, particularly polymers above $T_g$, elastic strain is not instantly reversible. Think about crumpling a plastic bag (a crisp packet for instance) and then releasing it; it unravels quickly at first and then progressively more slowly. This time-dependent behaviour is called **viscoelasticity**. Furthermore, under a constant stress such a material continues to deform. In such cases the elastic modulus cannot be defined unambiguously, and so a plastics beam could sag more and more under load. 'The material for a beam' could have discussed plastics like these, but the same theories do not apply; we consider how to take them into account in due course.

Another important time-dependent property occurs in plastics, metals and ceramics. Under a given stress they continue to deform plastically. The process is called **creep**. In crystalline materials it becomes important at temperatures above $0.4\,T_m$. Lead (melting temperature 600 K) creeps at ambient temperatures as Figure 6.8 demonstrates rather effectively. The reason why many glaciers move is that ice creeps. Even at 243 K ($-30°C$), ice is at about $0.9\,T_m$.

In metals and ceramics, elastic and plastic strain are easily distinguished (as SAQ 6.2 shows); in polymers both types of strain are time-dependent and the difference is far less discernible. Figure 6.9 illustrates one way of classifying the mechanical characteristics of materials. It covers the main types of strain response to an imposed stress.

Two points to notice about Figure 6.9:
• It is idealized and schematic. In real materials, especially in plastics, the demarcation between different types of behaviour is not sharp.
• Another type of response to an imposed stress — fracture — can occur at any point in the strain responses depicted.

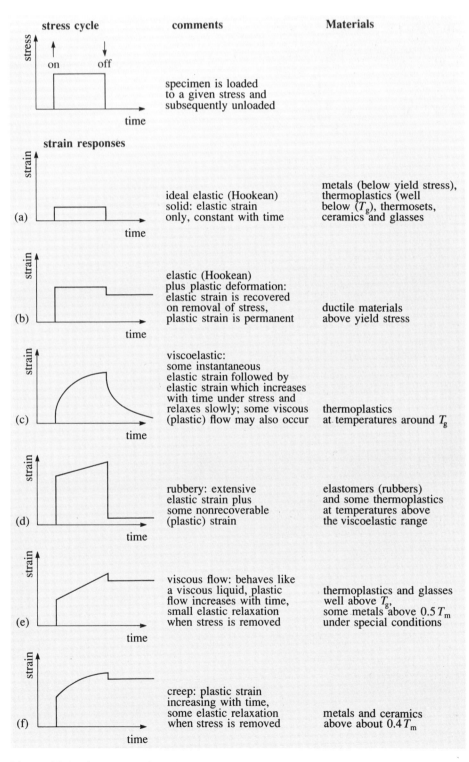

Figure 6.9 Various types of strain response to an imposed stress

We can conveniently group the responses in Figure 6.9 under two headings: **recoverable strains** and **nonrecoverable strains**.

The extent to which recoverable and nonrecoverable strains occur in materials under stress depends on their structure, their temperature and how the stress is applied. So, predicting the response of material to stress during manufacture and service is complex. Fortunately, for practical purposes not all responses are relevant to all materials. Consequently, as the third column in Figure 6.9 indicates, we need only concentrate on particular phenomena in different materials. Generally:

• In thermoplastics and elastomers, time-dependent recoverable strains are very important; they are fundamental to the performance these materials can achieve in service and are of significance in their processability.

• Ceramics are brittle; their viscoelastic and plastic regimes are negligible. An important consequence of their brittleness is that the standard methods of shaping products from large lumps of solid, forging and rolling for example, are not suitable; hence the powder route involving pressing and sintering introduced in Chapter 1. However, the situation is not quite black and white, as ▼**Stress systems and some consequences**▲ indicates.

• In metals and thermoplastics, plastic flow is extremely important. Familiar processes for shaping products, such as drawing, rolling and injection moulding depend for their feasibility on a significant (plastic) viscous regime. Whether such a regime exists depends on the propensity for plastic flow compared with that for fracture. ▼**To flow or to crack?**▲ addresses this.

We will tackle recoverable and nonrecoverable strains in turn.

# 6.4  Elasticity in atomic solids

For atomic solids — that is, solids composed of atoms held together by primary bonds — there is no need to develop the modelling of the elastic modulus beyond that covered in Chapter 3. You will remember that this was considered in terms of the force–separation curve for two bonded atoms.

SAQ 6.3  (Revision)
Given the force–separation curve of Figure 6.10:
(a) Over what range of the curve is Hooke's law obeyed?
(b) What differences in the shape of the curve would you expect for solids bonded by *i* primary and *ii* secondary forces?

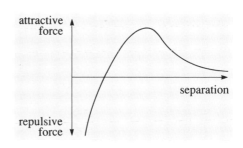
Figure 6.10 A force–separation curve

'The materials for a beam' gives an introductory idea of how Young's modulus features as a mechanical property for service. As far as processing is concerned it is of little consequence except, perhaps, for the phenomenon of **springback**. As you know, if during a tensile test the

# ▼Stress systems and some consequences▲

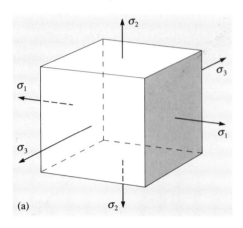
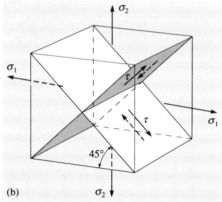
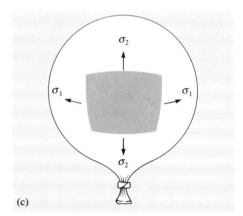

Figure 6.11 (a) The three principal stresses; (b) shear stresses created; (c) biaxial tension

First, a word about stress systems. In a tensile (or compressive) test the specimen is subjected to a very simple stress system — a uniaxial stress along the major axis. Now the shape of many products is such that during processing and in service they experience multi-axial stresses.

In general, any applied stress system can be considered in terms of three principal stresses acting in mutually perpendicular directions, Figure 6.11(a); these stresses are shown as tensile but they could be compressive.

Whenever any pair of principal stresses is unequal, there is a **shear stress**, which is a maximum on the two planes at 45° to the principal stresses, as shown in Figure 6.11(b). The magnitude of this maximum shear stress $\tau$ is half the difference between the principal stresses $\sigma_1$ and $\sigma_2$:

$$\tau = \frac{\sigma_1 - \sigma_2}{2}$$

Particular stress systems can now be described by the relative signs and magnitudes of each principal stress. Important examples are:

## Uniaxial tension/compression
Here $\sigma_1$ say, is the applied stress and $\sigma_2 = \sigma_3 = 0$. This is what happens in a tensile test and, for example, in ropes and in structural columns in buildings. So $\tau = \sigma_1/2$; the shear stress is half the applied stress. This is the conclusion reached in 'Shear deformation' in Chapter 2.

## Biaxial tension
How would you describe this in terms of the principal stresses?

Here only $\sigma_3$ (say) is zero. There is a special case called balanced bi-axial tension when two principal stresses are tensile and equal, the third is zero: say $\sigma_1 = \sigma_2$ and $\sigma_3 = 0$. Biaxial tension is important in shell-shaped products with a pressure difference between the inside and outside, for example gas cylinders and balloons, see Figure 6.11(c).

## Triaxial tension/compression
Here none of $\sigma_1$, $\sigma_2$ or $\sigma_3$ is zero. There is a special case when $\sigma_1 = \sigma_2 = \sigma_3$, which is usually called pure hydrostatic tension (or compression). Here shear stresses are not generated and so plastic shape changes are not possible. $\sigma_1 = \sigma_2 = \sigma_3$. Triaxial tension is extremely important because it occurs at the tips of cracks in solids under stress. Triaxial compression is becoming increasingly useful in the shaping of ceramics and other brittle material from the solid.

Can you see why?

If the principal stresses are unbalanced, shear forces can be generated without the use of tensile stress (which might nucleate and propagate cracks).

This is what happens in nature when rocks are extruded into crevices under extremely high pressures. These conditions can be created in hydrostatic extrusion.

As a process, extrusion was introduced in Chapter 1. Figure 6.12 illustrates typical equipment for conventional and hydrostatic extrusion. In the latter the ram is replaced by a fluid which generates hydrostatic pressure. However industrial application of this process is very expensive because the pressures needed may exceed 1.5 GPa and the production cycle time is very long.

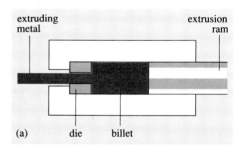
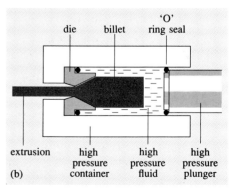

Figure 6.12 (a) Conventional extrusion; (b) hydrostatic extrusion

# ▼To flow or to crack?▲

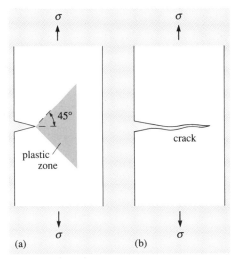

Figure 6.13 (a) Yielding and (b) fracture at a crack tip

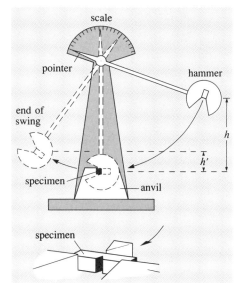

Figure 6.14 The Charpy impact test

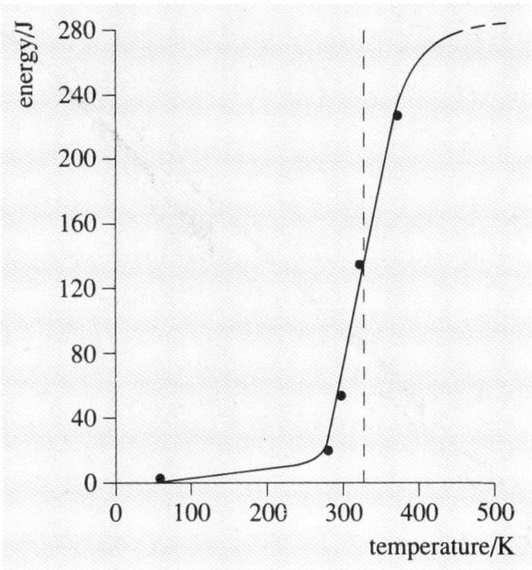

Figure 6.15 Ductile–brittle transition in a low carbon steel

Under an increasing tensile stress a material will, at first, deform in a recoverable way. Then at some point either it will yield, that is it will undergo plastic flow, or it will break, that is a crack will propagate through it. Which occurs all depends on what happens at stress concentrations such as crack tips. Figure 6.13 illustrates the two possibilities.

How do different types of material behave? Well, some unusual materials first; FCC metals are ductile at all temperatures and highly cross-linked thermosetting plastics are usually brittle. But most materials are ductile at high homologous temperatures and become brittle at some point as the temperature is lowered. They undergo a ductile–brittle transition.

It is common practice to determine the transition temperature using an impact test — the **Charpy impact test** being the most popular. It involves measuring the energy absorbed by a specimen broken by

a calibrated swinging pendulum (Figure 6.14). To assess the effect of temperature and to compare different materials specimens must have a standard size and shape. Figure 6.15 shows a typical result for a low carbon steel; far more energy is absorbed in breaking a ductile (tough)

specimen than a brittle one. The transition temperature usually taken as the midpoint of the transition region, about 320 K in this case.

Listed in Table 6.2 are the structural and operational factors that favour plastic flow or brittle fracture.

Table 6.2

| Favour plastic flow | Favour brittle fracture |
|---|---|
| Structure | |
| nondirectional bonding | directional bonding |
| mobile dislocations | immobile dislocations |
| (in crystals) at least 5 slip systems | fewer than 5 active slip systems |
| low yield strength | high yield strength |
| no cross-linking | extensive cross-linking |
| stereoregularity (see Section 5.4.2) | irregular molecules |
| Operation | |
| high homologous temperatures | low homologous temperatures |
| low rates of deformation (see Figure 2.16) | high rate of deformation |
| hydrostatic compression | triaxial tension |

specimen is unloaded at some stress above the yield stress it relaxes by a certain amount of elastic strain. This elastic strain is the origin of springback which can be significant in the plastic forming of metals, especially the working of wire, strip and sheet. In order to produce accurate shapes, allowance must be made for springback. When elastic strain is time dependent, as in polymers, matters get rather more complicated.

# 6.5 Elasticity in polymers

'The material for a beam' (Section 6.3) is an example of a major design aspect for many products that have to carry loads in service, namely the need to minimise the weight for a given elastic deflection. Metals and ceramics have the advantages that Young's modulus, which is determined by the atomic force–separation curve, does not vary with time and varies very little with temperature. It is one of the 'average energy' properties considered in Chapter 4.

On the other hand, as Figure 6.9(c) and (d) indicate, polymers are very different. In particular, the elastic modulus changes with time under load and with the rate at which it is loaded. It also changes markedly with temperature — by as much as a thousand times over a range of 200 K. This is extremely important since in many polymers the changes occur around ambient temperatures. Although in polymers a model for elastic modulus based on the atomic separation curve still applies, the more extensive movement of molecules provides other mechanisms for elastic strain. Let's explore them in more detail.

First, with all these uncertainties how can an elastic modulus be defined? Figure 6.16 shows the typical first stage of a polymer stress–strain curve together with three methods of specifying the modulus used in practice. The **initial modulus** is taken as the slope of the tangent to the curve at the origin. It is a convenient measure for purposes of comparison and we shall use it here. But it is rather difficult to determine, which is why the **secant modulus** is often used. This is defined as the slope of a line drawn from the origin to a point on the curve at some specified strain, usually 1.0% or 2.0%. However, for design purposes this may not be the most useful.

Can you see why?

As Figure 6.16 illustrates, the slope of the stress–strain curve, and therefore the modulus decreases with increasing stress and strain. To calculate deflections in components, values of the modulus are needed for the stresses or strains expected in service. For these purposes the **tangent modulus** is the most appropriate. More about this later in the section. To provide a basis for discussion, Figure 6.17 demonstrates the dramatic effect of temperature on the initial modulus of a typical amorphous polymer (PMMA and PS for instance); the values of modulus are only approximate but notice the log. scale. As shown, the curve can be divided into four viscoelastic regions. All polymers exhibit one or more of these regions; only amorphous polymers show all four. The regions are as follows.

• **Glassy region** — the polymer is rigid and behaves rather like an ideal elastic solid, as in Figure 6.9(a). As its name implies, the polymer is glassy and at sufficiently low temperature, brittle. This state exists well below $T_g$. The name is used because of the similarity to the behaviour of ceramic glasses at room temperature: brittle, with negligible

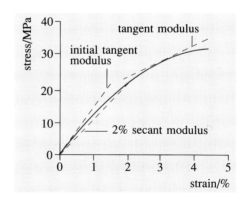

Figure 6.16 Ways of measuring the elastic modulus of a polymer

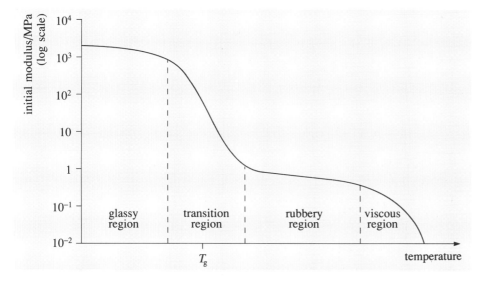

Figure 6.17 Temperature and modulus in an amorphous polymer

time-depenent behaviour. It follows that plastics used in this state have a $T_g$ higher than ambient; examples are PS (encountered in all sorts of containers, ballpoint pens and so on) with $T_g$ about 370 K and PMMA (in glazing and baths for example) with $T_g$ about 380 K.

• **Transition region** — contains $T_g$ and in it the modulus falls steeply with temperature, by as much as a hundred times or more. The polymer is sometimes said to be 'leathery'. It can be deformed and even folded but it recovers its shape relatively slowly. A good example is the plasticized PVC sold as 'leathercloth'.

• **Rubbery region** — the modulus is now very low and the main characteristic is the capacity for very large and rapidly reversible strains. In this form, it is familiar in rubber bands and bicycle inner tubes.

• **Viscous region** — this is, of course, non-recoverable flow and is included here for completeness. It is the region in which amorphous thermoplastics are processed to shape.

Of course, in practice, there is no clear demarcation between these regions, but in each one a different deformation mechanism is dominant. Remember, only amorphous thermoplastics exhibit all four regions. Which ones occur in partially crystalline thermoplastics, elastomers and thermosets and why, will become clear as we consider the four regions in more detail.

## 6.5.1 The glassy region

All polymers exhibit this physical state; at a low enough temperature they will be rigid, glassy, and often brittle. This is the habitual state for thermosets — those in GFRP and electrical fittings for instance — because the molecules are extensively cross-linked into a rigid framework. At low enough temperatures, rubber behaves like this too. An excellent demonstration of this is to cool a piece of rubber in liquid nitrogen and then to hit it hard. It shatters. In partially crystalline thermoplastics, this glassy state occurs well below $T_g$, where molecular motion is inhibited and all molecules are 'frozen' into position. It is important to appreciate that $T_g$ and the ductile–brittle transition temperature are not directly related.

In the glassy state, polymers are akin to metals and ceramics: their elastic modulus can be explained in terms of the force–separation model applied to secondary bonds between molecules.

EXERCISE 6.2   Using the schematic illustration of a simple amorphous thermoplastic in Figure 6.18, explain why in the glassy region, the thermoplastic has a much lower Young's modulus than diamond.

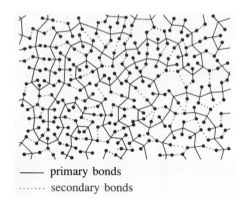

—— primary bonds
······ secondary bonds

Figure 6.18 An amorphous polymer (2D representation). The ● symbol represents a side group

The answer to Exercise 6.2 provides a strong clue to ways of increasing the elastic modulus of a polymer in the glassy state. If the modulus is determined by 'bond stretching', the greater the number of covalent bonds relative to secondary bonds lying along the stress axis the higher will be the modulus.

Can you think of two ways of achieving this in comparison with an amorphous polymer?

Increasing the degree of crystallinity and producing cross-links between molecules. The former leads to molecules aligning into regular arrays (see Figure 5.65) many of which will be parallel to the acting stress. A useful way of viewing the effect of cross-links is to compare a cross-linked polymer with diamond. In effect, diamond is the ultimate cross-linked polymer: each carbon atom is covalently bonded to four others, see Figure 3.23. So, the more cross-linking there is in a polymer, the higher is its elastic modulus.

One way of increasing the crystallinity of a thermoplastic is to deform it plastically by stretching. This causes the molecules to align with the stress axis. The process is called cold drawing and we consider it in Section 6.6.

## 6.5.2 Transition region

Clearly the glassy region exists below the brittle–ductile transition temperature. As the temperature is raised towards $T_g$ the increased thermal energy becomes sufficient to overcome the secondary bonds and to allow molecular movement by means of carbon–carbon bond rotation. Such effects are the source of viscoelasticity. Since they are thermally activated, the amount of strain produced by a stress depends on both temperature and time.

Which class of change with temperature discussed in Chapter 4 does the elastic modulus now fit into?

Accelerating. It can be modelled in terms of the fraction of atoms which have a much higher than average thermal energy.

It is important to appreciate that for most thermoplastics, $T_g$ is within 100 K or so of room temperature and so they are usually used either just in the glassy region or just in the transition region.

When a polymer is stressed in the transition region there are three kinds of contribution to the resulting strain:

(a) a small amount of strain expected from 'bond stretching'

(b) a large amount of time-dependent strain arising from bond rotation — called **rubber elasticity** or sometimes **long-range elasticity**

(c) viscous flow due to the sliding of molecules past one another — plastic strain.

On removing the stress, (a) is recovered instantly, (b) is usually recovered quite rapidly over a period of time, and (c) is not recovered — it is permanent strain. This behaviour is summarized in Figure 6.9(c).

We will consider viscous flow later. Here it is sufficient to point out that it happens because there is enough thermal energy to overcome the secondary bonding between molecules. Under stress they can slide over one another, rather like buttered spaghetti.

The new deformation mechanism to concentrate on is rubber elasticity. As you will expect from its name, it is the dominant one in the rubbery region but it is also significant in the transition region. Its main characteristic is the amount of instantly recoverable strain it can produce.

Why does a force–separation model not apply to rubber elasticity?

The strains produced are far too large. According to this model a solid cannot be strained beyond the maximum in the force–separation curve without breaking; this means quite small strains (less than 20%). The rubber elasticity originates in the many molecular conformations that can exist in a polymer. To explain this we need to use a thermodynamic model.

What is a conformation?

The shape of the backbone molecule. Since carbon–carbon bonds can rotate, carbon atoms can occupy different angular positions with respect to another. Recall the trans and gauche positions from Section 5.4.1.

So, in principle, bond rotation allows a vast number of different conformations, ranging from straight to complex kinked and twisted structures. In a thermoplastic or an elastomer (the general term for polymer rubbers) the molecules are not completely free. They will be restricted locally; in an amorphous thermoplastic primarily by physical entanglements (see Figure 5.64), in a partially crystalline polymer by being trapped between crystalline regions (see Figure 5.65) and in an elastomer by cross-links. In such systems, in which internal energy changes are small, the free energy decreases if the entropy increases.

307

Now, the configurational entropy of an unstressed polymer will be a maximum when the unconstrained molecule segments have random conformations, in other words when they are completely disordered. When the polymer is stressed, kinked segments can straighten and lengthen, thus contributing to the strain. Of course, this reorganization of the molecules into more ordered conformations means that the entropy decreases. So, when the stress is removed, the molecules wriggle back, over time, to their original lower energy positions. More about this model in the next section.

Since rubber elasticity is thermally activated, bond rotation may not happen immediately a stress is applied because each carbon atom is bonded to others in the chain. When the time required for the molecule to change conformation to accommodate the stress is long, the elastic strain will occur slowly. This is what happens in the transition region. At higher temperatures, typically 30 K or so above $T_g$ where the thermal energy is greater, high strains will develop almost instantaneously with stressing. This is what you experience when you stretch an elastic band and is what happens in the rubbery region of Figure 6.17 — more about this shortly.

As you would expect, the extent to which rubber elasticity and viscous flow contribute to the elastic strain depends very much on the structure of the polymer; in particular on the points of restriction along the molecules. For instance:

(a) Amorphous polymers with no cross-links or crystallinity have only entanglements as 'built-in' mechanisms to inhibit viscous flow or to enable rubber elasticity.

(b) In partially crystalline polymers the lengths of molecule packed into the orderly crystalline regions cannot contribute to rubber elasticity, and in addition viscous flow is difficult in these regions. So the higher the degree of crystallinity, the less the rubber elasticity and viscous flow.

(c) Even a few cross-links have a marked dampening effect on the viscous region and extend the rubbery region; we consider this in the next section. A lot of cross-links completely inhibit viscous flow and rubber elasticity, which is effectively what happens in a heavily cross-linked thermosetting plastic.

Because both rubber elasticity and viscous flow are thermally activated processes, they are dependent upon time as well as temperature. Figure 6.19 is an example of the effect, showing the modulus–temperature curve for two very different times of loading. Notice that the effect is most marked in the rubbery region.

Given this time dependence of deformation the stress–strain curve is not as useful as it is for, say, metals. In practice, the basic source of data for polymers is so-called **creep curves** in which strain is measured as a function of time for a given constant stress. Do not confuse this with the creep response described in Figure 6.9(f) in which the strain is

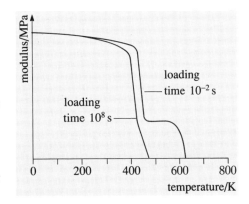

Figure 6.19 Modulus and temperature in PMMA

entirely plastic (permanent). In the case of polymers the strain is largely viscoelastic and therefore recoverable.

Figure 6.20(a) shows a set of creep curves for a variety of HDPE for a range of stress; they are typical of their kind. They show that the polymer continues to deform indefinitely. Of course under other conditions, for example at higher stress or in a harmful chemical environment, it might rapidly creep to failure. But you can see that for

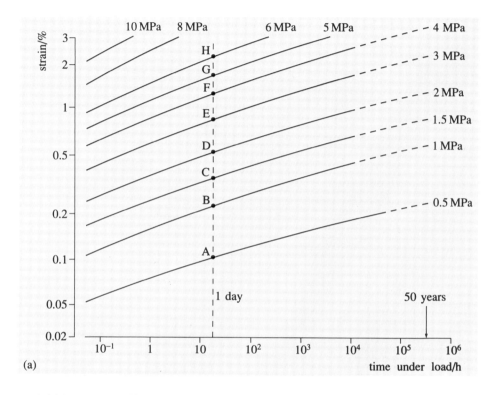

(a)

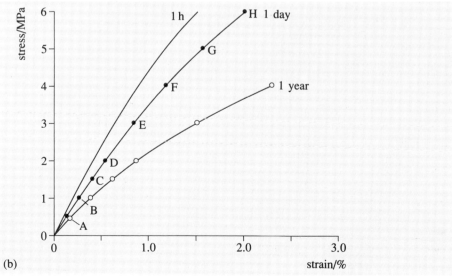

(b)

Figure 6.20  HDPE (a) tensile creep data (b) isochronous stress–strain curves

design purposes, data is required for very long test times. This is obviously a problem for polymers developed in the last few decades! So it is often necessary to extrapolate the creep curves to the time required by a design specification. This is the reason for plotting the data as log. strain against log. time — the curves are approximately linear and extrapolation is fairly straightforward.

For design purposes the creep data of Figure 6.20(a) is often plotted in other ways; Figure 6.20(b) shows one important example. It is a set of **isochronous** (meaning equal time) **stress–strain curves** obtained from Figure 6.20(a) as indicated by the points A–H for the test time of one day. Do not confuse these curves with stress–strain curves obtained in a tensile test. A tensile test involves completely different conditions; the sample is strained at a constant rate and the resulting stress is measured. ▼Thinking about a simple plastic beam▲ indicates how these curves are used.

## 6.5.3 The rubbery state

Most polymers become rubbery at some temperature. This includes many thermoplastics and the family of polymers called **elastomers** which exhibit very high recoverable strains. The only exceptions are heavily cross-linked thermosets such as phenol formaldehyde and epoxies. Nowadays the terms rubber and elastomer are used interchangeably; for historical reasons the word elastomer was coined to distinguish synthetic rubbers such as SBS from natural rubber (NR).

# ▼Thinking about a simple plastic beam▲

As I have already mentioned in 'The material for a beam', Equation (6.1), for an elastic material:

$$\Delta = \frac{Fl^3}{48EI}$$

You should also know that:

$$\sigma = \frac{yM}{I} \qquad (6.3)$$

where $\sigma$ is the stress at the top and bottom surfaces of the beam, $y$ is half of the depth of the beam and $M$ is the bending moment.

But how can a viscoelastic polymer be handled? A common approach is to assume, initially, that a plastic beam would behave elastically and then to adjust for the differences. It is called the **pseudo-elastic** method and is described here.

The important point to remember is that if you assume the polymer is elastic and

you select a modulus at the strain corresponding to the outer surfaces, you will predict a deflection larger than the actual value would be. The stress in the beam varies linearly from zero at the neutral axis to a maximum at the surface. So for a polymer which produces isochronous stress–strain curves like those in Figure 6.20(b), since the modulus decreases with increasing stress, the strain produced will be smaller at lower stresses. Consequently, the overall deflection of the beam will be less than for an elastic beam.

We will consider only one element in the design of a simple beam (like that shown in Figure 6.7), namely that the deflection $\Delta$ at the mid point should not exceed a certain amount in a specified time, say a year. The procedure then is:

(a) Calculate a suitable $I$ for the beam by putting a 'guesstimate' for the creep tangent modulus in Equation (6.1). That is, assume that the tangent modulus is a constant, like $E$.

(b) Using the value for $I$, calculate $\sigma$ in Equation (6.3).

(c) Calculate the tangent modulus for this value of $\sigma_{max}$ from the isochronous stress–strain curve for the specified design time, for example one of those in Figure 6.20(b).

(d) Substituting this value in Equation (6.1) gives the deflection $\Delta$.

(e) Since this deflection will be an overestimate, the value of the cross-section is reduced and the calculations repeated until the design deflection is achieved.

SAQ 6.4 (Objective 6.2)
A beam is made from the grade of HDPE which produced the isochronous stress–strain curves of Figure 6.20(b). If the stress on the beam is 4 MPa how much more will it deflect after one year than one day?

Would you expect $T_g$ to be above or below ambient for an elastomer?

Below ambient — rubber elasticity requires that the secondary bonding between molecules can be overcome by thermal energy.

In addition to the huge elastic strains possible, elastomers have other special characteristics. One is illustrated by the stress–strain curve of Figure 6.21. The whole of the strain is elastic and at high strains further deformation gets more difficult — the tangent modulus increases.

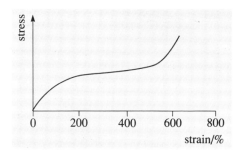

Figure 6.21 The stress–strain curve of a typical elastomer

> SAQ 6.5 (Objective 6.3)
> Use the earlier model of rubber elasticity to explain why
> (a) rubber elasticity occurs more readily at higher temperatures
> (b) the elastic modulus increases at high elastic strains.

The increase of modulus at high strains isn't the only surprising property of rubbers. In complete contrast to other materials, the modulus increases with temperature. Rubbers also get warm when stretched, as you can demonstrate for yourself by rapidly stretching a large rubber band held against your lip. ▼**The strange behaviour of rubber**▲ explains these peculiarities.

# ▼The strange behaviour of rubber▲

When rapidly compressed, most materials become warmer; and when expanded, they become cooler. The effect is particularly noticeable in gases (think of pumping up a bicycle tyre) but also occurs in liquids and in solids when they are elastically deformed. Do not confuse this with plastic deformation in solids, where 95% of the work done deforming the solid appears as heat. The effect I am concerned with here is reversible and due only to changes in the configurational entropy of the system.

If the order within any material decreases, as it does in most expansions, the configurational entropy increases and energy in the form of heat is needed to compensate. The system will supply the energy from within itself if the expansion is adiabatic, and it will cool down. In metals and ceramics, because there is a limit to elastic strain, the possible changes to configuration are small, hence the cooling and heating effects are small. The behaviour of polymers can be very different.

When a polymer in the rubbery region is stretched, the segments of molecules between points of restriction (cross-links

and so on) take up more linear conformations than the random ones in equilibrium. Thus, the polymer becomes more ordered; its entropy decreases and causes a temperature rise.

If the heat evolved were allowed to dissipate and then the stress on the polymer were relaxed, would you expect the polymer to cool down?

Yes, the straightened molecules would take up more random conformations, leading to an increase in entropy and a drop in temperature.

Now why does the elastic modulus increase with temperature? Consider the free energy equation from Chapter 4:

$$\Delta G = \Delta H - T\Delta S$$

Suppose a rubbery material is now stretched at constant temperature and pressure. How do these terms change as the material is deformed? That is, if $\Delta x$ represents the extension, what can we say about each of the terms in the following equation?

$$\frac{\Delta G}{\Delta x} = \frac{\Delta H}{\Delta x} - \frac{T\Delta S}{\Delta x}$$

The first term, $\Delta G/\Delta x$, represents the work done on the material per unit change of length as it is stretched, and if $\Delta x$ were made progressively smaller this ratio would come to represent the applied force.

Now, if the deformation occurs merely by reorienting and extending chains without major microstructural changes (such as crystallization), the internal energy $H$ will not change as the material is deformed. So, the second term, $\Delta H/\Delta x$, is zero.

Looking at the last term, recall that stretching the material reduces $S$. So $\Delta S/\Delta x$ will be negative as the material is stretched. The term

$$-\frac{T\Delta S}{\Delta x}$$

will therefore be positive. Thus the greater $T$, the greater will be $\Delta G/\Delta x$, which is related to the applied force. Provided temperature and pressure are constant during deformation, a greater force will be required to achieve the extension $\Delta x$ at a higher temperature than at a lower temperature. Hence the modulus has increased.

The critical structural feature of rubbery behaviour is clearly the number and effectiveness of the restrictive points along a molecule. In a commercial elastomer, the restrictive points are occasional cross-links between molecules, and for this reason they are often called lightly cross-linked thermosets.

With no cross-links, molecules slide readily over one another and viscous flow dominates. Rubber elasticity requires the presence of some cross-links but not too many. As the number of cross-links increases the segments of free molecule become so small that rubber elasticity is eventually inhibited altogether and the material becomes a heavily cross-linked thermoset. ▼**Vulcanization of natural rubber**▲ is a beautiful illustration of these changes.

However, not all elastomers rely on the presence of chemical cross-links. You will recall from 'Some copolymers in action' in Chapter 5 that in styrene-butadiene-styrene rubber (SBS) the butadiene rubbery segments are pinned at their ends by styrene blocks, see Figure 5.63.

# ▼Vulcanization of natural rubber▲

Natural rubber starts out as latex, the sap from rubber trees; it is a dispersion of polyisoprene particles in water. Its molecular structure is shown in Figure 6.22.

In its normal state at ambient temperature, after the water is removed, it is a solid which will readily creep because it has no cross links. It is made into a useful material by **vulcanization**, a process in which, essentially, rubber is heated with sulphur. The word vulcanization comes from Vulcan the Roman god of fire and metal working! The process was discovered by Charles Goodyear in 1839 and was vital to the development of the vehicle tyre industry.

Some of the sulphur atoms form cross-links by attacking the $CH_2$ groups neighbouring the double bonds in isoprene molecules (Figure 6.23).

By adjusting the processing conditions, the number of cross-links can be controlled: the more there are, the stiffer and harder the rubber. The natural rubber in a car tyre may contain 3–5 wt% sulphur, giving

a cross-link about every 500 carbon atoms along a molecule, whereas that in, for example, a traditional car battery case may have up to 40 wt% sulphur. The latter is extensively cross-linked and close

to being a glassy brittle thermoset. In fact, with time in service rubber can become brittle. This and related mechanisms of degradation will be discussed in Chapter 7.

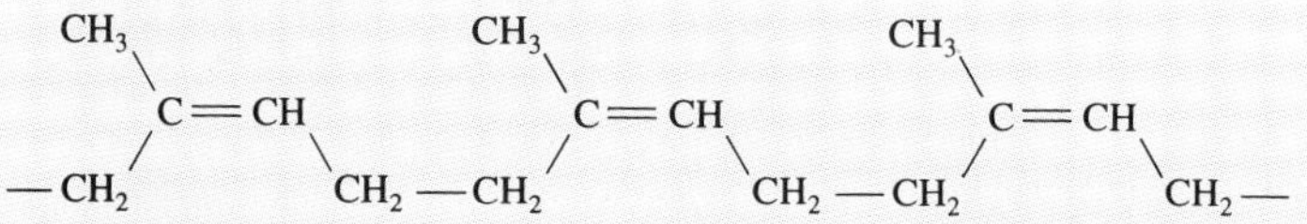

Figure 6.22 Molecular structure of polyisoprene

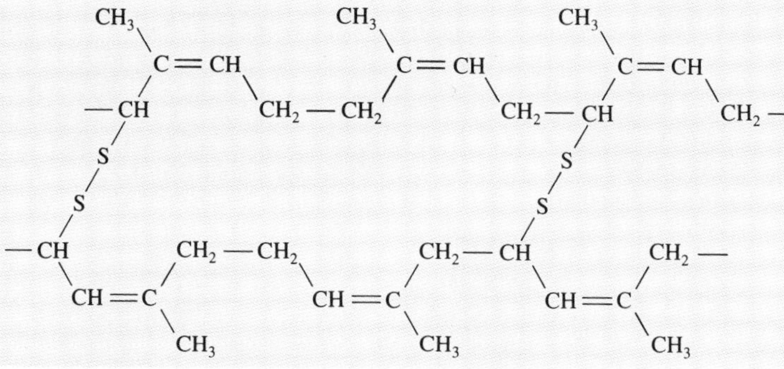

Figure 6.23 Cross-links in vulcanized rubber

## 6.5.4 Summary and comments

Many plastics in service at ambient temperature are fairly close to $T_g$ — the temperature about which their behaviour changes markedly. They operate at high homologous temperatures, roughly in the range 0.4–0.6 $T_m$. This means that thermal energy is very influential in determining properties; about as important in a polymer at ambient as it is in red hot steel.

Deformation mechanisms are thermally activated and are therefore both temperature and time dependent. The change in mechanical properties with increasing temperature is dramatic. For instance the initial modulus decreases from typically 2.5–5.0 GPa in the glassy state to 0.1–1.0 MPa in the rubbery state. In comparison, the Young's modulus of steel is 210 GPa at room temperature.

The detailed nature viscoelastic behaviour changes with temperature and can be divided into four regions: glassy, transition, rubbery and viscous. Different types of polymer behave differently.

- Amorphous thermoplastics show all four regions. Above $T_g$ physical entanglements between molecules provide the constraints between which segments can undergo some rubber elasticity. Without crystallinity and cross-links to provide strong intermolecular bonding, they flow viscously. This means they can be moulded well above $T_g$.

- In partially crystalline thermoplastics, the higher the degree of crystallinity, the higher the elastic modulus; that is, for a given stress the lower are the contributions to strain from rubber elasticity and viscous flow. In a fully crystalline material these strain mechanisms would not occur and the modulus would be fairly constant with temperature until $T_m$ was reached. However, the change to viscous behaviour is not sharp and at the usual processing temperatures a polymer melt retains some of its viscoelastic character. The reason is that molecules remain entangled to some extent in the melt and the result is a 'stiffer' melt that is more difficult to process. The persistence of entanglements depends on the length and configuration of the molecules. In fact, molecule length, more specifically relative molecular mass, plays a vital role in determining polymer properties, as you will see in the next section and in Chapter 7.

- In lightly cross-linked thermosets (elastomers or rubbers) occasional cross-links along a molecule allow extensive rubber elasticity and inhibit viscous flow. The greater the number of cross-links the less elastic (more rigid) the material becomes above $T_g$.

- In heavily cross-linked thermosets, the extent of cross-linking can be so great that the thermoset shows little rubber elasticity or viscous flow and does not melt when it is heated, but decomposes.

SAQ 6.6 (Objective 6.4)
Which of the schematic modulus–temperature curves 2–5 in Figure 6.24 matches to a highly (but not completely) crystalline thermoplastic. Explain your reasoning.

Figure 6.25 is another way of demonstrating the remarkable range of mechanical behaviour. It is a selection of (schematic) stress–strain curves, obtained in tensile tests, for a polymer at different temperatures. Curve (a) is in the glassy region and shows elastic deformation up to (brittle) fracture; curve (d) is in the rubbery region and like that of Figure 6.16 entirely elastic; curves (b) and (c) are in the transition region. In the plateau of curve (c) the specimen is undergoing extensive plastic deformation, brings us neatly to the subject of the next section.

# 6.6 Plastic flow in polymers

It is only necessary to consider partially crystalline polymers above $T_g$ here because they exhibit all of the observed effects.

When a partially crystalline thermoplastic is stretched continuously a point is reached beyond which (nonrecoverable) plastic strain sets in. ▼Pulling polyethylene▲ shows how you can observe what happens macroscopically for yourself.

Plastic strain leads to the appearance of a 'neck' in the specimen. However, the material within this neck does not keep on shrinking as it does in a metal. Compare specimen (a) in Figure 2.16 with that in Figure 2.18. We explore the reason for the difference in Section 6.9.3. Instead, once formed, the section of the neck remains constant and the neck simply grows longitudinally along the specimen. During this process, the stress–strain curve draws out to the shape shown as curve (c) in Figure 6.25. The neck first forms at point X and it grows longitudinally between Y and Z. The sample can extend by as much as three times.

Let's consider what is going on within the microstructure during cold drawing. Under small stresses, the amorphous regions deform by the molecules sliding over one another and this has the effect of (partially) aligning the crystalline regions (Figure 6.26).

At somewhat higher stresses, however, the packing of the lamellae (see Section 5.4.2) and even the lamellae themselves are unravelled and the molecules are rearranged to form **fibrils** as shown in Figure 6.27. The degree of crystallinity of the polymer is now greater than in the lamellar configuration.

At still higher stresses, the fibrillar structure may be broken down as the polymer chains are unfolded and extended to form a highly crystallized form of the polymer. The polymer is then in the cold drawn condition.

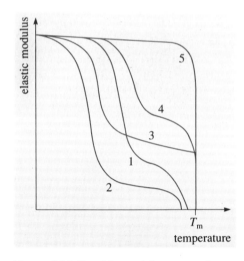

Figure 6.24 Possible modulus curves for a polymer

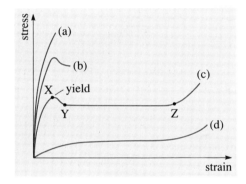

Figure 6.25 Stress–strain curves for a polymer with temperature increasing from (a) to (d)

# ▼Pulling polyethylene▲

You need to buy a pack of beer cans! What you are after is the polyethylene rings that hold the cans together.

Put two pencil marks about 3 cm apart on a segment of the strip and stretch it slowly with your hands. You will feel it give suddenly and then stretch easily. Watch carefully for any change in the dimensions of the sample as you stretch it. Unload and see if the length changes appreciably.

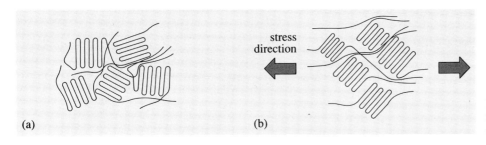

Figure 6.26 Orientation of crystalline regions in a polymer (a) no stress (b) small stress

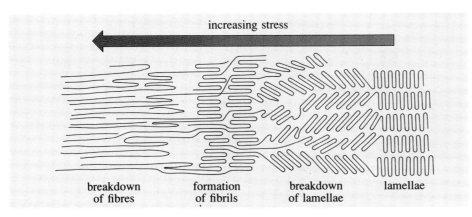

increasing stress

breakdown of fibres    formation of fibrils    breakdown of lamellae    lamellae

Figure 6.27 The effect of a gradually increasing stress on a partially crystalline polymer

It has a much higher strength and stiffness in the drawing direction because with the molecules extended in the direction of stressing, a much greater proportion of the load is being carried by the covalent bonds in the backbone of the molecules.

As you can see from Figure 6.27, during cold drawing the polymer changes from a partially crystalline solid with no preferred orientation of the crystalline regions to a highly oriented and much more crystalline solid in which a lot of the molecules are pulled out into a series of parallel strings lying along the direction of the stress. This is an irreversible process and the resultant cold drawn polymer, in property terms, is a very different material.

▼Drawing in action▲ shows that strengthening by drawing is used to good effect in the manufacture of strong fibres.

As with viscoelastic properties, the plastic response of polymers to tensile testing depends on time and temperature. At high strain rates cold drawing does not occur, compare (a) to (b) in Figure 2.16. Instead of drawing the polymer fractures. This is understandable when you consider that cold drawing involves a massive reorganization of the molecules in the solid. So if the test rate is too high, there is not sufficient time for the molecules to move past each other. The resulting higher stresses lead to fracture. You can see this for yourself if you repeat the test on a beer can loop, but pulling faster this time. Similarly since the rate at which the molecules can flow past each other is thermally activated, low temperatures produce the same effect as high testing rates.

# ▼Drawing in action▲

Drawing is used to good effect in the manufacture of polymeric, especially polypropylene, tape and twine, which have largely replaced steel wire and string. One consequence of drawing and its strong molecular alignment is that the properties are high anisotropic (directional). So the tape has high longitudinal strength and stiffness but low transverse strength and stiffness; it is easy to tear it along its length.

By far the widest application of drawing is in the production of synthetic textile fibres such as nylon and polyester. Textiles require the fibres to be strong and flexible, properties which can be achieved in partially crystalline thermoplastic by drawing them so that they have a fibrillar microstructure (see Figure 6.27) aligned along the fibres. Fibres, such as nylon, are produced as continuous filaments from a melt. Orientation of the molecules is then achieved by drawing the filaments to 4 or 5 times their original length as they are formed.

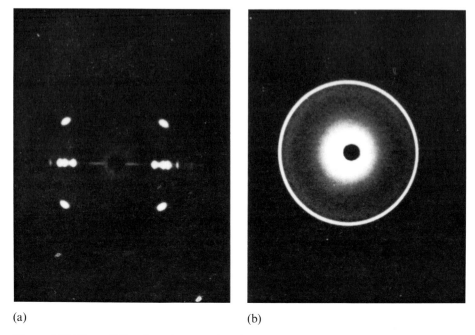

(a)                                      (b)

Figure 6.28  X-ray diffraction photographs of polypropylene

SAQ 6.7   (Objective 6.5)
Figure 6.28 shows X-ray diffraction photographs obtained from a
sample of PP before and after stretching. Identify (a) the unstretched
sample and (b) the stretched sample. Explain the difference between
the two photographs in terms of their microstructures.

Now, what about the behaviour of ductile completely crystalline
materials? Before considering them, we need to develop a few ideas
about plastic deformation and the nature of dislocations.

# 6.7  Some crystal plasticity

First, a recap:

SAQ 6.8   (Revision)
Figure 6.29 illustrates the crystal structure of MgO.

(a)  Which of the vectors **a** to **c** would you expect to be the slip
vector?
(b)  On which type of plane, A or B would you expect slip to occur?
(c)  Why must a ductile polycrystalline material deform on a number
of slip systems?
(d)  At what angle to a tensile axis will lie a plane on which the shear
stress is a maximum?

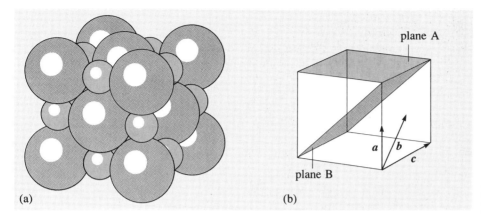

(a)     (b)

Figure 6.29 MgO (a) crystal structure (b) planes and vectors

Slip starts in a crystal when the shear stress reaches the critical level needed to produce dislocation movement. This **critical shear stress** determines the yield stress in a material and its magnitude depends on the resistance to dislocation motion presented by the crystal structure and microstructure of the material. The crystal structure provides an intrinsic resistance caused by the breaking of bonds between atoms or ions as a dislocation moves; the microstructure provides additional resistance in the form of obstacles to dislocation movement — solute atoms, grain boundaries, particles and so on.

The intrinsic resistance is higher in covalently bonded crystals such as silicon than in metals.

Why is this so?

Covalent bonds are highly localized strong stiff bonds which have to be broken in order for atoms to move relative to one another. Figure 2.63 gives a rough idea of what happens when an edge dislocation moves; imagine that the vertical lines are covalent bonds between neighbouring atoms. On the other hand, in most metals the electrons involved in bonding are highly delocalized — they orbit many atoms. So, in this case, the small relative movement of atoms that occurs when a dislocation moves only has a minor effect on the general bonding between delocalized electrons and metal ions (see Section 3.8.2).

As you would predict, this intrinsic resistance to dislocation motion may be overcome by thermal activation. Some examples of the consequences are shown in Figure 6.30 for single crystals of a range of materials with different types of bonding and crystal structures. Some crystalline materials that are completely brittle at ambient temperatures can flow plastically at elevated temperatures.

Notice, however, that Figure 6.30 is for single crystals. As you know from Section 2.6.1, in polycrystalline materials the situation can be very different. It can be shown (with difficulty) that, in order to conform to the deformation of its neighbours, an individual crystal in a

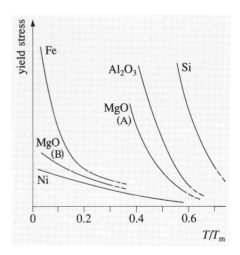

Figure 6.30 Yield stress and temperature in single crystal materials

polycrystalline aggregate has to slip on five independent slip systems. A system is independent if its slip vector cannot be matched by combinations of those for other systems.

For FCC metals, for instance, this presents no problems; of the 12 available slip systems, see Figure 6.31, five are independent. The same is true of silicon which slips on the same systems, but in this case it is brittle at room temperature because of its high intrinsic resistance. MgO is a different story. The temperature dependence of the yield stress for those systems involving the type B planes and the vector $b$, see Figure 6.29(b), is included in Figure 6.30. The intrinsic resistance is obviously not all that high. However, of the six available systems of this type only two are independent. So polycrystalline MgO is brittle at low temperatures. Notice, however, that plastic deformation becomes possible above about $0.6\,T_m$ (about 1900 K) when dislocations also become significantly mobile on planes of type A.

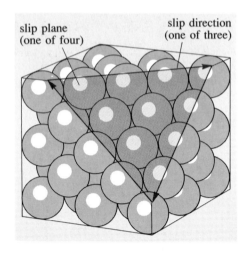

Figure 6.31 Slip systems in FCC crystals

EXERCISE 6.3   There are two geometric reasons why the tensile yield stress needed of a polycrystalline material is greater than the critical shear stress needed to move dislocations. What are they?

Of course, as you know from Chapter 5 in particular, other forms of resistance to dislocation motion — that is, strengthening — are provided by microstructural features such as solute atoms, grain boundaries and particles of second phase. More about these shortly, but first I need to consider the energy of a dislocation.

## The energy of a dislocation

As you know, the atoms around a dislocation are not in their normal equilibrium sites in the crystal. See Figure 2.63 for an illustration of this. It follows that these atoms have a higher energy than they would in the absence of the dislocation. This constitutes the energy of the dislocation. Very close to the dislocation (within, say 3 or 4 atoms) the crystal is highly distorted. Within this region, which is called the **dislocation core**, the atoms are a long way from their normal positions. However beyond the core the distortion is much less and the atom displacements are sufficiently small to be considered elastic and therefore subject to Hooke's Law. So, outside the core, the energy of the dislocation takes the form of elastic strain energy. In ▼**The elastic energy of a dislocation**▲ an equation for the elastic strain energy of a dislocation is derived and some significant consequences of the form of the equation are highlighted.

The fact that a dislocation has an associated strain field, and thereby an energy, has important consequences, in particular:

• The presence of a dislocation increases the internal energy of a crystal; the higher the dislocation density the higher the internal energy. The significance of this is revealed in Section 6.8.

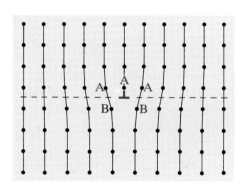

Figure 6.32 Atoms around an edge dislocation

# ▼The elastic energy of a dislocation▲

Outside its core, a dislocation generates an elastic strain field, and as you will see, many of the important characteristics of a dislocation are due to this strain field.

A screw dislocation produces a simple strain field and so it is fairly straightforward to derive an equation for its elastic strain energy. Consider the

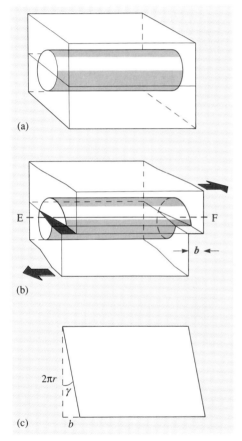

(a)

(b)

(c)

Figure 6.33 Calculating the elastic strain energy of a screw dislocation

screw dislocation EF in Figure 6.33(b). To remind yourself of the configuration, compare it with Figure 2.56. In the present case the screw dislocation runs right through the crystal. Now imagine a very thin-walled cylinder, radius $r$, of material in the undeformed crystal, as illustrated in Figure 6.33(a). When this cylinder surrounds the screw dislocation it is distorted into the helix of Figure 6.33(b); the deformation is equal to the shear displacement $b$ produced by the dislocation. If the cylinder is opened out in its deformed state it will look like the strip shown in Figure 6.33(c).

The shear strain in the strip is

$$\gamma = \frac{b}{2\pi r}$$

The shear stress is

$$\tau = \mu\gamma$$

where $\mu$ is the shear modulus.

Consider now that the cylinder of radius $r$ has wall thickness $dr$ and unit length, and that the task is to calculate the elastic strain energy within it. From Hooke's law the elastic strain energy per unit volume is

$$\tfrac{1}{2} \times \text{stress} \times \text{strain} = \frac{\mu\gamma^2}{2}$$

If you are unsure about this, see Chapter 2 'Work of deformation' and 'Shear deformation'.

The volume of the cylinder is $2\pi r\,dr$, so the elastic energy is given by

$$\frac{\mu\gamma^2}{2} \times 2\pi r\,dr = \frac{\mu}{2}\left(\frac{b}{2\pi r}\right)^2 2\pi r\,dr$$

$$= \frac{\mu b^2}{4\pi}\frac{dr}{r}$$

The total elastic energy per unit length of the dislocation is then the sum of that

within all cylinders from the core radius $r_0$ to the outer radius of the crystal $r_1$. This is

$$\frac{\mu b^2}{4\pi}\int_{r_0}^{r_1}\frac{dr}{r} = \frac{\mu b^2}{4\pi}\ln\left(\frac{r_1}{r_0}\right)$$

Note the following points stemming from this equation:

(a) The vector $b$, magnitude $b$, is used to describe the slip displacement produced by the screw dislocation. This applies generally to all dislocations; it is called the **Burgers vector**. For simplicity, in Chapter 2 both the shear produced by slip and that produced by a dislocation were characterized by the slip vector $s$. In the study of dislocations, the Burgers vector $b$ is used because only in simple cases does $s = b$. In some crystal structures, FCC and CPH metals for instance, $s$ is achieved by a combination of dislocations ($b_1 + b_2 = s$). In addition, dislocations with $b < s$ are responsible for other types of shear transformation — for example the martensitic transformation discussed in Chapter 5. Here we will be assuming $s = b$.

(b) The energy of a dislocation is proportional to $b^2$ and this affects the way that dislocations are likely to combine or divide.

*i* Two dislocations with Burgers vectors $b_1$ and $b_2$ will not tend to join together to form a dislocation with Burgers vector $b_3$ unless $b_3^2 < b_1^2 + b_2^2$.

*ii* A dislocation with Burgers vector $b_3$ can reduce its energy if it can split into two components with Burgers vectors $b_1$ and $b_2$ for which $b_1^2 + b_2^2 < b_3^2$.

(c) The strain field around an edge dislocation is very different from that of a screw dislocation but the equation for the elastic energy has the same form.

• Through the influence of their strain fields, dislocations interfere with the movement of one another and with any other structural feature such as solute atoms that also produce a strain field.

It is the interaction between dislocations moving on intersecting slip systems and the increase in their density which gives rise to work hardening. Work hardening is extremely important in the cold working of metals. You will recall from Chapter 2 that the term cold working is used for shaping processes such as rolling carried out at low temperatures — below about $0.3\,T_m$. Figure 6.34 demonstrates the effects of work hardening. The basic stress–strain curve is that of a specimen pulled continuously until it fractures. As you know, for Hookean solids such as metals at stresses up to the yield stress ($\sigma_A$ in Figure 6.34), deformation is elastic; if at any point the specimen is unloaded, on reloading the original stress–strain plot is followed. Exactly the same thing happens at stresses beyond the yield stress. Thus if the specimen is unloaded at B it relaxes elastically to C, leaving plastic strain of $\varepsilon_C$. On reloading the elastic region CB is retraced and the specimen has a new yield stress $\sigma_B$. The previous plastic deformation has work hardened the material and the yield stress has been increased from $\sigma_A$ to $\sigma_B$. However, the ductility of the material has been decreased.

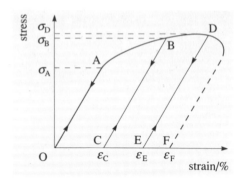

Figure 6.34 Loading and unloading cycles in a tensile test

In terms of % strain what would be the ductility of specimens (a) tested directly to fracture and (b) reloaded after having been unloaded at D?

For (a) the ductility would be $\varepsilon_F$% and for (b) it would be ($\varepsilon_F - \varepsilon_E$). In other words at cold working temperatures a ductile material has a certain capacity for plastic flow and no more.

Because of work hardening, yield stress increases with plastic strain until at some point the stress required to maintain dislocation motion becomes so high that crack initiation and propagation occur instead — the material breaks. Work hardening is an example of an important general relationship between yield stress and ductility: as the yield stress increases, ductility decreases. The reasons underlying this relationship will become clear in the next section. For the moment, note that it is a prime example of conflict between property requirements for processing to shape and for performance in service. Shaping processes such as forging and rolling require high ductility, accompanied by a low yield stress to minimize the size and power of the processing machinery. In service the need is usually for a high yield stress. Essentially the conflict arises because a high yield stress stems from inhibiting dislocation movement and high ductility requires mobile dislocations. In the next three sections we explore the conflict and start by considering ways of increasing the yield stress.

# 6.8  Achieving strength for service

Because their intrinsic resistance to dislocation movement is very low, metals are naturally soft. Indeed some of the metals we use are fairly soft, for example copper wire and aluminium foil, yet we know from common experience that most metals in service have high yield stresses.

Why, basically, are high yield stresses desirable?

Because the higher the yield stress the smaller is the cross-section required to carry a given mechanical load. So artefacts become more practicable, lighter, more convenient to use, require less material and so on.

Let's assume for the moment that we are concerned only with increasing the yield stress of a material to improve its performance in service. The strategy is to hinder the movement of dislocations in the crystal lattice and there are four distinct ways of doing this.

- work hardening
- reducing the grain size
- solid solution strengthening
- precipitation/dispersion hardening

We will address each in turn.

## 6.8.1 Work hardening

As you know from the previous section, the work hardening that occurs during the cold working of an artefact produces an increase in yield stress — the aluminium or steel in drinks cans is much stronger than the unworked form. Work hardening can also be useful in service. For example, if locally in a component the stress exceeds the yield stress, the resultant plastic flow and work hardening will leave that region more resistant to the next overload. With metal components this effect is called 'shakedown'. It is not possible with brittle materials such as ceramics.

Work hardening occurs because the density of dislocations increases rapidly during plastic flow and because dislocations moving on intersecting slip systems tangle with one another. As the dislocation density increases the tangling gets worse. ▼Dislocation density▲ gives an idea of the numbers involved.

What actually happens when dislocations meet depends on whether they are edge or screw. However, in general, the interactions are determined by the elastic strain fields of the dislocations. We can get an idea of what is involved by looking at a simple example.

Figure 6.35 shows two edge dislocations lying on the same slip plane. First of all, notice that immediately above each dislocation the atoms are closer together than the normal spacing and therefore in compression. See also Figure 6.32. Below the dislocations the atoms are

# ▼Dislocation density▲

The dislocation density increases rapidly with the amount of plastic strain — compare, for example, the specimen in Figure 2.67 which contains about 1% plastic strain with that in Figure 2.58 in which plastic flow has barely begun. It is usual to describe the number of dislocations in a crystal in terms of a dislocation density $\rho_D$ which is normally taken as the number of dislocations intersecting a unit area (usually $m^2$) of the crystal. So, for instance, what very roughly is the $\rho_D$ of the sample shown in Figure 2.58?

There are about 60 dislocations intersecting the surface of the sample. Now the area of the photograph is about $24\,cm^2$ and the magnification is $20\,000$, so the actual area of the sample is

$$\frac{24}{20\,000^2} \times \frac{m^2}{10^4} \approx 6 \times 10^{-12}\,m^2$$

60 dislocations in an area of $6 \times 10^{-12}\,m^2$ gives $\rho_D$ as

$$\frac{60}{6 \times 10^{-12}} = 10^{13}\,m^{-2}$$

In extensively deformed metals such as heavily rolled copper $\rho_D$ can reach values of $10^{16}\,m^{-2}$ and higher.

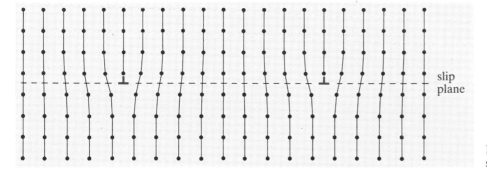

slip plane

Figure 6.35 Two edge dislocations of the same sign on the same slip plane

in tension. (It may help you to visualize how the tension and compression arise by imagining that the dislocation is produced by cutting the crystal down to the slip plane and then inserting the 'half-plane' of atoms). The two dislocations in Figure 6.35 are of the same sign (that is with the extra half plane on the same side of the slip plane).

Will the dislocations attract or repel one another?

Hint: Consider what would happen to the elastic energy if the two dislocations were brought together to form a single dislocation.

They will repel one another. If each of the two original dislocations has a Burgers vector $b$, and the two combine, they would create a new dislocation with a Burgers vector of $2b$. Since the elastic strain energy per unit length is proportional to $b^2$ (see 'Elastic energy of a dislocation') the combined dislocation will have an elastic energy twice that of the sum of the energies of the original two ($4b^2$ compared with $2b^2$). Hence a repulsion. In order to move the dislocations closer together, a greater stress must be applied to the crystal, so dislocation interactions like this are one source of work hardening. Figure 2.58 is an example of what can happen. The dislocations in the row are all of the same sign and they are 'piled up' against an obstacle, in this case a grain boundary, a situation we consider in Section 6.8.2. To produce more slip, a higher stress is needed in order to overcome the repulsive force between the dislocations and move them closer together and towards the obstacle.

SAQ 6.9 (Objective 6.6)
Will the two dislocations in Figure 6.36 attract or repel?

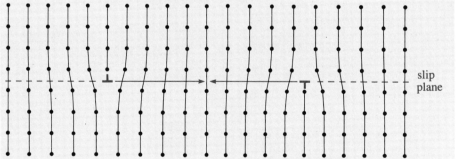

Figure 6.36 Two edge dislocations of opposite sign on the same slip plane

Another important work hardening mechanism depends on the spacing between the dislocations in a crystal and the ease with which they can bend between obstacles.

Since a dislocation has an elastic energy, its energy will be a minimum when it is straight, that is, when it is short as possible. So, when a dislocation is forced into a curved shape there will be a restoring force acting to straighten it. The restoring force is called the **line tension** and is equal to the dislocation energy per unit length.

The flexibility of a dislocation is important for the following reason. Consider a dislocation which is moving across a slip plane and encountering obstacles such as a 'forest' of dislocations moving on intersecting planes, as shown in Figure 6.37. If the dislocation cannot move through these obstacles it must bend between them. In order to bulge through a gap, the dislocation must bend to a semicircle (thereafter its radius increases again). A greater force is required to bend the segment of dislocation to a semicircle between the obstacles at A and B than between C and D because AB < CD. A useful analogy (from M. Ashby and D. Jones, *Engineering Materials — an Introduction to their Properties and Applications*, Pergamon Press, 1980) is that it is rather like blowing up a balloon against the bars of a birdcage. You can imagine that a hard blow is needed to bulge the balloon between the bars after which further expansion is easy. The closer together the bars, the harder you would have to blow. So the closer together the dislocation 'trees' in the 'forest' the more effective the barrier.

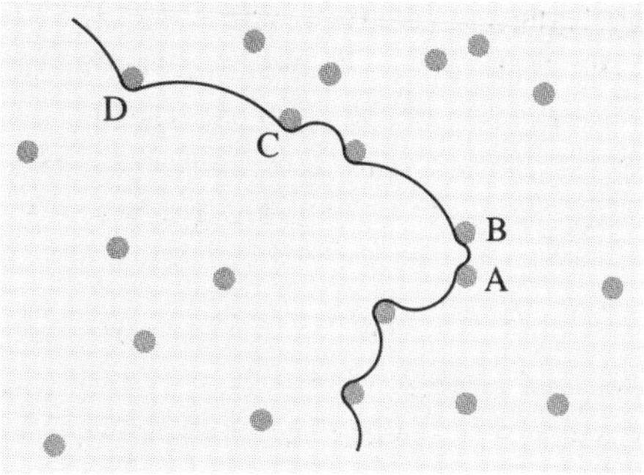

Since the dislocation density increases with plastic strain the average spacing between dislocations decreases. This means that an ever-increasing shear stress is required to bend dislocations moving between the others in the tangle. So the yield stress depends on the dislocation density.

Figure 6.37 A dislocation meeting obstacles such as forest dislocations

## 6.8.2 Reducing the grain size

Grain boundaries are effective obstacles to dislocations because they interrupt the continuity of the slip planes in a crystal. Figure 5.1 illustrates this. A grain boundary is a region of atomic disorder and a dislocation simply cannot pass through it. Since each grain in a polycrystal is surrounded by a grain boundary, a dislocation can move only within the grain in which it was created. The more grain boundaries there are in a material (the smaller the grain size) the more difficult it is to plastically deform the material.

## 6.8.3 Solid solution strengthening

As you know from Chapter 5, to form solid solutions the atoms or ions of the two components must be fairly similar. However, they are not identical and the differences are sources of increased resistance to dislocation movement. The main effect is due to the difference in size between the solvent and solute atoms. Like dislocations, they produce elastic strain fields which interact.

EXERCISE 6.4 Consider the atom sites A and B in Figure 6.32. In to which sites would you put (a) a large substitutional solute atom and (b) a small substitutional solute atom in order to minimize the elastic energy?

These are attractive interactions between solute atoms and dislocations which, because they reduce the overall elastic strain energy, make the dislocation 'reluctant' to move on and leave the solutes behind. A larger force is needed to move the dislocation away from the solute.

## 6.8.4 Precipitation and dispersion hardening

This form of strengthening is based on the obstruction of dislocation movement by small hard particles of a second phase. To be most effective, the particles should be uniformly distributed, very small and close together.

Suggest a way of producing a fine distribution of particles.

By precipitating them from a supersaturated solid solution; this method and the precipitation hardening it produces were considered in Section 5.3.2. Of course particles can be produced in other ways, such as mixing powders of a metal and a ceramic ($Al_2O_3$ say) and then compacting and sintering them. Such dispersions give rise to what is called **dispersion hardening**. Of course, the strengthening mechanisms are the same as for precipitation hardening.

In general, the particles of ceramic or intermediate compound have complex crystal structures for which the intrinsic resistance to dislocation motion is very high. So, dislocations cannot move through them. In order to keep moving, the dislocations must bulge between the particles — it is the balloon in a birdcage problem again. There is a limit to the strengthening that can be obtained: when the particles are too close together the stress to bow dislocations between them becomes so high that either the particles or the matrix crack and the material fails.

These, then, are the four main methods of manipulating structure to produce an increase in yield stress. ▼**Strengthening of copper**▲ provides some indication of their relative effectiveness. So far, we have considered ways of increasing the yield strength in order to improve performance in service. This invariably leads to a reduction in ductility.

Further increases in yield strength could be achieved at the expense of ductility. Why is this undesirable?

To maintain toughness. By inhibiting crack growth plastic flow is a most effective toughening mechanism. It also provides a safety mechanism for artefacts that are overloaded in service. By yielding it can shed its load onto other components.

We now look at the other side of the coin — how to encourage extensive plastic strain. This is obviously a prime property requirement for the shaping of products from the solid.

# ▼Strengthening of copper▲

Each of the strengthening mechanisms is used to produce commercial copper alloys, so copper provides a good basis for comparison — so do aluminium, nickel, iron and others incidentally. Table 6.3 gives typical values of yield stress and ductility at room temperature.

Using alloy *ii* as a basis for comparison, the data reveal the following important points:

• Solid solution hardening is a useful source of strengthening (alloy *iv*); work hardening is very potent (alloy *iii*) and precipitation hardening is the most effective (alloy *vii*).
• Work hardening is roughly additive with solid solution hardening (compare alloy *ii* with alloy *iii*, and alloy *iv* with alloy *v*) and supplements precipitation hardening (alloys *vii* and *viii*)
• The general observation that increasing the yield stress decreases ductility is borne out. Notice however that even the strongest commercial alloy has some ductility.

Table 6.3

| Composition | | Condition | Mechanism | Yield stress /MPa | % Elongation | Typical engineering uses |
|---|---|---|---|---|---|---|
| i | 99.99% Cu | specially grown single crystal | intrinsic resistance | 10 | 100 | not commercial |
| ii | 99.99% Cu | annealed average grain size ≈ 0.01 mm | grain size | 60 | 55 | electrical conductors of all sorts |
| iii | 99.99% Cu | cold worked | work hardening | 325 | 4 | electrical contacts |
| iv | 75% Cu–25% Ni | annealed | solid solution hardening | 140 | 40 | boiler tubes |
| v | 75% Cu–25% Ni | cold worked | solid solution and work hardening | 390 | 15 | 'silver' coinage |
| vi | 98% Cu–2% Be | solution treated | solid solution hardening | 185 | 45 | |
| vii | 98% Cu–2% Be | precipitation hardened | precipitation hardening | 1100 | 5 | |
| viii | 98% Cu–2% Be | solution precipitation hardened and cold worked | precipitation and work hardening | 1260 | 2 | springs and pressure tubing |

SAQ 6.10   (Objective 6.7)
Which of the following statements is incorrect?

(A) An interstitial carbon atom in FCC iron (austenite) would prefer to be on the tension side of an edge dislocation.
(B) Two negative edge dislocations moving on the same slip plane will repel one another.
(C) Within a grain boundary a dislocation loses its identity.
(D) Large widely separated strong precipitates will be more effective strengtheners than small closely spaced strong ones.

# 6.9  Ductility for processing

## 6.9.1  What limits the extent of plastic strain?

EXERCISE 6.5   What does limit the extent of plastic strain?
Hints: Think of the consequences of strengthening outlined in the previous section.
Compare Figures 2.16(a) and 2.18.

It is important to appreciate that, as well as a large ductility a low yield stress is also desirable for processing purposes. It has considerable implications for the size, power and cost of rolling mills, presses, extruders and so on.

The mechanical working of metals can be carried out hot or cold; the distinction and the main types of process were discussed in Chapter 1. As you will see shortly, cold working has considerable advantages compared with hot working, but the amount of plastic strain possible is limited by the work hardening that occurs. However, the situation can be remedied by annealing.

## 6.9.2 Annealing

The effects of cold working can be removed by heating the metal to a sufficiently high temperature. Figure 6.38 shows a typical tensile test for a ductile material that work hardens. If the specimen is unloaded at the point B, the material contracts in Hookean fashion along BC and if it is retested it does not yield until point B. However if, after unloading, we subject the sample to a heat treatment at a temperature of about $0.7\,T_{m}$ then retest the sample we obtain the following results.

The material yields at point D which is lower than B and at the same stress level as A, the original yield stress, and then it work hardens as before. The final fracture will occur at point E. Although it represents a strain to fracture of 0.65, this is in addition to the strain of 0.1 achieved before heat treatment. The overall deformation (shape change) of the specimen that it was possible to achieve has been increased. We could repeat this heat treatment cycle at a point such as F and produce even more deformation without exhausting the ductility. This technique is used in wire drawing, rolling and so on for shaping metal components. It allows large shape changes to be accomplished if heat treatments are repeated frequently. This technique is known as **annealing**.

In the annealing of severely cold worked metals, it is possible to identify three temperature ranges in which different phenomena occur — recovery, recrystallization and grain growth.

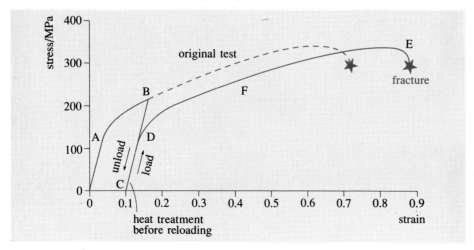

Figure 6.38 The effect of heat treatment on the stress–strain curve

## Recovery

Figure 6.39 shows the effect of temperature on three properties of cold worked copper; hardness, electrical resistivity and density. As the copper is heated progressively up to 580 K (about $0.4\,T_m$) there is a gentle increase in density and a gentle fall in resistivity and hardness. The process is called recovery; the properties recover towards their original values with no visible change in the microstructure as viewed by optical microscopy.

The source of recovery is spontaneous reduction in the dislocation density by mutual annihilation. This process is caused by the attractive elastic interaction between dislocations of opposite sign. You met an example of dislocation annihilation in SAQ 6.9. However, this can happen by slip alone only when two dislocations of opposite sign lie on precisely the same atomic plane.

Edge and screw dislocations behave differently in this respect. A dislocation can move by slip only on a plane that contains both its line and its vector. For an edge dislocation this defines a specific plane. A screw dislocation is not constrained in this way; since this line is parallel to its vector, it can slip on any plane on which it lies. If you are uncertain about this, look back at Chapter 2 where the geometry of slip was considered.

So recovery by the annihilation of screw dislocations of opposite sign can occur readily — they always have a common slip plane — but edge dislocations are different. Edge dislocations can annihilate if they happen to be slipping on the same plane. At low temperatures such occurrences will not be frequent. However at higher temperatures the number of encounters will become much greater because thermal energy is sufficient to allow edge dislocations to move out of their respective slip planes onto a common plane on which they can slip towards one another and cancel out. ▼Reducing dislocation density by climb and slip▲ describes what happens. Not all of the dislocations are eliminated during recovery. A minority, typically about $10^{14}\,\mathrm{m^{-2}}$, survive but this is still a very large dislocation density compared with that of an undeformed metal.

## Recrystallization and grain growth

If you glance back at Figure 6.39 you will see that when the annealing temperature is raised to 673 K there is a sharp fall in the hardness and electrical resistivity of copper. This is accompanied by a major reformation of the grains. New grains differing in size, shape and orientation appear in place of the old grains. These grains can be seen by optical microscopy, Figure 6.41. Whereas the old grains were highly distorted by slip, the new grains are more equiaxed and contain a much smaller density of dislocations, of the order of $10^{11}\,\mathrm{m^{-2}}$. This process is aptly known as **recrystallization** and it occurs when cold worked metals are annealed at temperatures above about $0.4$–$0.5\,T_m$. The incentive or

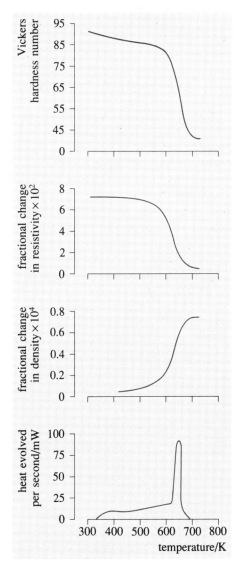

Figure 6.39 Some physical properties of cold worked copper as a function of temperature

# ▼Reducing dislocation density by climb and slip▲

Dislocation climb is the movement of an edge dislocation in a direction normal to its slip plane, that is the extra half-plane of atoms gets longer or shorter. In particular, if the central A atom in Figure 6.32 is replaced by a vacancy the dislocation line moves up (climbs) by one atomic spacing, but only in the plane of atoms shown.

How can this removal of atoms be achieved?

By self diffusion. As you know from Chapter 4, it occurs by the migration of vacancies. As temperature increases, the equilibrium concentration of vacancies increases and so does diffusion. Every time a vacancy arrives at the end of the half plane, the dislocation climbs.

Figure 6.40 shows schematically, how two edge dislocations of opposite sign can annihilate one another by a combination of climb and slip. Each dislocation undergoes climb until both lie on the same slip plane. Since edge dislocations of opposite sign attract, they move together by slip and annihilate one another. The process of climb is restricted to edge dislocations. Screw dislocations are not associated with 'extra material' and so vacancy migration cannot help the dislocations to move.

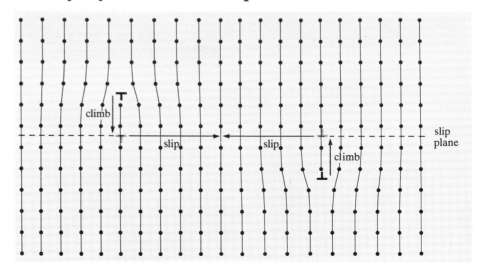

Figure 6.40 Annihilation of an edge dislocation by slip and climb

Usually, the dislocations in crystals are of the mixed type; each consists of some segments which are predominantly edge (with the slip vector perpendicular to the line of the dislocation) and segments which are mainly screw (with the slip vector parallel to the line of the dislocation). Generally, edge dislocations are less mobile than screw dislocations because they can only move by slip on one plane.

It is the segments of edge dislocation which restrict the mobility of a mixed dislocation. Raising the temperature to the range in which self-diffusion occurs readily permits the edge segments to move by climb with a consequent increase in the mobility of the whole length of the dislocation. You will see the significance of this when we consider the creep of crystalline materials in Section 6.10.

'driving force' for recrystallization is the energy of the dislocations that are annihilated by the process. The evolution of this energy as heat can be detected during recrystallization using a sensitive calorimeter (see bottom of Figure 6.39). In effect the materials is now 'as good as new'. This is the explanation for the decrease in the yield stress produced by annealing of the specimen illustrated in Figure 6.41.

(a)  0    500 µm

(b)  0    500 µm

Figure 6.41 Micrographs of copper specimens (a) cold worked (b) recrystallized

If heating of the metal is continued after recrystallization has occurred, the average size of the grains increases. This is known as **grain growth**. The incentive for this to happen comes from the decrease in the total area of grain boundary. Because the atoms near a grain boundary are displaced from their equilibrium sites, a boundary has a higher internal energy; grain growth leads to a decrease in this energy. The same sort of thing happens in a foam of bubbles, in order to reduce the total surface energy, some bubbles grow at the expense of others.

## Working and annealing

We have considered cold working and annealing as two separate but complementary operations: cold working hardens a metal and annealing softens it again.

So, what is hot working?

It is the mechanical working of metals at temperatures at which the annealing mechanisms occur concurrently with plastic deformation; this is typically at about $0.5\,T_m$. When processing conditions settle down into steady state flow there is a dynamic balance between the rate of work hardening and the rate of softening due to recovery and recrystallization. This balance enables metals to be hot worked to large strains using only modest stress. So, which is best, hot working or cold working? ▼Hot working or cold working?▲ summarizes the arguments.

SAQ 6.11 (Objective 6.8)
Distinguish between the two ways by which dislocations can move in crystals.

So, there is a means of extending the ductility of cold worked metals, what about the problem of necking? One approach is to try to prevent necking when the material is deformed in tension (drawn); the other is to produce the same shape using compression — by extrusion say. Let's first consider what happens when a material is drawn.

## 6.9.3 Drawing

Drawing can be very straightforward: a glass blower can make a fibre simply by pulling apart the two ends of a hot glass rod. Thermoplastic melts can also be stretched extensively without breaking, rather like chewing gum. If a small 'neck' happens to exist in one part of the rod, it will not shrink in cross-section any more rapidly than the rest of the rod and so the material can be stretched without the neck developing rapidly into a fracture.

# ▼Hot working or cold working?▲

The main difference between cold working and hot working is work hardening: it occurs in cold working and not in hot working. Consider each of the processes.

## Hot working

At temperatures above about $0.5\,T_m$, a metal is soft and doesn't harden with plastic flow. So large shape changes are possible. That is why it is very popular for the initial processing stages for many products. A big advantage of hot working is that, because metals are soft at these temperatures, the forces required to deform them are low; hence the presses, mills and so on can be small and relatively inexpensive.

The main disadvantages of hot working are to do with the surface of the product. At hot working temperatures most metals oxidize readily. This means a poor surface finish and dimensional accuracy in the product; due to degradation of both the product and the machine tool. Also, of course, since no hardening occurs, the product is still soft.

## Cold working

The rolls, dies and presses have to exert large forces in order to overcome the increasing work hardening and yield stress. So the machines are big and expensive. On the other hand, the product has a good surface finish and has been strengthened by work hardening. The metal can then be annealed and the process of recrystallization controlled to give a very fine grain size. Further cold work then produces material strengthened by both work hardening and small grains. For these reasons cold working is a popular finishing technique.

Other important limitations are that the product obviously has a lower ductility. Also to produce further shape changes, the metal has to be annealed. This 'work then anneal' sequence is not always well suited to continuous production processes.

This behaviour is a consequence of the way that hot glasses and polymers flow. To understand this, we begin by defining the strain $\varepsilon$ as being equal to the fractional change in length $l$.

$$\delta\varepsilon = \frac{\delta l}{l} \tag{6.8}$$

Since the volume of a material is not changed by flow, the strain is also equal to the fractional change in cross-sectional area $A$.

$$\delta\varepsilon = -\frac{\delta A}{A} \tag{6.9}$$

The minus sign is included because as the length increases the area must decrease. If the length increases by 1% say, then the cross-sectional area must decrease by 1% in order to keep the volume unchanged. From equation 6.9, the rate of change of strain with respect to time $t$ is

$$\frac{d\varepsilon}{dt} = -\frac{1}{A} \times \frac{dA}{dt} \tag{6.10}$$

If a rod of hot material is stretched, it is found that the rate of strain produced in the rod is proportional to some power $n$ of the longitudinal stress $\sigma$.

$$\frac{d\varepsilon}{dt} = k \times \sigma^n \tag{6.11}$$

where $k$ is simply a constant of proportionality. By definition $\sigma$ is the force $F$ along the rod divided by the cross-sectional area $A$. Substituting this into Equation (6.11) and combining it with Equation (6.10) we get:

$$-\frac{1}{A} \times \frac{dA}{dt} = k\left(\frac{F}{A}\right)^n \tag{6.12}$$

From this we conclude that the rate at which the cross-section of the rod shrinks during stretching is

$$\frac{dA}{dt} = -k \times \frac{F^n}{A^{n-1}} \tag{6.13}$$

In hot glass the rate of straining varies linearly with applied stress and the power $n$ is approximately one. In this case $dA/dt$ does not depend at all on $A$ and all cross-sections of the rod shrink at the same rate; necks do not tend to shrink preferentially. This is also approximately true for thermoplastics melts.

When a typical metal rod is stretched, the strain rate depends usually on about the fourth power of the stress ($n = 4$). In this case you can see that the rate at which the cross-section shrinks must be proportional to $A^{-3}$, so narrow cross-sections will shrink much faster than wide ones and necks will shrink at an ever increasing rate until fracture occurs. For this reason metals cannot usually be drawn extensively in tension.

However, as earlier stress–strain curves have shown, metals can be stretched to small strains before a neck forms. Why does a neck not develop as soon as the metal yields? The following argument shows that this is due to work hardening by the metal.

Suppose that a metal rod is being stretched and one part of the rod stretched slightly more than the rest. A small neck appears in that part of the rod and as its cross-section shrinks so the stress within the neck increases above that of other cross-sections. At small strains, the strength of the metal increases sufficiently (work hardens) to match this increase in the applied stress and the neck is stable; the rest of the specimen length thins too. However, as the strain in the rod increases, so the work hardening rate of the metal decreases (as shown by the slope of the stress–strain curve) and a point is reached at which an increment of stress in the neck exceeds the increment in work hardening. The neck then deforms at an accelerating rate until it breaks.

## 6.9.4 Optimizing processing and service performance

The yield and flow of materials highlights very well the conflict between properties for performance and for processing. Essentially, for easy shaping a low yield stress and high ductility are desired. On the other hand, to perform well in service a high yield stress is usually necessary together with some ductility (for toughness). One of the reasons why heat treatments like annealing are so vital is that they can often resolve the conflict.

The general approach can be summarized as:

(a) Treat the material so that it has a microstructure that is easily worked to shape — as soft and ductile as practicable.
(b) Shape it.
(c) Heat treat it to obtain the microstructure with the desired properties for service — low electrical resistivity, high hardness, high magnetic remanance and so on.

Of course, the approach is useless for brittle materials, but the same philosophy is at work. For instance:

• Fine powders of clay, oxides, tungsten etc. are utilized by forming a slurry which is poured into a mould and then fired (heat treated) to remove the liquid.

• The thermosets in GFRP are not hardened until the composite has been shaped as required.

Here are two examples involving metals. First, look back at 'A hacksaw blade' in Chapter 2. The blades are formed and the teeth cut and set with the steel in a soft and ductile condition. It is then heat treated — the processes of quenching and tempering discussed in Chapter 5 — to produce hard (tempered martensite) teeth. A nice point about this example is that it also demonstrates the idea of treating the material to have the properties (microstructure) where it is needed: hard in the teeth, but still tough round the fixing pins.

Second, the main final shaping process for the copper wires used in electrical power conductors is cold drawing. They are then annealed at about 700 K. (The melting temperature of copper is 1335 K.)

EXERCISE 6.6   (a) Explain why the cold drawing and annealing steps are used. (b) Sketch a PPPP tetrahedron that summarizes the links for annealing at 700 K.

SAQ 6.12   (Objective 6.9)
Consider alloy c in Figure 5.40.

(a) Which method of strengthening has the most potential?
(b) At which stage of the heat treatment would you work the alloy to its required shape? Explain your answer.

So far in this section on yield and (plastic) flow I have largely ignored the effect of time under stress. It is clearly important in determining the behaviour of polymers and also in the plastic behaviour of crystalline materials. This is the subject of the next section.

# 6.10 Understanding, avoiding and utilizing creep strains

The permanent (plastic) strain produced in crystalline materials with time under constant load is called **creep**. Don't confuse this with the (recoverable) viscoelastic strain considered in Section 6.5.2. However, the two phenomena do have something in common: they are both thermally activated. They become important at temperatures above about $0.4\,T_m$, hence, for example, thermoplastics creep at ambient temperatures and so does lead, see Figure 6.8, but special alloys of nickel can operate for many hours in jet engines at more than 1100 K (see 'Hot material for jet engines' in Chapter 3) and ceramics such as alumina in furnace linings can do even better. ▼A creep experiment▲ shows how you can observe creep for yourself.

## 6.10.1  Creep in service

Figure 6.43 illustrates the typical creep curves observed in metals for a range of stress and temperature. Generally, the strain–time curves show four distinct regions:

(a) **Instantaneous strain** — this is the strain that occurs immediately the stress is applied.

(b) **Primary creep** — the period during which the creep rate ($d\varepsilon/dt$) decreases.

## ▼A creep experiment▲

You need a coil of 'electrical' solder — the sort used to solder electrical contacts. It is based on the eutectic Pb–Sn alloy and has a melting temperature of about 455 K. Ambient is greater than $0.6\,T_m$; Figure 5.25(c) shows the phase diagram.

Coil the solder about six times round a stick (such as a broom handle) into a tight spiral, remove it and suspend it from one end. Observe the lengthening of the spiral over a period of several hours. If you would like to measure the change, a ruler will do. I suggest you put a marker about two coils down the spiral and measure the movement of this point as well as the bottom end.

Figure 6.42 shows my results. The self-weight of the spiral is sufficient to develop the modest stresses needed to produce permanent creep strain which opens out the spiral with time. Afterwards, if you lay the extended spiral horizontally it will not contract, as it would if the deformation were viscoelastic.

Since the length of wire in each coil is the same, the stress at each point is proportional to the number of coils below it. So you can also explore the effect of stress on creep rate.

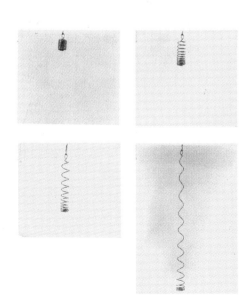

Figure 6.42 Creep in solder spirals. This creep was observed over several hours

(c) **Secondary creep**—here $d\varepsilon/dt$ is a constant; it is often called **steady-state creep.**

(d) **Tertiary creep** — the creep rate accelerates and leads to fracture.

Secondary creep is of most interest by far because it is the region in which products operating in creep conditions spend the greatest part of their life. It is this region which is of most concern in design. So, what determines the rate?

At elevated temperature the yield stress decreases with increasing temperature. This is because dislocation motion is thermally activated, which is what you would expect from 'Reducing dislocation density by slip and climb.' Above $0.3\,T_\mathrm{m}$, suffcient thermal activation is available for dislocation movement to occur at stresses below the normal yield stress. What is more, this deformation will continue for a long period at a constant value of stress: metals do not appear to exhibit the usual work hardening behaviour. The fact that creep strain continues indicates that dislocations are moving and multiplying without hindrance. The answer is that work hardening is offset by recovery. Dislocations move, interact and tangle (the work hardening process), but the rate of entanglement is matched by a rate of disentanglement (the recovery process of dislocation annihilation).

As creep is a thermally activated process it should be possible to determine the activation energy for secondary creep and to check whether it correlates with this mechanism.

Which atomic event would you expect the activation energy to depend on?

Self-diffusion. This is the process on which dislocation climb, hence recovery, relies.

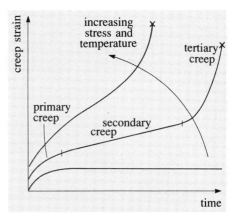

Figure 6.43 Typical creep curves for different stresses and temperatures

EXERCISE 6.7 Specify the steps you would follow in order to determine the activation energy for secondary creep. Hint: Thermally activated processes were considered in Chapter 4.

In fact, the activation energy does turn out to be close to that for self-diffusion, so recovery is the critical mechanism.

Of course, for products which creep in service, materials with low creep rate are needed. So what strategies are there for producing them? There are two approaches: limit dislocation motion and/or limit self-diffusion.

SAQ 6.13 (Objective 6.10)
For each of these two approaches suggest structural/property features that would restrict secondary creep at a given temperature.

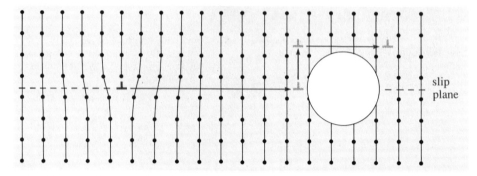

Figure 6.44 Dislocation climb over particles

Mainly because they are larger, precipitates are better than solutes which is why precipitation hardening systems are commonly used for high temperature alloys. But precipitates are not completely successful because dislocations can still climb over them, albeit taking a long time to do it. Figure 6.44 illustrates very schematically what happens. It is similar to dashing down a motorway to the next traffic jam! Just like other properties we have discussed in this chapter, whilst creep can produce problems in service, it can also provide processing opportunities. Let's briefly explore an example of growing importance.

## 6.10.2 Superplastic forming of metals

Superplasticity is the ability of some metal alloys to undergo extremely extensive (100% and more) uniform plastic flow without necking; they behave like thermoplastics and glasses in the viscous state.

The shaping method (Figure 6.45) has a strong affinity with the vacuum forming of thermoplastics — see Section 1.6.2. A sheet of 'superplastic' metal is clamped over a mould, heated to the appropriate temperature and slowly forced into the shape of the mould using air pressure. Although this is a common method of shaping thermoplastics, until recently it was very unusual for metals. This is because metals exhibit superplasticity only under very stringent conditions of grain size, deformation rate and temperature.

The most important of these is the deformation rate; the applied stress, grain size and temperature requirements are the means of providing this rate. In Section 6.9.3 the necking of metals in tension was explained in terms of the parameter $n$ in Equation (6.13); typically for metals $n = 4$, whereas $n \approx 1$ for glass and thermoplastics. Now, metals treated such that the grain size is extremely small ($< 10\,\mu m$) and deformed extremely slowly (about $10^{-3}\,s^{-1}$) have values of $n$ of about 0.9. Note just how slow this rate is compared with, say, forging, in which a sheet 1 m long will stretch by 1 mm per second.

Clearly, to give this unusual viscous-like behaviour the deformation mechanism must be rather different from the dislocation movement we have been considering. In fact, the important agents are vacancies not dislocations, as ▼**Diffusion creep**▲ explains.

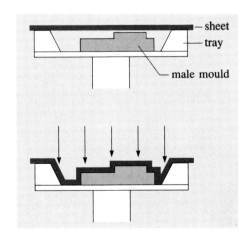

Figure 6.45 A simple version of superplastic forming

# ▼ Diffusion creep ▲

It is possible for a crystal under stress to steadily change its shape with time (that is, flow plastically) by the movement of atoms. This is called diffusion creep. Figure 6.46 represents a grain in a polycrystalline material under simple tension. Atoms migrate preferentially so that the crystal elongates and the applied stress is relieved. The flow of atoms in one direction is equivalent to a flow of vacancies in the opposite direction.

Two diffusion paths can be involved: normal diffusion through the body of the crystal, and diffusion along the grain boundary.

This sort of creep becomes important when creep by dislocation movement is negligible, in ceramics for instance, and in metals at stresses below those for slip. Also, of course, it is significant only at temperatures at which diffusion is reasonably rapid; typically $0.4\,T_m$. The rate of creep depends on the diffusion coefficient for the mechanism(s) in operation.

Would you expect grain boundary diffusion to dominate bulk diffusion at lower temperatures?

Yes, because diffusion is quicker through the disordered grain boundary regions. But it isn't quite this simple because there are many more 'diffusion paths' through the volume of a crystal than over the area of a grain boundary.

Is the grain size important?

Yes, the smaller the grain the shorter the distance atoms have to diffuse. This is why superplasticity requires such small grains.

Finally, with all grains in a polycrystalline material elongating in the same way, holes would be created at grain boundaries. To maintain contact some readjustment of grains is needed. The phenomenon of superplasticity depends on diffusion creep and in this case the grains accommodate their relative changes in shape by sliding round one another.

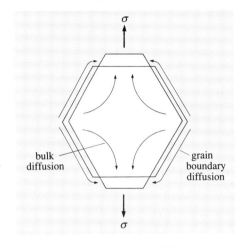

Figure 6.46 Mechanisms of diffusion creep

Popular commercial superplastic alloys are based on aluminium and titanium and they are used in products as diverse as camera housings, door and body panels for road vehicles and covers for plane undercarriages. They have some important advantages over plastic rivals, particularly that they can have very deep drawn sections and sharp corners. However, they are produced by creep and should only be used at temperatures well below those at which creep is a problem, that is $\ll 0.5\,T_m$. Ambient is no problem for aluminium and titanium superplastic alloys.

So far we have considered the mechanical behaviour of materials both for processing and for performance in service. I want to end the chapter with a brief description which is exclusive to service, namely the failure of materials by fatigue.

## 6.11 Fatigue failure in service

Consider two identical tensile test specimens, one of which is subjected to a constant stress of $\frac{1}{2}\sigma_y$ (that is half the yield stress) and the other to a stress which alternates continuously between stress values of 0 and $\frac{1}{2}\sigma_y$. Which will break first?

Common sense would suggest the former, because the latter spends most of the time under stress less than $\frac{1}{2}\sigma_y$. But, the latter breaks first. Apparently there is a loss of performance or 'fatigue' that comes from using a fluctuating stress. The effect is called **fatigue**. The majority of failures in structures and machines are probably due to fatigue. Fatigue occurs in most materials but here we will concentrate on metals.

335

Measuring the response of a metal to fatigue is notoriously difficult because it is affected by so many variables, particularly the form of the stress cycle, the surface condition of the material and the size and shape of the specimen. ▼Assessing fatigue behaviour▲ describes the usual approach.

Fatigue failure requires a fluctuating tensile stress and is a consequence of highly localized plastic deformation which, with its attendant work hardening continues cycle after cycle until a short sharp crack is formed. The cracks are usually initiated at a stress concentration on the surface and propagate into the material in a series of discrete jumps. Each cycle may advance the crack a little further. Some of the positions at which the crack is arrested are readily seen on the fracture surface by optical microscopy, Figure 6.49, and are usually called **beach markings** because they resemble the ripple marks on a beach. Scanning electron microscopy reveals that the ostensibly smooth regions between these markings in Figure 6.49(a) are similarly marked on a finer scale — they are called **striations**. At some stage the crack reaches the critical length

# ▼Assessing fatigue behaviour▲

Fatigue tests are usually carried out by applying known cyclic stresses to specimens of standard size and shape and counting the number of cycles required to produce failure. Various types of stress cycle can be used, from a simple sinusoidal applied in a bending configuration to a complex programmed push-pull sequence of different amplitudes to simulate the anticipated service stress, Figure 6.47.

When a cyclic stress of constant amplitude is used for each test, the resulting data can be plotted as a graph of stress amplitude $S$ versus number of cycles to failure $N$, as indicated in Figure 6.48.

Figure 6.48(a) is a typical $S$–$N$ curve for a ferrous material. Notice that it flattens out to give a value of stress amplitude below which fatigue failure is unlikely to occur no matter how many cycles are applied. This stress is called the **fatigue limit**. It is used as a design criterion both in determining the intended overall stress level in a component and in consideration of geometrical features which produce stress concentrations: holes, corners, keyways and so on.

Figure 6.48(b) is typical of a nonferrous material, where the curve continues to fall, flattening out gradually but not completely. Such materials do not have a clearly defined fatigue limit. So the design criterion adopted is the fatigue strength at

a given number of cycles which is unlikely to be exceeded in the lifetime of the component, usually $10^8$. In other words, with nonferrous metals, instead of using the fatigue limit, a designer would use the stress level at which failure within $10^8$ cycles is unlikely.

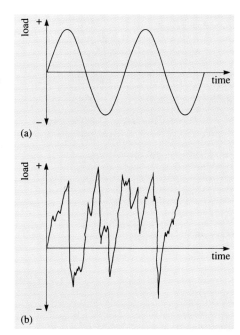

Figure 6.47 Loading spectra (a) simple sinusoidal; (b) random

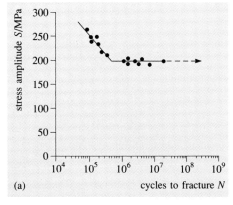

(a)

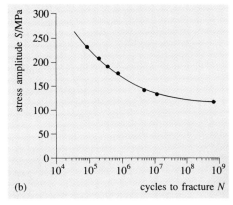

(b)

Figure 6.48 Typical $S$–$N$ curves for (a) mild steel; (b) cold worked copper

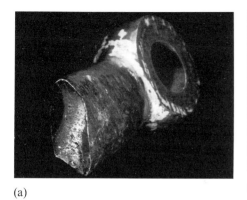

(a)

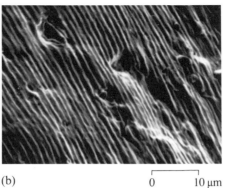

(b)

0        10 μm

Figure 6.49 (a) Beach markings on the fatigue fracture of a crankshaft; (b) scanning electron micrograph of a region between beach markings

at which rapid fracture occurs. In practice, the fractured surface consists of two areas: one area bearing beach markings — often disfigured by the corrosion that occurs whilst the crack propagates — and the remaining area with either a relatively smooth surface (brittle failure) or a rough, torn surface (ductile failure) produced when the area can no longer sustain a load.

Although fatigue in metals is associated with slip, it is of such a localized nature that very little overall plastic deformation is produced during development of a crack. Consequently, when the two halves of a fatigue fracture are placed together, they reveal no more distortion than occurs with a brittle fracture. It is this lack of visible distortion that makes fatigue cracks so difficult to detect in service prior to the final catastrophic failure. All that shows is a faint hair-like mark which may well be obscured by surface dirt or rust.

## 6.11.1 Prevention of fatigue

From this description of fatigue failure, three strategies for improved fatigue resistance can be surmised: preventing plastic deformation, removing stress concentrations and inhibiting crack nucleation and/or growth. Let's look at each in turn.

How can the extent of plastic deformation under a given stress be reduced?

By increasing the yield stress (by the mechanisms discussed in Section 6.8). Of course, if the increase is accompanied by an enhanced susceptibility to cracking, no improvement may result.

Stress concentrations can arise from the following sources: microstructural features of the material such as pores and sharp particles of second phase; the surface condition of a component, for instance scratches, a rough surface produced by machining, pits caused by corrosion; and the overall design of the product, of which holes, corners, oil holes and keyways are particularly important. Clearly all should be avoided.

Crack nucleation and growth only occur under tension, so they can be inhibited if the material is in a state of compression.

A general way of achieving this is to induce a structural volume expansion in the surface of a component — the important region. It is particularly applicable to steels. For instance, there is a volume expansion when austenite is quenched to form martensite. Suppose the surface of a component is heated into the austenite region and then quenched. On quenching, the underlying material restrains the surface expansion that should occur when the surface layer transforms to martensite. So high residual compressive stresses are created in the surface layer and balancing tensile stresses within the 'core'. This technique is widely used in components such as drive shafts in motor vehicles.

SAQ 6.14   (Objective 6.11)
List possible ways of achieving greater resistance to fatigue failure under the headings (a) component design and (b) materials selection.

## Objectives for Chapter 6

After studying this chapter you ought to be able to:

6.1 Decide, giving your reasons, whether a product is (a) microstructure dominated, (b) shape dominated or (c) dominated neither by microstructure nor by shape (SAQ 6.1).

6.2 Specify a method for coping with the viscoelastic behaviour of a polymer in the design of a load bearing component (SAQ 6.4).

6.3 Describe a model for rubbery elasticity which accounts for changes in modulus with temperature and strain, and for the effects of cross-linking (SAQ 6.5).

6.4 Explain the influence of polymer structure on the temperature dependence of the elastic modulus (SAQ 6.6).

6.5 Describe what happens to the microstructure of a thermoplastic on cold drawing (SAQ 6.7).

6.6 Predict the result of interactions between two given dislocations (SAQ 6.9).

6.7 Describe in terms of dislocations the basis of strengthening by other dislocations, grain boundaries, solute atoms and particles of another phase (SAQ 6.10).

6.8 Explain how dislocation climb affects recovery (SAQ 6.11).

6.9 For specified material conditions explain how the microstructure can be modified to suit processing and product performance (SAQ 6.12).

6.10 Specify, and explain, the structural features that inhibit secondary creep (SAQ 6.13).

6.11 Suggest ways of increasing the fatigue resistance of a component (SAQ 6.14).

6.12 Define or distinguish between:

viscoelasticity
creep in polymers
initial/tangent/secant modulus
rubber elasticity
cold drawing
fibrils
independent slip system
elastic energy of a dislocation
dislocation density
dislocation climb
cold work/hot work
dislocation line tension
annealing
recovery
recrystallization
grain growth

primary/secondary/tertiary creep
superplasticity
principal stress
uniaxial/biaxial tension
hydrostatic pressure (triaxial
   compression)
triaxial tension
ductile–brittle transition
vulcanization
Burgers vector of a dislocation
diffusion creep
fatigue
fatigue limit
S–N curve
beach markings

# Answers to Exercises

**EXERCISE 6.1** The values I obtained for the merit index $\rho/E^{1/2}$ are:

| | |
|---|---|
| wood (spruce) | $190\,\mathrm{kg\,m^{-3}\,GPa^{-1/2}}$ |
| CFRP | $137\,\mathrm{kg\,m^{-3}\,GPa^{-1/2}}$ |
| aluminium | $320\,\mathrm{kg\,m^{-3}\,GPa^{-1/2}}$ |

On this single merit index, CFRP has a clear advantage. But it is very expensive!

**EXERCISE 6.2** Essentially there will be two very different types of force–separation curve. One will represent the primary covalent bonds between carbon atoms within a molecule. It will be of the type shown in Figure 6.50(a). The other will represent the secondary bonds between side groups on neighbouring molecules. This will have a low gradient and give rise to a low modulus, as in Figure 6.50(b).

So the covalent bonds are stiff and the secondary bonds are floppy. Under a given stress, the strain in a thermoplastic originates mainly from the floppy secondary bonds. In diamond, the stiff covalent bonds determine the modulus.

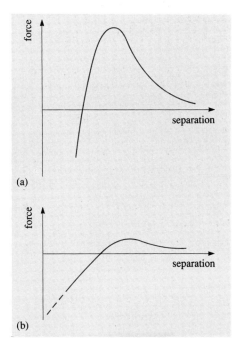

(a)

(b)

Figure 6.50 Force–separation curves for (a) primary bonding (b) secondary bonding

**EXERCISE 6.3** Firstly, from 'Shear deformation' (Section 2.6) we know that the shear stress $\tau$ produced by a tensile stress $\sigma$ is a maximum on planes at 45° to the tensile axis. Here $\tau = \sigma/2$. So if the critical shear stress is $\tau_c$, the tensile yield stress should be $2\tau_c$.

Secondly, although the critical shear stress is the same for all of the slip systems required to operate in each grain, not all will be ideally oriented for maximum resolved shear stress. So the acting stress has to be increased to achieve the critical shear stress on the least favourably oriented system. In fact, this means a factor of about 1.5. Overall then, taking these two factors into account:

$$\sigma_y \approx 3\tau_c$$

**EXERCISE 6.4** The A atoms in Figure 6.32 are closer together than the equilibrium spacing and so in that locality the crystal is in compression. The B atoms are further apart and the crystal is in tension.

(a) A solute atom that is larger than the solvent atoms expands the crystal structure. It will decrease the overall strain if it substitutes for a B atom.

(b) A small solute atom will reduce the overall strain around the dislocation if it substitutes for an A atom.

**EXERCISE 6.5** There are really two aspects to the question.

*i* Strenthening reduces ductility. This links back to 'To flow or to crack?' and to the consequences of increasing the yield stress of a ductile material. Essentially, if the stress required to initiate and propagate cracks is less than that required for plastic flow, fracture ensues. So increasing the yield stress takes the material nearer to the stress level for fracture after necking.

*ii* Necking reduces ductility. The common forming processes usually involve tensile stress in one way or another. So in metals, the production of a neck and the subsequent rapid failure is a limitation. In materials in which a neck is stable, such as polyethylene, extensive plastic strain is possible.

**EXERCISE 6.6**

(a) Cold drawing produces a wire with high dimensional accuracy, and therefore a uniform conductivity. This is of prime importance. However, cold drawing produces a high dislocation density in the wire. Being sites of disorder, dislocations lower the electrical conductivity. Annealing causes recrystallization to occur and therefore a dramatic decrease in dislocation density. The annealed wires have high conductivity and ductility (and hence are easily bent to shape), both highly desirable properties for general electrical wiring.

(b) My tetrahedron would be that in Figure 6.51.

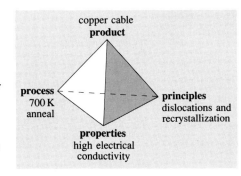

Figure 6.51 PPPP for annealing copper wires

**EXERCISE 6.7** As you know from Section 4.5.2, if the process is thermally activated the creep rate $d\varepsilon/dt$ is given by:

$$\frac{d\varepsilon}{dt} = \text{constant} \times \exp(-E_a/kT)$$

where $E_a$ is the activation energy, $k$ is the Boltzmann's constant and $T$ is the temperature in Kelvin. So the steps are:

(a) Determine $d\varepsilon/dt$ at a number of temperatures for a given stress.

(b) Plot $\log_e d\varepsilon/dt$ against $1/kT$. This should yield a linear plot. (Remember that although the slope of the line is negative, $E_a$ is positive).

(c) Determine the slope of the line. This is the activation energy.

If you had problems with this exercise, reread Section 4.5.2, particularly 'Using Arrhenius plots.'

# Answers to self-assessment questions

## SAQ 6.1

(a) and (f) are shape dominated: in order to work properly both must have specific dimensions.

(d) and (e) are microstructure (property) dominated: the particular materials — rubber and, say, uranium dioxide ($UO_2$) — are the source of the products. Without such materials, the product would not be possible.

(b) and (c) are not dominated by either shape or microstructure. They can be made in different shapes and of different materials, and indeed are.

## SAQ 6.2

(a) The elastic strain at A is OE

(b) The elastic strain at B is DC

(c) The plastic strain at B is OD

(d) Young's modulus is given by EA/OE, and also CB/DC which is equivalent to EA/OE, since Young's modulus does not change with plastic strain.

## SAQ 6.3

(a) The elastic modulus is directly proportional to the slope of the force–separation curve. Hooke's law is obeyed for the range of separation over which the slope (that is, Young's modulus) is constant.

(b) i Primary bonds are indicative of high Young's modulus and bond energy. So the force–separation curve would have a steep slope and a large area beneath it — as you will recall from Chapter 3, the bond energy is proportional to this area.

ii Secondary bonds lead to low modulus and have low bond energies. Hence a force–separation curve with lower slope. Figure 6.50 exaggerates the differences.

## SAQ 6.4  From Equation (6.1)

$$\Delta_d = \frac{Fl^3}{48E_d I} \quad \text{and} \quad \Delta_y = \frac{Fl^3}{48E_y I}$$

where the subscripts d and y denote values for one day and one year respectively.

Since $F$, $l$ and $I$ are fixed

$$\frac{\Delta_y}{\Delta_d} = \frac{E_d}{E_y}$$

When I calculated the tangent moduli from the one day and one year curves of Figure 6.20(b), my results gave:

$$\frac{\Delta_y}{\Delta_d} \approx \frac{275}{108}$$

After a year the deflection is about 2.5 times that after one day.

## SAQ 6.5

(a) Rubber elasticity occurs in the low temperature transition region as well as in the rubbery region. In constant stress (creep) tests, it occurs more slowly at the lower temperatures because the thermal energy available to induce changes in molecule conformations is much lower, so the frequency of change is much smaller.

(b) The source of rubber elasticity is the straightening and recoiling of segments of molecules. At high elastic strains, more and more molecules will become fully extended in the direction of the applied stress. The modulus will then begin to increase towards that for the normal bond stretching strain; the stress must now distort primary covalent bonds.

## SAQ 6.6  Curve 4 is the likely one. The important factors in determining this choice are:

(a) Crystallinity decreases the transition region. The crystalline regions are stiffer (because the molecules are packed closer together), so the glassy region extends to higher temperatures — that is, $T_g$ increases and the modulus is increased in the viscoelastic region relative to the amorphous polymer (curve 1). This rules out curve 2.

(b) The polymer is not completely crystalline; there will be amorphous regions which will exhibit rubber elasticity. This rules out curve 5. These regions will also undergo viscous flow which rules out curve 3.

## SAQ 6.7  Photograph (b) is the unstretched sample and (a) is the stretched sample.

In the unstretched sample there is no preferred orientation of the crystalline regions. So the crystalline regions appear as a distinct diffraction ring. The

amorphous regions appear as a diffuse halo; compare Figure 3.37.

Stretching of the polymer leads to alignment of the crystalline regions in a preferred direction (see Figure 6.27) and to the alignment of molecules in the amorphous regions. The preferred directions give rise to diffraction spots or streaks on part of the halo.

## SAQ 6.8

(a) The expected slip vector is the shortest between equivalent atom/ion sites in the structure. Hence **b**,

(b) Three points here:
The slip plane must contain the slip vector, both plane A and plane B do, since equivalent vectors can be drawn in each plane.

Other things being equal, the favoured slip plane is the one which the atoms/ions are most closely packed together. In MgO planes of type A are more closely packed than type B.

However, type A planes slipping in the direction of a vector **b** brings like ions into closer contact than the equilibrium spacing. The resulting high electrostatic repulsion militates against slip on this plane. The same restriction does not apply to type B planes — the preferred slip plane.

(c) In a polycrystalline material, the individual crystals lie in different orientations with respect to one another. So in order to maintain contact during plastic deformation they need to undergo a general change of shape. Since slip on one system produces a very specific change of shape, deformation on a number of slip systems is necessary.

(d) For a given tensile stress, the resolved shear stress is a maximum ($\tau = \sigma/2$) on a plane at 45° to the tensile axis.

If you are unsure about these answers, study again Section 2.6.1 and 'Shear deformation' in Chapter 2.

## SAQ 6.9  The two dislocations are of opposite sign, they will attract one another and cancel each other out. On meeting, their half-planes line up to leave perfect

341

crystal. The elastic energy of the two original dislocations is reduced to zero. This is a mechanism for reducing the dislocation density; you will see its significance in Section 6.9.2.

SAQ 6.10 A is correct. An interstitial carbon atom distorts the FCC structure of iron, see 'Arrangements in solid solutions' in Chapter 5, so it creates less distortion if it goes into the expanded region below the half-plane of the dislocation.

B is correct. If the two dislocations came together, the combined dislocation would have an elastic strain energy twice that of the separate dislocations ($4b^2$ compared with $2b^2$).

C is correct. The character of a dislocation is determined by its vector which, in turn, depends on the crystal structure. Locally within a grain boundary the orderly crystal structure is lost, see Figure 5.1.

D is wrong. Given that both types of precipitate are strong, the important factor is the distance between the precipitates. The smaller the distance, the greater the stress needed to move a dislocation between the precipitates.

SAQ 6.11 An edge dislocation can move by slip and climb.

In slip, movement is on a specific slip plane as dictated by its line and vector. It is driven by a shear force arising from either an external force or an internal force generated by elastic strain fields in the solid.

In climb, movement is normal to the slip plane. It is thermally activated and is due to the diffusion of vacancies to the edge of the half-plane. The rate of climb is linked to the rate of diffusion.

SAQ 6.12

(a) Precipitation hardening by quenching the alloy to ambient after a solution treatment in the $\alpha$ phase; and then ageing at a temperature below the solvus. See Section 5.2.5.

(b) In the solution treated condition. At this stage the alloy is a solid solution and in the softest and most ductile state.

SAQ 6.13

(a) To restrict dislocation motion I would:
i Use a material with a high intrinsic resistance to dislocation motion, that is

one with directional (covalent) bonding. Of course this may mean deleterious consequences for other properties, particularly ductility and toughness.
ii Use effective obstacles to dislocation motion — as much solute and precipitation as possible. The precipitates would need to be stable at the service temperature.

(b) To restrict diffusion I would use a material with a high melting temperature. The lower on the homologous temperature scale ($T/T_{\mathrm{m}}$) is the service temperature, the lower will be the effect of thermal energy.

SAQ 6.14

(a) Component design — redesign with a view to
i reducing the overall operating stress
ii reducing stress concentration sites, for example, sharp corners, section changes.

(b) Materials selection — select an alternative material with

i a higher yield stress
ii the potential of heat treating to produce a residual compressive stress in the surface of the component
iii no intrinsic large stress concentrators such as sharp particles or voids resulting from the processing of the component.

# Chapter 7 Chemical properties for processing and use

## 7.1 Introduction

SAQ 7.1 (Revision of Chapter 4)
What, in thermodynamic terms, grants permission for a change to take place spontaneously?

Some changes, in Chapter 4 terminology, happen 'at a critical temperature'. What, thermodynamically, is special about the critical temperature?

When a change is 'allowed', it does not happen instantaneously. Why? What controls the rate it happens at?

This chapter employs ideas from SAQ 7.1 to help us understand the chemistry of processing and service of many materials. Winning metals (from ore), making plastics, and setting concrete are all chemical processes; so are many forms of degradation. Although each material has its own chemistry, there are fundamental principles which apply to all materials. In particular, temperature is an important factor in controlling chemical change, because it governs thermodynamic permission.

Chemical reactions which go in either direction,

reactants $\rightleftharpoons$ products

must have $\Delta H$ positive for one direction and negative for the other. Similarly $\Delta S$ has opposite sign for the two directions. This is just the same as a reversible *physical* event like boiling a liquid to vapour or condensing a vapour to liquid:

liquid $\underset{\Delta H \text{ and } \Delta S \text{ negative}}{\overset{\Delta H \text{ and } \Delta S \text{ positive}}{\rightleftharpoons}}$ vapour

| | |
|---|---|
| molecules bound tighter; muddle slight | molecules bound more weakly; muddle great |

The free energy equation is merely expressing observed facts: *below* a critical temperature liquid is the stable form; above that temperature vapour is stable. During this chapter we shall find *chemical* examples of the same thing. Two examples from the field of organic polymers will set up the ideas.

The first example is the joining of many small monomer molecules into few large polymer molecules.

$$n \text{ monomer} \quad \rightleftharpoons \quad \text{polymer}$$

many unjoined molecules = high entropy; weaker double bonds so high enthalpy

few molecules gives low entropy; stronger bonds means lower enthalpy

For the forward reaction (towards polymer),

$$\Delta G = \Delta H^{(-)} - T\Delta S^{(-)}$$

$\Delta G$ can only be negative if the temperature is not too high. There is a temperature 'ceiling' above which the polymer cannot be made or used. There are at least two polymers (PMMA and PTFE) where the mode of thermal degradation is *de*polymermerization, in which monomer molecules are detached from the ends of polymer chains. Usually some other mechanism obtains, as the second example shows.

Heating PVC produces HCl (which is bad news for the waste disposal business):

$$\cdots CH_2-CHCl \cdots \longrightarrow \cdots CH=CH \cdots + HCl$$

This degradation is of the form

$$\text{solid} \longrightarrow \text{solid} + \text{gas}$$

Gases have high entropy (many molecules, widely distributed, offer scope for disorder). So in

$$\Delta G = \Delta H^{(+)} - T\Delta S^{(+)}$$

$\Delta G$ becomes negative when the temperature is high enough. We shall be looking out for reactions of this sort throughout the chapter.

The actual changes in $H$ and $S$ which would be caused by a reaction, if it happened, are more or less independent of temperature. So, within that approximation, a plot of $\Delta G$ versus $T$ will be a straight line cutting $\Delta G = 0$ at the critical temperature. The gradient is determined by $\Delta S$. Compare the equations

$$\Delta G = \Delta H - T\Delta S$$

$$Y = C + XM$$

The gradient $M$ of the standard straight line equations maps to $-\Delta S$ in the free energy equation. We will get a positive gradient if $\Delta S$ is negative (that is, the reaction goes to its low-entropy side). Figure 7.1 shows this together with the mirror image line describing the oppositely

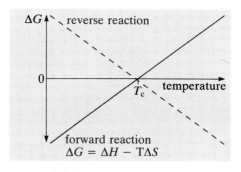

Figure 7.1 Simple Ellingham diagram

directed reaction (going to the high entropy side). Such a graph of $\Delta G$ against $T$ for a reaction is known as an **Ellingham diagram**. We shall be using them shortly.

One final point requires attention. You know that the model as it stands, with the reaction happening completely in one direction or the other and flipping at $T_c$, is not realistic. The state of equilibrium is actually a dynamic balance between the two opposing reactions. The equilibrium composition of the mixture of reactants and products can be described by

$$K \approx \exp\left(\frac{-\Delta G}{RT}\right)$$

The 'equilibrium constant' $K$ is built up as a suitable function of the concentrations of all the ingredients. If $K$ is very large, the reaction goes one way; if small, the reaction goes the other way. We will not pursue the details of the theoretical chemistry here. But by similar arguments to those used in Chapter 4 for other thermally activated phenomena, while $\Delta G$ changes around zero by a few $RT$ per mole (and therefore a few $kT$ per molecule), the value of $K$ changes from very large to very small.

If the lines in Figure 7.1 slope steeply, then the unrealistic model of the reaction as a process that switches direction at $T_c$ is a reasonable approximation. The range of temperatures for which $\Delta G$ is close to zero is narrow (Figure 7.2a). However, if the slopes are gradual, then this range of temperatures is wider (Figure 7.2b). Within this range where both reactions are significant, factors such as pressure can have a strong influence on which reaction is 'winning'.

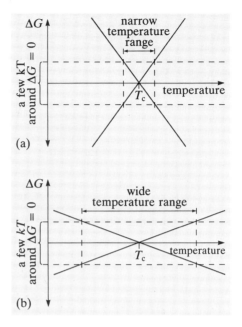

Figure 7.2

# 7.2 Oxides: alloys and ores

Because oxygen is both an abundant element in Earth's crust and a voracious grabber of electrons (Chapter 3), it is not surprising that almost all the other elements are found as oxides. In oxides, the bonds (ionic or covalent) are strong, so oxides are very stable compounds. Hence oxides are used when great stability is required. Glasses and ceramics are examples of mixed oxides which are corrosion resistant and heat resistant. (Concrete is another example, although it is not refractory because it contains water.) Oxides are also used as sources of metals, that is as ores. Here their chemical stability is a problem to be overcome. And because oxygen is a major constituent of the atmosphere, metals, once extracted from ore, will tend to revert to their thermodynamically stable oxide form. This is corrosion.

In this section we shall consider the chemistry of oxides both as engineering materials and as ores. We shall be much occupied with reactions of the form

metal + oxygen $\rightleftharpoons$ metal oxide

and Figure 7.3 gives Ellingham diagrams for a range of metals. ▼**Notes on Figure 7.3**▲ gives some comments. ▼**Oxidation and reduction**▲ explains some of the jargon we shall use.

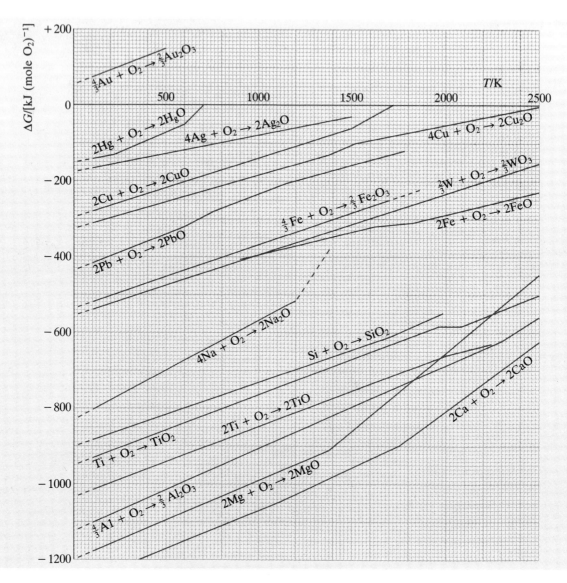

Figure 7.3 Ellingham diagrams

346

# ▼Notes on Figure 7.3▲

The lines all refer to reactions of the sort

metal(s) + oxygen(g) ⟶ metal oxide(s)

   high entropy side         low entropy side

**SAQ 7.2** (Objective 7.1)
Check that the gradients should be positive (as drawn) for the reactions in the direction arrowed.

The entropy of a gas is dependent upon pressure (that controls the space the molecules have available to move in) and of course the entropy must be related to the number of molecules of gas. To make the information comparable between different metals, all the lines are for reactions under a pressure of one atmosphere and one mole of oxygen molecules. This is why funny numbers like $\frac{4}{3}$ Al are used on Ellingham diagrams.

Read this as $\frac{4}{3}$ of a mole of aluminium, not as $\frac{4}{3}$ of an atom of aluminium.

**EXERCISE 7.1** How many atoms of aluminium react with one mole of oxygen?

Why are all the lines roughly parallel?

Because each line represents the loss of one mole of gas. In any reaction in which gas is consumed or liberated, the entropy change associated with the gas is usually the dominant entropy change for the reaction. Slight variations in slope are due to different entropies of the solids, and kinks indicate a phase change (such as melting).

The intercept at $T = 0\,\text{K}$ represents the chemical bond energy of the oxides. This can be checked by putting $T = 0$ in the free energy equation, so $\Delta G = \Delta H$. At this temperature $\Delta G$ represents the chemical potential energy lost when metal and oxygen atoms bond together (as oxide) rather than bonding to their own species (as elements).

Figure 7.3 confirms many of the things we know about metals:

• Gold does not tarnish. The line shows $\Delta G > 0$ for oxide formation at room temperature.

• Mercury can be extracted simply by roasting its ore. $\Delta G$ goes positive at modest temperatures.

• Other oxides do not decompose even at fierce furnace temperatures. They are refractory materials not easily attacked chemically. This thermal stability of many metal oxides reflects their strong bonding.

# ▼Oxidation and reduction▲

What all metals have in common are some s-state electrons at their outer surface which can easily be removed or used for bonding. Loss of these electrons leaves a positively charged ion:

$$M - ne^- \longrightarrow M^{n+}$$

A common way for the electrons to be lost is to oxygen, for example:

$$2M + O_2 \longrightarrow 2M^{2+} + 2O^{2-}$$

By extension, **oxidation** is the term used to describe any reaction in which electrons are lost from a substance. Another common oxidation reaction, not involving oxygen, happens when a metal dissolves in acid:

$$M + 2H^{2+} \longrightarrow M^{2+} + H_2$$

Yet another sort of oxidation can happen in electrochemistry. Metal can be oxidized, sending electrons into an electrical circuit. The metal ions pass into solution in the electrolyte. (These electrochemical terms will be explained shortly.)

The opposite of oxidation is **reduction**. Any reagent receiving electrons, for example hydrogen ions in an acid reaction, is said to have been reduced. In the reduction of iron oxide to metallic iron, the iron ions receive two virtually massless electrons as the much heavier oxygen ions are disconnected. (This decrease of mass during reduction is apparently the origin of the term.)

Again in electrochemistry we shall find reduction reactions, where electrons are donated to metal ions which then come out of solution as atoms (**electroplating**).

By implication, whenever something is reduced, something else is oxidized. **Redox** is the term applied to reactions where electrons are swapped in this way.

## 7.2.1 Pottery and Portland cement

Chapter 5 showed that the components of alloy systems can be non-metallic. For example, Figure 5.30 is a phase diagram for the $SiO_2$–$Al_2O_3$ system. Both bricks and cement use such oxide-alloy systems. The ways in which they change from a mouldable wet condition to a rigid solid are interestingly different.

The main ingredient of clay is the mineral kaolinite $Al_2O_3.2SiO_2.2H_2O$. The delightful plastic quality of wet clay is well accounted for by the crystal structure of kaolinite. The unit cell of the crystal structure is a hexagonal assembly of thin ionic and covalently bonded layers (Figure 7.4a). The crystals seen in Figure 7.4(b) are several times thicker than this cell because the cells bond to each other by hydrogen bonding. Both surfaces of the crystals can also hydrogen bond to water, so with water added to the clay the platy crystals slide over each other easily and the potter can mould shapes. Firing clay first involves carefully drying off the free water. Then as the temperature rises to around 800 K the kaolinite crystal structure is destroyed as combined water is removed. You can see this as

$$solid \longrightarrow solid + gas$$

so the decomposition represents a big entropy increase and $-T\Delta S$ overwhelms the binding enthalpy of the crystal as the temperature gets high enough. But what becomes of the residual solid? You can answer this question by reference to Figure 5.30.

EXERCISE 7.2   The residual solid $Al_2O_3.2SiO_2$ is 44% alumina by mass. What are the thermodynamically stable phases of this mixture? Up to what temperature will these phases be stable?

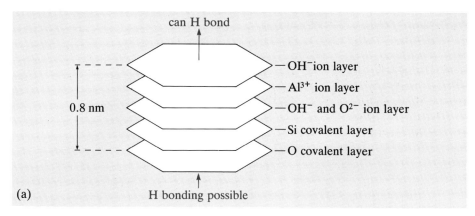

(a)

(b)          0      1 μm

Figure 7.4 (a) Unit cell of kaolinite. (b) Microstructure of kaolinite

(a)

(b)

├──────────────┤
0        10 μm

Figure 7.5 (a) Samian pot. (b) Micrograph of pot

Fired clay begins to hold together by sintering if the temperature is taken to about 1100 K, and earthenware or low quality bricks are so fired. However, much stronger ceramic can be made if the sintering can be accelerated.

How can this be done? Remember (Chapter 4) that sintering is a diffusion process.

Two ways: higher temperature and the presence of a liquid phase.

The melting temperature of silica is brought down sharply by small amounts of impurity. Only 5% of alumina is required for the eutectic on the phase diagram you have been looking at. Clay is rarely pure kaolin so impurities can react with the silica to lower the melting temperature. Alkali oxides do this very effectively, so incorporating them into the clay mixture offers a way to making stronger ceramic. Firing to around 1400 K then produces 'stoneware', which is strong and watertight. 'Engineering bricks', used prolifically in Victorian railway engineering, are strong by virtue of liquid-assisted sintering. The Romans valued highly beds of 'illitic' clay, which has significant potassium content. From these they made the famous Samian ware. Figure 7.5 shows a Samian pot and a scanning electron micrograph of a fracture surface of this ware. The toffee-like appearance and the sharp edged flats on this picture bear witness to liquid assisted sintering. This was strong pot in spite of its extensive porosity (black parts of the micrograph).

Portland cement, our other oxide-alloy system, is an alloy of calcium, silicon and aluminium oxides, plus minor constituents. It is made by roasting a mixture of about one part clay ($Al_2O_3.2SiO_2.2H_2O$) and three parts limestone ($CaO.CO_2$) to around 1900 K. This produces an alloy whose main phases have compositions near the stoichiometry of the first three entries in Table 7.1.

Table 7.1 Phases of Portland cement

| Name | Chemical formula | Mass proportion % |
|------|------------------|-------------------|
| tricalcium silicate | $3CaO.SiO_2$ | 55 |
| dicalcium silicate | $2CaO.SiO_2$ | 20 |
| tricalcium aluminate | $3CaO.Al_2O_3$ | 12 |
| tetracalcium alumina ferrite | $4CaO.Al_2O_3.Fe_2O_3$ | 8 |
| hydrated calcium sulphate | $CaSO_4.H_2O$ | 3.5 |
| alkali oxides and other constituents | $K_2O$ and $Na_2O$ | 1.5 |

With kaolinite we were interested in what happened when the water was removed. Now we are interested in what happens when water is added. A complex series of chemical reactions, rate controlled by diffusion, leads to the matted microstructure depicted in Figure 7.7(c). ▼Setting cement▲ outlines the reactions which lead to this structure. The important feature of these reactions is that, from the moment the concrete is mixed (Portland cement, 'aggregate' and water), setting is delayed long enough for the concrete to be transported to the site, poured into prepared shuttering and consolidated by vibration. (Consolidation ensures that surplus water and most of any entrapped air rises to the surface.)

Chapter 2 commented on the role of the aggregate in toughening and strengthening the concrete composite. But the quantity of water in the concrete also affects its quality. There are three ways in which water can be held in concrete.

• Combined chemically within the hydrated compounds. This water is about 25% by mass of fully reacted cement.

• In the gel pores as **gel water**. Since the pores are so small, typically only an order of magnitude greater than the diameter of a water molecule, secondary bonding of this water to the gel particles contributes to the strength of the gel. About 15% by mass of the reacted cement is gel water.

• As free water in capillary pores. Capillary pores are full of water if the concrete is immersed, but in dry conditions many of the pores dry out.

This list suggests a minimum water/cement ratio of 40% is needed: 25% as chemically bound water and 15% as gel water. Any water in excess of this amount will establish the capillary pores within the setting cement. Thus the water/cement ratio chosen for the mix influences the overall porosity of the concrete, and hence its strength. Figure 7.6 shows how the compressive strength varies with water content for a typical concrete. In practice rather more water than that minimum is used both to aid workability and to avoid holes due to inadequate compaction.

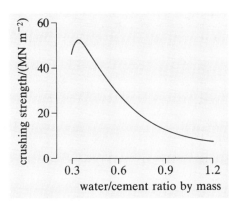

Figure 7.6 Compressive strength versus water/cement ratio

# ▼Setting cement▲

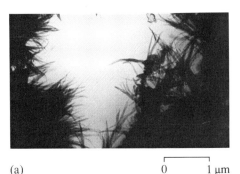

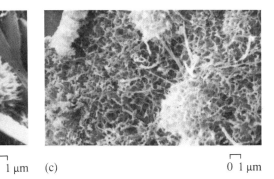

(a)      0   1 μm   (b)      0   1 μm   (c)      0   1 μm

Figure 7.7 Cement setting. (a) Gel fibres. (b) and (c) Growth of gel fibres from the cement particles and larger crystals. The markers are 1 μm long.

The composition of Portland cement is shown in Table 7.1. When water is added, the first reactions rapidly produce $OH^-$ ions in solution. In these reactions, calcium oxide dissolves from the major ingredients

$$CaO + H_2O \longrightarrow Ca^{2+} + 2OH^-$$

and the 1.5% alkali oxides react:

$$K_2O + H_2O \longrightarrow 2K^+ + 2OH^-$$

The pH rises and soon reaches 12.6. Crystals of calcium hydroxide then precipitate. This is the first new solid to appear.

$$Ca^{2+} + 2OH^- \longrightarrow Ca(OH)_2(solid)$$

To promote this precipitation and so increase the initial setting rate, extra $Ca^{2+}$ ions can be added in the form of calcium

chloride dissolved in the mixing water. But the chloride ions jeopardize the corrosion resistance of reinforcing steel in concrete, so it has its snags.

The next stage of setting is a polymerization reaction. As the calcium ions are stripped from the surface of cement particles, the silicate ions join by covalency:

$$2SiO_4^{4-} + H_2O$$
$$\longrightarrow Si_2O_7^{6-} + 2OH^-$$

This is the first step as two silica tetrahedra join corner-to-corner, ejecting an oxygen atom. Of course this happens repeatedly and randomly. A 'gel' of amorphous silicate network quite quickly coats the dry interior of each cement particle. Further reaction is slowed because it requires water to diffuse through the coating. As water diffuses in

towards the solid it releases calcium ions, which can move out. The silicate ions on the other hand cannot diffuse out through the network. The excess silicate ions attempt to combine as before but as the pH within the trapped solution rises, due to production of $OH^-$ ions, the reaction is choked. Every so often the pressure inside the gel builds to such a level that the coating bursts, squirting the inside solution into the mixing water. Instantly, at the lower pH of the mixing water, silicate ions react making new gel at the interface between the two liquids. Complex shapes such as curled sheets, fibres and even hollow tubes are formed. Figure 7.7 shows the resulting tangle which holds the cement together. Because the mechanism is controlled by diffusion, cement takes some time to set. Indeed, the process continues for months.

351

## 7.2.2 Winning metals

How may metals be won from their oxides if using heat alone requires impractical temperatures? The Ellingham diagram (Figure 7.3) suggests a way. For any two reactions on the diagram, the one with the lower line can be used to reduce the higher one. Suppose at some temperature

$$A + O_2 \longrightarrow AO_2 \qquad \Delta G = -a\ lot$$

$$BO_2 \longrightarrow B + O_2 \qquad \Delta G = +a\ little$$

Add them together,

$$A + O_2 + BO_2 \longrightarrow AO_2 + B + O_2 \qquad \Delta G = -some$$

The mole of oxygen need not be involved; it is on both sides of the reaction. (This is why data are specified per mole of oxygen on the diagram.) So

$$A + BO_2 \longrightarrow B + AO_2 \qquad \Delta G = -some$$

Since $\Delta G$ is negative, this is a thermodynamically feasible reaction. But three questions need to be answered.

1 What A should be used as a reducing agent?
2 At what temperature?
3 How can the reactants be got together?

We'll look at two sets of answers to these questions in the context of iron production. The first, ▼**Iron from ore**▲, uses carbon monoxide to reduce the iron. The second, the ▼**Thermit welding process**▲ (used for joining railway lines on site), employs solid aluminium as a reducing agent. Aluminium is much more expensive than iron because of the difficulty of extracting it from its oxide. Clearly, reduction by aluminium is not a practical way to produce iron in bulk

## ▼Iron from ore▲

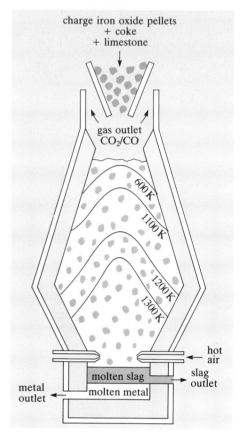

Figure 7.8 Blast furnace temperature distribution

Choosing carbon monoxide as a reducing agent solves the problem of getting the reactants in contact, but what metal oxides will carbon monoxide reduce by the following oxidation?

$$2CO + O_2 \longrightarrow 2CO_2$$

In particular how can it be used to get iron from its oxides in a blast furnace?

The air blast at the bottom burns coke to produce the temperature profile shown in Figure 7.8. Figure 7.9 shows that blowing air over hot coke produces CO rather than $CO_2$ if the temperature is above 1000 K (line 6 is below both 5 and 3 at temperatures above 1000 K), so the gas rising up through the furnace is mostly CO.

The reaction we wish to promote is

$$Fe_2O_3 + 3CO \longrightarrow 2Fe + 3CO_2$$

but it doesn't happen in a single jump. We have to see how CO reacts in a sequence of reactions

$$Fe_2O_3 \longrightarrow Fe_3O_4 \longrightarrow FeO \longrightarrow Fe$$

as the ore moves down the furnace, getting progressively hotter.

Line 3 being below line 1 indicates that CO can 'steal' oxygen from $Fe_2O_3$ at any temperature; and above about 800 K it can do the same for $Fe_3O_4$ (line 3 drops below 2). But the final stage of the reaction,

$$FeO(s) + CO(g) \rightleftharpoons Fe(s) + CO_2(g)$$

apparently won't go to the right because the FeO line (4) is below the CO line (3). However the slopes of the lines are very close. We are therefore in the sensitive regions I mentioned at the start of the chapter. The equilibrium condition of the reaction is delicately controlled by temperature and the concentrations of the reagents. Back in Chapter 4 in 'Statistical games for $W$' we saw that mixing two species produces entropy. The forward reaction above (producing Fe and $CO_2$) generates a mixture of gases, $CO_2$ and CO. So this reaction produces a bit more entropy than the Ellingham lines for FeO and CO imply and the system's free energy will be reduced as the mixture is made. This effectively moves the cross-over point to a higher temperature. Figure 7.10 shows the ratio of CO to $CO_2$ for equilibrium at increasing temperatures lower in the furnace. But the upward flow of gas in the furnace carries away the $CO_2$ and ensures there is excess CO ('typical' in Figure 7.10). So the reaction keeps trying to make $CO_2$ and continues reducing the iron oxide.

Another clever feature of the blast furnace is its ability to separate iron from all the other rubbish in the ore. It does this because the coke is strong enough not to crush to powder and become compacted under the load above it. Thus gas can flow up the furnace and liquid products can trickle downwards. The iron, having dissolved an appreciable amount of carbon, is liquid above 1500 K and the slag (mixed oxides of Si, Al, and Ca) can be liquefied at a similar temperature. At the base of the furnace the low-density slag floats on the more dense iron so it can be run off separately.

In many places around the world there are vast deposits of virtually pure iron oxide ores. Coking coal on the other hand is comparatively scarce. Given such pure ore, the complexity of the blast furnace, with its ability to remove impurities as slag, is unnecessary. A reducing gas is all that's required. One technique for doing this is the Midrex process. It uses a mixture of CO and $H_2$ for the reduction. (This reducing gas is produced separately by passing steam over coke of any grade.) The ore is directly reduced to solid iron by the gas. Energy is not 'wasted' melting the iron or its impurities, so the process is more energy efficient than the blast furnace.

SAQ 7.3   (Objective 7.1)
The Ellingham line for
$2C + O_2(g) \rightarrow 2CO(g)$ has negative gradient and therefore crosses metal oxide lines at various critical temperatures. Explain, in terms of gas generation, why this line slopes negatively.

SAQ 7.4   (Objective 7.2)
In the lower part of the blast furnace, liquid slag trickles over the hot coke. The slag contains $SiO_2$, $Al_2O_3$ and CaO. Is the temperature high enough to allow reductions of the form

$$oxide + carbon \longrightarrow element + CO$$

for the oxides in the slag?
Hint: Transfer line 6 from Figure 7.9 to Figure 7.3.

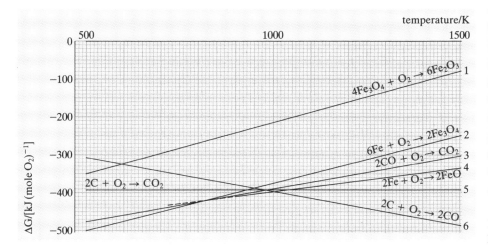

Figure 7.9 Ellingham diagrams for iron and carbon oxides

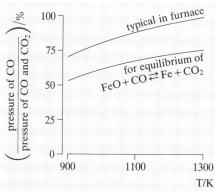

Figure 7.10 Ratio of CO and $CO_2$ partial pressures in the furnace

# ▼Thermit welding process▲

On Figure 7.3 the aluminium line lies below the one for iron at *all* temperatures, so apparently aluminium can be used to reduce iron oxide at any temperature. The necessary intimate contact of reagents is achieved by mixing aluminium powder and iron oxide powder ($Fe_2O_3$). But simply mixing the powders at room temperature does not activate the reaction. A magnesium flare is used to cause a local rise in temperature, giving a high initial reaction rate. Because the reaction is strongly exothermic, the high temperatures and rates are self-maintaining until all the reactants are consumed. The result is molten iron and aluminium oxide slag.

To see the thermodynamics in action we note the overall reaction as

$$Fe_2O_3(s) + 2Al(s)$$
$$\longrightarrow Al_2O_3(s) + 2Fe(l)$$

which can be split into two parts:

$$\tfrac{4}{3}Al + O_2 \longrightarrow \tfrac{2}{3}Al_2O_3$$

and

$$\tfrac{2}{3}Fe_2O_3 \longrightarrow \tfrac{4}{3}Fe + O_2$$

To estimate the overall free-energy change, look at the Ellingham diagram for these two reactions. On Figure 7.3 the $Al_2O_3$ and $Fe_2O_3$ lines are almost parallel, so the free-energy change in bonding aluminium with the oxygen instead of iron is the same at all temperatures. Take the 1000 K numbers to see how things work out. For 1 mole of $O_2$ reacting with Al the graph shows $\Delta G$ is about $-900\,kJ$, while for iron the equivalent value is $-370\,kJ$. Thus *reduction of* the iron oxide requires $+370\,kJ\,mol^{-1}\,O_2$.

Add the equations and the $O_2$ cancels out:

$$\tfrac{4}{3}Al + O_2 + \tfrac{2}{3}Fe_2O_3$$
$$\longrightarrow \tfrac{4}{3}Fe + O_2 + \tfrac{2}{3}Al_2O_3$$

Multiplying by $\tfrac{3}{2}$ gives an equation for one mole of $Fe_2O_3$, so the energy yield is about

$$\tfrac{3}{2}(900 - 370)\,kJ\,mol^{-1}\,Fe_2O_3$$
$$\approx 900\,kJ\,mol^{-1}\,Fe_2O_3$$

The intense heat would actually vaporize the iron if conduction through the rails did not limit the temperature rise. Thermit welding is suitable for railway work because it requires no bulky equipment (power packs, gas cylinders) and untrained operatives can get good results.

SAQ 7.5 (Objective 7.2)
Figure 7.11 shows Ellingham lines for the formation of PbS, PbO and $SO_2$. From antiquity the smelting of lead from its sulphide ore has been easy. The furnace was first fired open to the air, during which some ore was oxidized:

$$PbS + \tfrac{3}{2}O_2 \longrightarrow PbO + SO_2$$

Then the ore was raked over to mix oxide and sulphide, and roasting continued with the furnace sealed. The reaction

$$PbS + 2PbO \longrightarrow 3Pb + SO_2$$

is thermodynamically possible above a critical temperature. Examine Figure 7.11 and by adding reactions and free energies discover the free energy changes required at 500 °C and 1000 °C. Hence estimate the critical temperature above which the reaction will go forward.

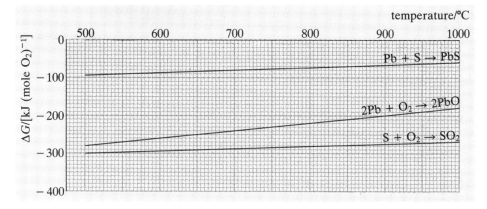

Figure 7.11 Ellingham diagrams

## 7.2.3 Hot oxidation

Metals are often required to serve at high temperature, for example in boilers and engines. The Ellingham diagrams show that at practical temperatures all the important metals will corrode in air. What determines the corrosion resistance of some metals? Naturally we fall back on rate effects to account for the failure of a system to take immediate advantage of thermodynamic permission, but in high temperature applications fast reactions are to be expected.

The first obvious limitation of rate is that only the surface of a metal object is exposed to corrosive attack; but a clean metal surface exposed to oxygen must respond instantly to its thermodynamic instincts. Almost immediately, there is layer of oxygen molecules tacked onto the surface, diffused a little way in, ionized and ionically bonded to metal atoms. Within minutes this should have grown to a prototype oxide skin a few atoms thick. In so far as any real crystal structure can belong to so thin a layer, we should expect local nuclei to have formed and grown sideways until the metal is covered.

Thus a barrier is formed which separates the metal from oxygen. What happens subsequently is entirely decided by the rates at which reacting species can diffuse through this layer. Figure 7.12 shows some possibilities.

In Figure 7.12(a) electrons, having diffused from the metal to the free surface, react with oxygen to produce oxygen ions and these diffuse back to the metal surface to form more oxide.

What mechanisms might be involved in such a cycle? See Section 4.5.5.

The electrons could diffuse by 'hopping' between cations of variable valancy, such as $Fe^{2+}$ and $Fe^{3+}$. For oxygen ions to diffuse to the metal there must be vacancies in the lattice.

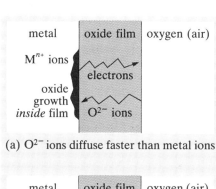

(a) $O^{2-}$ ions diffuse faster than metal ions

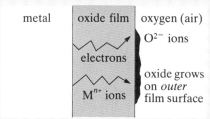

(b) Metal ions diffuse faster than $O^{2-}$ ions

Figure 7.12 (a) and (b) Two possible mechanisms of corrosion

If oxide growing at the metal–oxide interface has a greater volume than the metal from which it is formed, stresses will be generated in the oxide film. The film may then rupture, exposing clean metal for further direct oxidation.

Figure 7.12(b) shows a different cycle. Here both electrons and metal ions are seen to diffuse down the concentration gradients from the metal to the outer surface. When the oxygen ionizes, the metal oxide is formed on the outer surface, where stresses are easily relieved. The metal is then soon protected from further corrosion as the diffusion length through the film becomes sufficient to impede growth.

The volume ratio of metal oxide to metal is known as the **Pilling–Bedworth ratio**. Its first measurers (in the 1920s) wanted to know why some metal oxide films spall off whereas others are stable. Table 7.2 gives some values. Since most are significantly greater than unity, the corrodibility of the metal is decided not by this volume change but by which diffusion mechanism dominates in the growth of the oxide.

The two mechanisms just described can be demonstrated by oxidizing wires of titanium and nickel in hot oxygen. Titanium wire ends up as powder. A hollow oxide tube is the result with nickel.

Which metal has its ions diffusing through its oxide?

Nickel. The metal has moved to the outer surface to oxidize and mechanical integrity results. Titanium oxide allows faster diffusion of $O^{2-}$ ions and the oxide layer is repeatedly fractured. In jet engines, the lighter titanium alloys are favoured at the cooler compression end of the engine, while heavier nickel alloys must be employed in the hot combustion and turbine sections.

Even in a simple metal–oxygen system, corrosion is more complex than described here. The different processes will change in predominance with temperature and with the chemical accuracy (stoichiometry) of the oxide. Grain boundaries may provide fast diffusion paths for ions, or indeed for molecular species such as $O_2$. Stresses caused by temperature fluctuations can do just as much damage to a protective oxide film as the chemically induced stress. Finally of course, hot corrosion is rarely just oxidation. Fuels used in engines or power stations, for example, contain many corrosive substances. Alloys of several metals with chromium, aluminium, and yttrium (MCrAlY, said 'emcrally'!) have been devised as coatings for metal parts subject to such environments. They form protective corrosion products. They are reserved for critical situations, however, because of the expense of sputtering them as coatings onto large surfaces.

Table 7.2   Pilling–Bedworth ratios for common metals

| Metal | Pilling–Bedworth ratio |
| --- | --- |
| sodium → $Na_2O$ | 0.6 |
| calcium → $CaO$ | 0.6 |
| magnesium → $MgO$ | 0.8 |
| aluminium → $Al_2O_3$ | 1.3 |
| titanium → $TiO_2$ | 1.5 |
| nickel | 1.6 |
| copper → $Cu_2O$ | 1.6 |
| iron → $FeO$ | 1.6 |
| iron → $Fe_2O_3$ | 2.1 |
| chromium → $Cr_2O_3$ | 2.1 |

## Summary

Oxides are chemically stable because oxygen makes strong bonds. Ultra-stable oxides (often as oxide-alloys systems) are favoured for refractory and corrosive environments. Ellingham diagrams express free-energy changes for oxidation and reduction.

Reversion of metals to their thermodynamically more stable oxides is controlled by diffusion. Relative diffusion rates of metal and oxide ions through the oxide skin determine whether the skin sticks on or falls off.

# 7.3 Electrochemistry of metals

Electrochemistry is distinguished from other sorts by the feature that reactants get the energy necessary for chemical reactions from electrical sources.

Electrical initiation of reactions requires two conditions. First, the reactants must be charged particles so that electrical forces can act on them to change their energy. Ions and electrons are such. Second, the medium for the reaction must allow the charged particles to meet and react. Ions in solution (in water) or in molten ionic compounds can move, and metals allow electrons to move. So electrochemical reactions can occur at the interfaces between metals and solutions or molten salts.

Electrochemistry covers such things as batteries and fuel cells, the electro-refining and electroplating of metals, electrolysis for winning reactive elements (such as chlorine and aluminium) from raw materials, and wet corrosion of metals. It also covers some of the ways of combatting corrosion. Electrolytic capacitors are a direct application of electrochemistry to produce an electronic device.

The basic electrochemical 'machine' is depicted in Figure 7.13. It shows an electrical circuit comprising an electron-transport section (metals $M_1$ and $M_2$), and an ion-transport section (electrolyte, which may be a solution or a molten salt). At the two interfaces (electrodes) reactions must occur to allow the current to flow. Note that an electrochemical circuit must have at least two electrodes. There cannot be just a single metal/electrolyte interface. This seemingly obvious fact needs to be stated in view of later ideas concerning electrode potentials.

The e.m.f. in the circuit lets us decide which way current will flow, which reactions are possible and what rate they shall go at. Even when we study the wet corrosion of a single metal the same circuit will apply, except that $M_1$ is the same as $M_2$ and the e.m.f. is zero. You'll notice two jargon words, **anode** and **cathode**, on Figure 7.13. These are the conventional names for the electrodes in contact with the electrolyte. The anode is electrically connected to the positive side of the e.m.f., so it is sending electrons *to* the e.m.f. The cathode is connected to the negative side of the e.m.f., so it is receiving electrons *from* the source. The reactions at the electrodes are anodic or cathodic. Anodic reactions must result in free electrons in the anode to be sent up the wire. Here

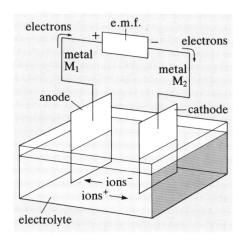

Figure 7.13 Basic electrochemical machine

are some anodic reactions.

$$M \longrightarrow M^{n+} + ne^-$$

$$2Cl^- \longrightarrow Cl_2 + 2e^-$$

$$4OH^- \longrightarrow 2H_2O + O_2 + 4e^-$$

Are the reagents on the left-hand side of these statements oxidized by the reaction or reduced?

They are oxidized because in each case electrons are taken away from them. In the first, atoms in the anode are ionized and pass into solution. In the second and third, ions become atoms (or molecules) and come out of solution. If all three reagents are present, any combination of the reactions might happen. Naturally we should like to be able to control what happens.

Typical cathode reactions might be

$$M^{n+} + ne^- \longrightarrow M$$

$$2H^+ + 2e^- \longrightarrow H_2$$

$$2H_2O + 2e^- \longrightarrow H_2 + 2OH^-$$

$$2H_2O + O_2 + 4e^- \longrightarrow 4OH^-$$

And again what *actually* happens is the outcome of competition between these possibilities.

Note: *a*node = oxid*a*tion; *c*athode = redu*c*tion

Our way forward, therefore, is to establish some measure of 'thermodynamic permission' for any electrochemical reaction and to find a way to control the rates of anodic and cathodic reactions. Consider first the equilibrium condition.

## 7.3.1 Equilibrium between a metal and a solution

'Solubility and free energy' in Chapter 5 explained the equilibrium balance of positive $\Delta H$ and $\Delta S$ as a solid solute disperses into solution. We can put this piece of theory to the hypothetical case of a piece of metal in a vacuum, Figure 7.14. Imagine a vapour of ions forming around the metal with the surplus electrons held by the metal. The reaction

$$\dot{M}(solid) \longrightarrow M^+(vapour) + e^-(in\ solid)$$

would have $\Delta H$ positive (energy needed to break the metallic bonds) and $\Delta S$ positive as the ions disperse. However, there is now an extra dimension. The separation of charges also requires energy, so $\Delta H$

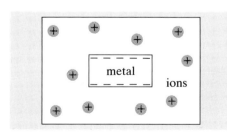

Figure 7.14 Metal in vacuum showing cloud of ions

becomes *more* positive as the reaction proceeds. Eventually at equilibrium there is an electrostatic potential difference between the metal and the ion vapour surrounding it.

If, rather than being in a vacuum, the metal is immersed in water, two further effects join the scene.

• The metal ions 'hydrate', that is they attract polar water molecules around them. The extra chemical potential energy needed to ionize the metal atoms is then offset to some extent by the ions' bonding to the water molecules, so $\Delta H$ is smaller and *more* ions can be formed before $\Delta H = T\Delta S$.

• The water can be a solution of a salt of the metal. Then the concentration effect which reduces the entropy increase for further dissolution is enhanced so *fewer* ions can be formed before $\Delta H = T\Delta s$.

This progression of ideas therefore leads us to some general statements.

1 At the condition of equilibrium between a metal and a solution of its ions an electrical potential step exists at the interface between metal and solution (this is the **contact potential**).

2 The magnitude of that voltage depends on the strength of the bonding of metal atoms in the metal and of metal ions in the solution.

3 The voltage also depends on the concentration of the metal ions in solution, which can be varied by balancing positive ions with negative ions in the solution.

We cannot measure the contact potential of a single electrode, because to do so would require that the electrode be part of an electrical circuit, and, as we have seen, that would require a second electrode, which would introduce its own contact potential into the circuit. In practice, the contact potential of hydrogen is taken as zero and other contact potentials are measured relative to it. ▼**Electrode potentials**▲ describes how it's done.

The equilibria we have been thinking about are *dynamic*. That is to say, within the boundary layer between metal and electrolyte, thermal activation is making things happen, be they ions detaching themselves from the metal and entering solution or, in the other direction, ions separating from water molecules and being adsorbed onto the metal. At equilibrium both these reactions are happening at an interface, and proceed at the same rate. Since they involve charge transport, they amount to equal and opposite electric currents, or **exchange currents** ($i_e$) as they are called. Catalysts and barriers, which alter activation energies, are very important in deciding the relative rates of competing reactions.

Tables 7.3 and 7.4 give some values of $i_e$ (as current densities) for one molar solutions at 25°C. Table 7.3 is for laying metals onto themselves, while Table 7.4 shows the rates for hydrogen evolution, often a competing process.

But equilibrium is, technologically speaking, useless. To achieve anything electrochemically one reaction has to go faster than the other. The external source of e.m.f. in the basic machine enables us to intervene.

Table 7.3   Exchange currents $i_e$ for metal/ion reactions

| Reaction | Onto | Exchange current $i_e$/(A m$^{-2}$) |
|---|---|---|
| $Ag^+ + e^- \rightleftharpoons Ag$ | Ag | $10^4$ |
| $Cd^{2+} + 2e^- \rightleftharpoons Cd$ | Cd | $10^3$ |
| $Cu^{2+} + 2e^- \rightleftharpoons Cu$ | Cu | $10$ |
| $Zn^{2+} + 2e^- \rightleftharpoons Zn$ | Zn | $10^{-3}$ |
| $Fe^{2+} + 2e^- \rightleftharpoons Fe$ | Fe | $10^{-4}$ |
| $Ni^{2+} + 2e^- \rightleftharpoons Ni$ | Ni | $10^{-5}$ |

Table 7.4   Exchange current densities for hydrogen evolution from PH 0 acid on various metals

| Cathode material | Exchange current $i_e$/(A m$^{-2}$) |
|---|---|
| Pt, Pd | $10$ |
| Ni, Au | $10^{-1}$ |
| Fe, Cu | $10^{-2}$ |
| Sn, Ti | $10^{-4}$ |
| Zn, Al | $10^{-6}$ |
| Cd, Mn | $10^{-7}$ |
| Pb, Hg | $10^{-8}$ |

# ▼Electrode potentials▲

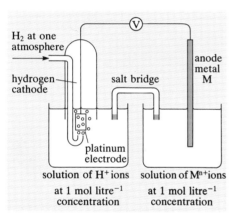

Figure 7.15 Battery with hydrogen cathode used in determining standard electrode potentials.

Table 7.5   Standard electrode potentials

| Material | Anode reaction | Standard electrode potential/V |
|---|---|---|
| Au | $Au \rightarrow Au^{3+} + 3e^-$ | $+1.43$ |
| Ag | $Ag \rightarrow Ag^+ + e^-$ | $+0.80$ |
| Cu | $Cu \rightarrow Cu^{2+} + 2e^-$ | $+0.34$ |
| H | $H \rightarrow H^+ + e^-$ | $0$ |
| Pb | $Pb \rightarrow Pb^{2+} + 2e^-$ | $-0.13$ |
| Sn | $Sn \rightarrow Sn^{2+} + 2e^-$ | $-0.14$ |
| Ni | $Ni \rightarrow Ni^{2+} + 2e^-$ | $-0.25$ |
| Cd | $Cd \rightarrow Cd^{2+} + 2e^-$ | $-0.40$ |
| Fe | $Fe \rightarrow Fe^{2+} + 2e^-$ | $-0.44$ |
| Zn | $Zn \rightarrow Zn^{2+} + 2e^-$ | $-0.76$ |
| Ti | $Ti \rightarrow Ti^{2+} + 2e^-$ | $-1.63$ |
| Al | $Al \rightarrow Al^{2+} + 2e^-$ | $-1.66$ |
| Mg | $Mg \rightarrow Mg^{2+} + 2e^-$ | $-2.38$ |
| Na | $Na \rightarrow Na^+ + e^-$ | $-2.71$ |

Electrode potentials cannot be determined absolutely; they must be measured relative to another electrode potential. Arbitrarily, the electrode potential of hydrogen under standard conditions is used. (Standard conditions are atmospheric pressure, a temperature of 25 °C and a concentration of 1 mol litre$^{-1}$.) To measure electrode potentials of other materials relative to it, a special battery, with a standard hydrogen cathode, is used (Figure 7.15).

Hydrogen at one atmosphere pressure is bubbled into a solution containing 1 mol litre$^{-1}$ of hydrogen ions (that's quite a concentrated acid, pH = 0). A metal anode is separately immersed in a solution of its ions. The electrolytes are connected by a salt bridge, which prevents them from mixing significantly. The electron

part of the circuit is completed by a high impedance voltmeter, so virtually no current flows; the reactions are at equilibrium. The voltmeter then measures the electrode potential $\varepsilon$ of the metal anode relative to the standard hydrogen cathode. Naturally its value will depend not just on the metal used, but also on the conditions around the anode. If the conditions around the anode are standard (which means 25 °C, atmospheric pressure, and a metal-ion concentration of 1 mol litre$^{-1}$), the electrode potential so measured is the **standard electrode potential** (SEP, $\varepsilon^{\ominus}$).

The values obtained for different metals are ranked in Table 7.5. What does this ranking mean?

Elements whose atoms most easily lose

electrons develop the largest negative SEP (Ti, Al, Mg, Na). Some elements lose electrons less easily than hydrogen. These have positive SEPs. Note the similarity of ranking of SEPs to the sequence of lines on the Ellingham diagram of Figure 7.3. It makes little difference whether ions are dispersed into solution or bound to oxygen, the elements show the same relative ease of shedding electrons.

SAQ 7.6   (Objective 7.4)
A Daniell cell consists of zinc and copper electrodes in molar solutions of zinc and copper. There is no applied e.m.f., the cell itself acting as a battery. What is its open-circuit voltage? Which metal will corrode away in this cell when it delivers a current?

## 7.3.2 Reaction rates

If the contact potential at an electrode is displaced from its equilibrium value by applying a voltage from an external source, the activation of one reaction is electrically assisted, and that of other impeded. Suppose at equilibrium the two reactions are

$$M^{n+} + ne^- \longrightarrow M$$

$$M \longrightarrow M^{n+} + ne^-$$

Their activations are governed by an Arrhenius rate equations. Let their rate equations be

$$R_r = R_1 \exp \frac{-E_r}{kT}$$

for the reduction reaction and

$$R_o = R_2 \exp \frac{-E_o}{kT}$$

for the oxidation, where $E_r$ and $E_o$ are the respective *thermal* activation energies. Associated with each reaction is a flow of current, which is a measure of the rate of the reaction. In equilibrium conditions this is the exchange current $i_e$. So

$$R_r = R_o = i_e$$

As soon as the contact potential is displaced from its equilibrium value, by applying an **overpotential** from an external source, one reaction becomes easier than the other. Let the overpotential be $\eta$. Thus the extra work in adding or removing an electron is $e\eta$, where $e$ is electronic charge. If the reduction is the favoured reaction,

$$R_r = R_1 \exp \frac{-E_r + ne\eta}{kT} = i_e \exp \frac{ne\eta}{kT}$$

and

$$R_o = R_2 \exp \frac{-E_o - ne\eta}{kT} = i_e \exp \frac{-ne\eta}{kT}$$

The net current is $R_r - R_o$,

$$i = i_e \left[ \exp \frac{ne\eta}{kT} - \exp \frac{-ne\eta}{kT} \right]$$

This is plotted in Figure 7.16(a). Because $kT \approx \frac{1}{40} eV$ at room temperature, only a small overpotential soon makes these terms *very* different, and

$$i \approx i_e \exp \frac{ne\eta}{kT}$$

or

$$\log_e i \approx \log_e i_e + \frac{ne\eta}{kT}$$

Thus

$$\eta \propto \log_e i - \log_e i_e$$

so the curve in Figure 7.16 straightens out. Extrapolating the straight part back to $\eta = 0$ allows $i_e$ to be estimated, hence the values in Tables 7.3 and 7.4.

In this simple model, as soon as one reaction is favoured ($|\eta|$ *just* greater than zero) the net current that flows is suddenly at least $i = i_e$. It's as though one reaction has been 'switched on'. Of course, all that's happened is that one of the exchange currents has been suppressed and the favoured $i_e$ can now be detected flowing through the external circuit. Changing the polarity of $\eta$ 'switches on' the reverse reaction.

> EXERCISE 7.3 What rate of hydrogen production (tonnes/day) would be achieved with a nickel cathode of area of $700\,m^2$ and the reaction just 'switched on'? (Such a large cathode is feasible industrially by using a finely divided deposit of nickel on a substrate.) Atomic mass of hydrogen $= 1$ and 1 faraday $\approx 10^5$ coulombs. (A faraday is the total charge of an Avogadro number of electrons, or a mole of $M^+$ ions.)

A graph of $\eta$ against $\log i$ (to any base) is a **Tafel plot** (Figure 7.16). The steepness of slope, according to the ultrasimple model, depends only on the number of electronic charges $n$ on the ion. Actually it is not so simple because as ions pass through the boundary layer there are energy losses from, for example, viscous drag. To be properly useful the line has to be plotted by experiment or derived from more subtle theory. Also, with reaction rates dependent on so many factors (solution concentrations, association between ions, their electrical mobility and, of course, temperature), a straight-line model may be too naive. So in examples which follow the arguments are correct merely in principle.

I now want to combine the equilibrium potential and the overpotential data on the same diagram. The new diagrams are called **polarization curves**. By plotting data for more than one reaction onto a single diagram, the rates of competing reactions can be assessed. Figure 7.17 gives the general idea. The voltage zero on the graph is the standard hydrogen electrode potential. As the currents fall to zero (remember that the current is a logarithmic scale, so zero current is not plotted) the

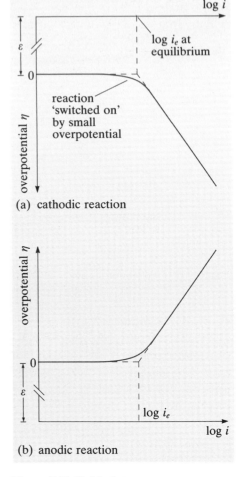

(a) cathodic reaction

(b) anodic reaction

Figure 7.16 Tafel plot

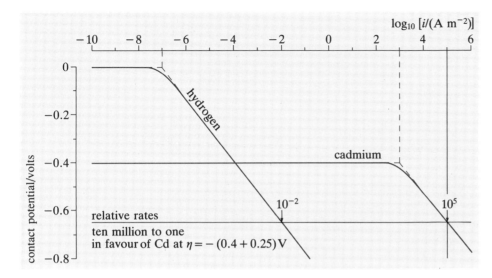

Figure 7.17 Polarization curves. Contact potential is relative to standard hydrogen cathode

electrodes will adopt their equilibrium potentials $\varepsilon$ relative to the hydrogen reference. Displacements of electrode potentials $\varepsilon$ by applying overpotentials will cause currents to flow. The convention is that anodic reactions require a positive potential displacement (curves sloping upwards for increased current) and that cathodic curves slope downwards.

Let us use a polarization diagram to look at metal reduction (cathodic). Suppose cadmium is to be plated onto itself using a molar solution acidified to pH 0 to improve solution conductivity. Will hydrogen evolution spoil the plating? In Figure 7.17, $\varepsilon$ for *both* elements is plotted (relative to standard hydrogen electrode so $\varepsilon = 0$ for hydrogen). Now the problem is that the hydrogen generation reaction is running with an overpotential of 0.4 V before the cadmium plating reaction even switches on. But since $i_e = 10^{-7}\,A\,m^{-2}$ for H on Cd while it is $10^3\,A\,m^{-2}$ for Cd on Cd, as soon as the cadmium reaction can occur its rate exceeds the hydrogen rate. A small overpotential suffices. In this case with $i_e$ so grossly different, poor knowledge of the Tafel slopes won't be too important.

EXERCISE 7.4 If the circuit is run with a cathode overpotential for the cadmium reaction of $\frac{1}{4}$ volt, how thick will the coating be after 1 minute? Use data from Figure 7.17. Atomic mass of cadmium = 112, density = $8650\,kg\,m^{-3}$ and 1 faraday $\approx 10^5\,C$.

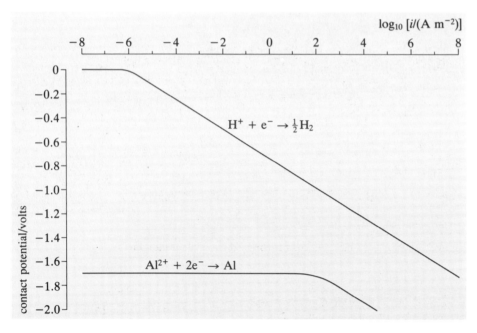

Figure 7.18 Polarization curves. Contact potential is relative to standard hydrogen cathode

A complete contrast is provided by aluminium smelting. If we tried to win aluminium from a molar acid aqueous solution, as Figure 7.18 shows, the hydrogen production rate is a million times greater than the aluminium rate. Nor is there salvation in using an alkaline solution to suppress the population of $H^+$ ions. Aluminium is soluble in alkali. Commercial ▼**Aluminium smelting**▲ uses a molten salt electrolyte.

SAQ 7.7   (Objective 7.4)
In a student exercise on electrolytic winning of copper, a quantity of ore is leached with boiling sulphuric acid and the solution electrolysed. The solution was still and a copper cathode was used. Most of the current was used depositing hydrogen. Two experts came. One advocated using titanium cathodes, the other stirring the solution during electrolysis. The first did not work. The second did; copper was laid down with only slight hydrogen generation. Explain.

## 7.3.3 Wet corrosion of metals

Metals, as we have seen, are vulnerable to oxidation (by various agents including oxygen). If the environment is moist, corrosion may proceed by electrochemical reactions. This is **wet corrosion**. Corrosion may be halted if the metal's surface is sealed, and self-sealing metals (such as aluminium) certainly have an advantage over continuously corroding metals, of which iron is one of the worst offenders. Indeed most widely used metals have this self-sealing property. Iron is so abundant that we tolerate its corrosion for the sake of its other advantages. But notice

# ▼Aluminium smelting▲

Aluminium is a very reactive metal and ionizes readily. It also has a strong affinity for oxygen. Hence, although it is a common element in the earth's crust, most of it is tied up in clay and rocks — impure aluminium oxide formed as the earth was created.

We've already eliminated reduction by carbon at over 2000 °C as a feasible method of converting $Al_2O_3$ (alumina) into aluminium, so what alternatives exist?

Reduction by magnesium or calcium are feasible at any temperature (see Figure 7.3) but first you then have to produce magnesium and calcium, hardly a feasible approach. Because aluminium cannot be deposited by electrolysis of aqueous solutions, the approach is to electrolyse a fused salt bath. The major problems are the melting temperature of alumina (2300 K) and the fact that any fused salt would contain large quantities of $Fe^{2+}$, $Ti^{2+}$ and $Mg^{2+}$ as impurities.

Why do these ions present problems?

Iron and titanium are more noble and their ions could be preferentially discharged at the cathode of any cell. Magnesium is just below aluminium in Table 7.5, so its ions should not be discharged if the cell voltage is controlled precisely to be between that to discharge aluminium ions and that for the discharge of magnesium ions.

The impurity problems can be sidestepped by purifying the ore before electrolysis using the Bayer process. The material is leached under pressure with strong caustic soda solution at a temperature of 430 K to produce a concentrated solution of sodium aluminate which, on cooling, forms a precipitate of aluminium hydroxide in a solution of sodium hydroxide. The aluminium compound is separated from the solution by filtration and when heated converts into clean aluminium oxide ($Al_2O_3$).

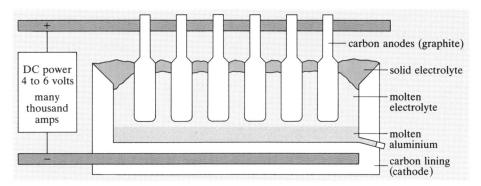

Figure 7.19 The Hall–Heroult process for extraction of aluminium

The problem of electrolysing this at high temperature was overcome simultaneously by Hall in the United States and Heroult in France. They discovered that the mineral cryolite, $Na_3AlF_6$, which melts at 1300 K, can dissolve $Al_2O_3$ up to 10 to 15%, and with further additions of $CaF_2$ the fusion temperature can be lowered to 1150 K. Note that sodium and calcium ions introduced into the melt are more electronegative than aluminium, and therefore will not be discharged during electrolysis. Fluorine is more strongly ionized than oxygen, therefore oxygen is discharged at the anode, although it is possible for some fluorine to be produced, particularly when the alumina content of the melt falls below 1%. Figure 7.19 shows the construction of a cell using their discovery.

As with any electrolysis process, the resistivity of the melt must be overcome on top of the theoretical voltage needed to split $Al_2O_3$ into its components, and cells are operated typically at 4 to 6 V.

Reactions taking place in the cell are summarized as follow:

Dissociation of alumina

$$Al_2O_3 \longrightarrow 2Al^{3+} + 3O^{2-}$$

Reaction at the cathode

$$2Al^{3+} + 6e^- \longrightarrow 2Al$$

Reaction at the anode

$$3O^{2-} \longrightarrow \tfrac{3}{2}O_2 + 6e^-$$

The carbon anode also burns giving

$$C + O_2 \longrightarrow CO_2$$

The latter chemical reduction of oxygen by the graphite actually saves energy in the process which would otherwise have to be found from the electrolysis current. Oxygen is not fully reduced, however, and the exit gas typically consists of a mixture of 20% CO and 80% $CO_2$.

The minimum potential drop needed to sustain these reactions is 1.6 V therefore the voltage efficiency is no more than 40%. Total power requirements have decreased from 25 MWh tonne$^{-1}$ in 1950 to less than 14 MWh tonne$^{-1}$ today due to better understanding of the process. It is nevertheless a huge power requirement and aluminium smelters must be located near to a cheap source of power such as hydroelectricity.

that sodium and calcium, two *very* abundant metals, are not useful in engineering — partly because they corrode even faster than iron. Here is a list of metals around my home which seem to be stable in ordinary conditions.

| | |
|---|---|
| aluminium | window frames |
| lead | roof flashings |
| zinc | galvanized corrugated iron sheet |
| copper | (as bronze/brass) fittings on horse tackle (slow tarnish) |
| chromium | plating on car parts |
| tin | plating on food cans |
| silver | jewellery (tarnishes) |
| gold | jewellery |
| stainless steel | cutlery |

Meanwhile with wire brush and red lead primer I annually maintain the steel oil tank, the gate hinges and of course the car. But roof slates shower down at every gale as old iron nails finally corrode away, and the cast iron guttering has now been replaced.

To understand the electrochemical actions, first let us note that the metal is destroyed by oxidation:

$$M \longrightarrow M^{n+} + ne^-$$

and recognize this as an *anodic* reaction. If the corrosion is to proceed by an electrochemical mechanism the electrons generated at the anode must move to a cathode where a corresponding amount of reduction takes place. In general there will be such a low concentration of metal ions in the electrolyte that some alternatives to plating out $(M^{n+} + ne^- \rightarrow M)$ will predominate. The most important cathode reactions in wet corrosion are

$$2H^+ \text{ (solution)} + 2e^- \text{ (cathode)} \longrightarrow H_2 \text{ (gas)}$$

$$2H_2O + O_2 \text{ (dissolved)} + 4e^- \text{ (cathode)} \longrightarrow 4OH^- \text{ (solution)}$$

The first is available in acid environments, which may of course include rainwater, as well as places where service in acid is required. The reaction of dissolved oxygen to generate hydroxyl ions ($OH^-$) is ubiquitous because a film of moisture covering a metal will have oxygen dissolved in it to saturation. When this reaction occurs, then both the cathode and the anode reaction produce ions in solution in the electrolyte. What happens next depends on the chemistry of the particular metal. If the ions stay in solution the metal can continue to corrode, especially when the water is flowing — bringing fresh oxygen and carrying away the ions as they form. Under the right conditions, though, the hydroxide may precipitate and dehydrate:

$$M^{2+} \text{ (aq)} + 2OH^- \text{ (aq)} \longrightarrow M(OH)_2(s)$$

$$M(OH)_2 \longrightarrow MO + H_2O$$

Variations on this involving other ingredients of the environment are often seen:

$$Fe(OH)_3 \longrightarrow FeO \cdot OH + H_2O \text{ (rusting)}$$

$$Fe(OH)_3 + Cl^- \longrightarrow FeO \cdot Cl + OH^- \text{ (marine corrosion)}$$

$$2Cu(OH)_2 + CO_2 \longrightarrow CuO \cdot CuCO_3 + H_2O \text{ (green patina on}$$
$$\text{copper sheathed roofs and on old bronze)}$$

$$PbO_2 + SO_2 \longrightarrow PbSO_4 \text{ (white deposit on lead in sulphur}$$
$$\text{pollution)}$$

The metal is 'passivated' by these solid reaction products, implying that the coating of products impedes further reaction. *If* the solids adhere well to the metal, further reaction is indeed impeded (for example, aluminium, chromium), but when further reactions cause volume changes and spalling (for example, in rust on iron) or when cyclic mechanical stress repeatedly disrupts an adherent film, corrosion continues. Passivation is not always a reliable anti-corrosion method.

Polarization diagrams can also be applied to corrosion to give at least a qualitative understanding of the electrochemical reactions. Let us suppose that a chunk of metal could support corrosion by an anodic reaction:

$$M \longrightarrow M^{n+} + ne^-$$

and that, nearby, hydrogen ions could be discharged by a cathodic reaction:

$$2H^+ + 2e^- \longrightarrow H_2(g)$$

We know that under standard conditions the metal/solution potential for this cathode reaction at pH 0 is 0 V (because we effectively have a hydrogen cathode) and that any corrodible metal at which the anode reaction takes place has a negative standard potential. Generally we won't know either of these potentials because concentrations are not necessarily 'standard', but our would-be anode will be more negative than the would-be cathode. Figure 7.20 shows the polarization diagram. The metal, being a conductor, cannot simultaneously be at two potentials. It has some uniform potential, denoted $E_{corr}$ on the diagram, which will be somewhere between the two electrode potentials. $E_{corr}$ is more negative than the cathode potential and more positive than the anode potential. Thus the electrode contact potentials provide overpotentials for each other. Both reactions are 'switched on' and polarized to some common current $i_{corr}$. The rates of reaction as measured by anode and cathode currents must be equal and so $E_{corr}$, $i_{corr}$ define the conditions of the circuit. The variations of current with overpotential at each electrode are plotted nominally as straight lines. Schematic though it is, this diagram does suggest how we can influence corrosion rate. To slow corrosion, the current must be reduced.

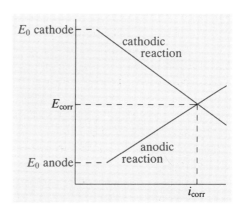

Figure 7.20 Anodic and cathodic polarization curves

367

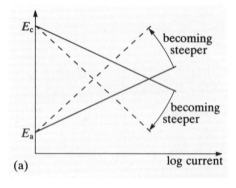

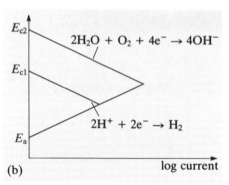

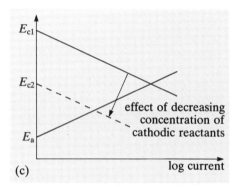

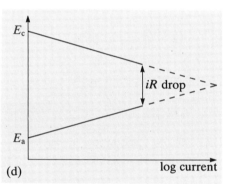

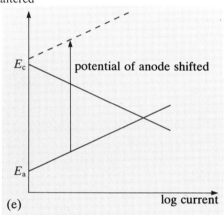

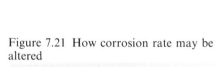

Figure 7.21 How corrosion rate may be altered

Remember it's a log $i$ scale, so modest changes of the intercept point may represent a large change in $i_{corr}$. Figure 7.21 shows some strategies, and the following comments apply respectively to the diagrams.

(a) If either reaction can be made more difficult to activate the slopes will be increased (more extra volts for more reaction). That's the role of a passivating layer.

(b) If the dissolved-oxygen cathode-reaction joins in, $i_{corr}$ is liable to increase.

(c) If the solution is starved of $H^+$ (that is, it is alkaline) and of dissolved oxygen, the cathode reactions are suppressed: $i_{corr}$ falls.

(d) If the electrolyte has poor conductivity even a small current will drop a significant voltage: $iR$ drop limits $i_{corr}$.

(e) If the metal potential can be brought to be cathodic relative to any other cathodic reactions there is no corrosion.

Thus the routes to corrosion protection are to cut down the rate of *either* reaction. For example:

• 'Anodized' aluminium has been deliberately corroded to give a thick alumina coat. This passivates the metal. Since alumina grows as a cellular structure under the right conditions, these pores can hold dye to give pretty colours.

- 'Primer' paints of red lead oxide, chromates, phosphates, zincates and so on all give stable passive layers on steel.

- Stainless steel is steel alloyed with sufficient chromium (12%) to provide a chromia ($CrO_2$) layer instead of rust. At less than 12% Cr the chromia layer is pervious, being poorly stoichiometric and allowing oxygen ion diffusion.

- Zinc plating on iron makes a barrier. If the barrier is breached, the zinc corrodes rather than the iron since the zinc becomes the anode. Hence galvanized iron.

- Tin plate is *merely* a protective coat using the passivity of tin. But if breached, tin becomes cathodic (with large area) while iron becomes anode. Fast, deep corrosion with high current density at the scratch is seen.

A parallel example to the last mentioned is at the junction of copper water pipes with iron radiators. If a leak develops, oxygenated water corrodes the iron. The copper becomes the cathode, setting $OH^-$ into solution. Breaking the metal circuit with an insulator can help and is common practice in chemical plant where pipework of different metals join. Inside the radiator *de*oxygenated water provides only limited cathodic action and black $Fe_3O_4$ is found. See ▼**An experiment with iron**▲

# ▼An experiment with iron▲

Both air and water are needed for iron to rust. Steel legs of oil rigs corrode in their splash zones (just below the water line), rather than at their deepest immersed parts; and there is a famous iron pillar in Delhi which has not rusted in that dry climate during hundreds of years exposure to air alone.

We have seen that when water containing oxygen is in contact with metal, the following reaction can occur.

$$2H_2O + O_2 + 4e^- \longrightarrow 4OH^-$$

The metal supplies the electrons and becomes ionized. Rust is the result when the metal is iron. Rust is a mixture of iron compounds, $Fe(OH)_3$, $FeO \cdot OH$, $Fe_2O_3$. Its non-adherent, flaky texture owes much to the variability of its composition. Its crystal structure is imprecise and contains internal stresses. Because of this the metal

is repeatedly exposed to further attack.

Ferroxyl indicator is a mixture which reveals $Fe^{2+}$ ions and $OH^-$ ions. One ingredient gives a blue colour with $Fe^{2+}$ ions, another turns pink when $OH^-$ are present. When a drop of brine containing ferroxyl corrodes clean steel, random pink and blue spots are seen at first. Later the centre of the drop becomes blue ($Fe^{2+}$), and the edge, into which oxygen diffuses faster because of its bigger ratio of surface area to volume, shows pink.

The first reactions occur because of microstructural irregularities causing small, local variations of voltage which are sufficient to drive a corrosion current. Later, with the centre of the drop depleted of oxygen, the metal oxidizes and its ions pass into solution.

A small pit develops, which further impedes oxygen transport, confirming this point as the anode of the cell.

SAQ 7.8 (Objectives 7.4 and 7.6) Why do cars rust particularly at spot-welded seams (Figure 7.22)?

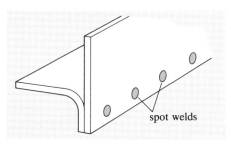

Figure 7.22 Spot-welded seam

## Summary

A metal in an electrolyte comes to chemical equilibrium with a contact potential between itself and the electrolyte. The equilibrium is dynamic, with equal and opposite 'exchange currents' flowing across the metal/electrolyte interface.

An overpotential which displaces the contact potential from its equilibrium value will cause a net current to flow. A polarization curve relates this current to the electrode contact potential (measured relative to a standard hydrogen cathode). Where there are competing reactions at an electrode, the polarization diagram can be used to predict the outcome.

Currents flowing at the two electrodes of a cell must be equal. Polarization curves for the anode and cathode must cross at that current. Wet corrosion occurs when, in the presence of an electrolyte, one area of a metal is anodic relative to an electrically connected cathodic area.

SAQ 7.9   (Objective 7.4 and 7.6)
The area where a bronze propellor joins the steel drive shaft at the back of a ship is susceptible to corrosion. Explain why. Protection is afforded by bolting blocks of magnesium onto the hull close to the corrosion zone. Why does this work?

SAQ 7.10   (Objectives 7.4 and 7.6)
Buried steel pipes are coated to protect them from corrosion. However, joins have to be coated on site and are therefore vulnerable. A secondary defence is therefore provided by an electric current, using the pipeline and lumps of scrap iron as electrodes (Figure 7.23). Which way round should the supply be connected. How would each of the following factors affect the current supplied?

(a) Saline ground water.
(b) Ground water lacking dissolved oxygen.
(c) Joint left uncoated.

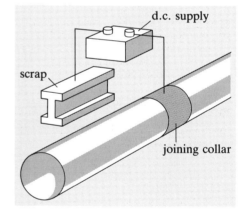

Figure 7.23

# 7.4 Organic polymers

## 7.4.1 Properties and molecular mass distribution (MMD)

Earlier chapters have ascribed the properties of polymers to the structure of their repeat unit. Chapter 5 further suggested that a given polymer could exist in different grades, and that these different grades could differ usefully in their mechanical properties (for example LDPE and HDPE). The different grades of a polymer are distinguished by the size of their molecule chains — by the 'polyness' of the polymer.

Molecular mass is a measure of 'polyness', so an appreciation of how a polymer's properties vary with its molecular mass, and of how the mass may be controlled, is essential to polymer engineers.

Apart from so-called monodisperse polymers, in which every chain is of the same size, it is unlikely that a polymer's molecules will be uniformly long. The molecular masses might therefore be widely distributed, as shown for example in Figure 7.24. (Note the molecular mass scale is logarithmic.)

The techniques and conditions of polymerization affect both the average chain length and the distribution of lengths. ▼**Molecular masses of polymers**▲ defines what 'average' and 'distribution' mean in this context, and ▼**GPC**▲ describes how they are measured.

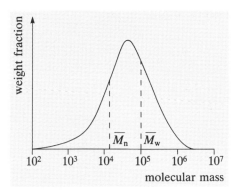

Figure 7.24 Molecular mass distribution for HDPE. The dispersion is 7.64

# ▼Molecular masses of polymers▲

The molecular mass of a polymer can have a wide distribution, so it is useful to be able to speak of its average molecular mass. There are two sorts of average molecular mass. Their derivations are very similar to those used in the law of mixtures in Chapter 2.

Suppose a polymer is made up from two species having individual molecular masses $M_1$ and $M_2$. Let there be a weight (properly, mass) $w_1$ of the first species and a weight $w_2$ of the second. The **weight fraction** of each species is

$$\frac{w_1}{w_1 + w_2} \text{ for the first and } \frac{w_2}{w_1 + w_2}$$

for the second. The **weight average molecular mass** $\bar{M}_w$ is

$$\bar{M}_w = \frac{w_1}{w_1 + w_2} M_1 + \frac{w_2}{w_1 + w_2} M_2$$

If there are several constituent species, then

$$\bar{M}_w = \text{sum of (weight fraction} \times \text{molecular mass)}$$

$$\text{or } \bar{M}_w = \frac{\Sigma w_i M_i}{\Sigma w_i}$$

For example, suppose 1 kg of polymer consists of equal parts by weight of species with individual molecular masses $10^4$ and $10^5$. The weight fraction for each is 0.5. Figure 7.25 shows the molecular mass distribution. So

$$\bar{M}_w = (0.5 \times 10^4) + (0.5 \times 10^5) = 55\,000$$

Notice that this average is weighted in favour of the heavier species.

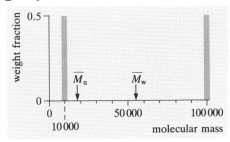

Figure 7.25 MMD for equal parts by mass of polymer mixture of molecular masses $10^4$ and $10^5$

The other average molecular mass is in terms of the number of molecules of each species. This is the **number average molecular mass**, $\bar{M}_n$. The number of molecules is counted as the number of moles of each species, so

$$\bar{M}_n = \text{sum of (mole fraction} \times \text{molecular mass)}$$

$$\text{or } \bar{M}_n = \frac{\Sigma n_i M_i}{\Sigma n_i}$$

The number of moles $n_i$ has to be calculated from $w_i$. The molecular mass of this species is $M_i$, so a mole of it weighs $M_i$ grams, or $M_i \times 10^{-3}$ kg. So an amount $w_i$ kg of a particular species comprises

$$n_i = \frac{w_i}{M_i \times 10^{-3}} = \frac{w_i \times 10^3}{M_i} \text{ moles}$$

*provided weights are in kg and moles are in grams.*

For the polymer of Figure 7.25,

$$n_1 = \frac{0.5 \times 10^3}{10^4} = 5 \times 10^{-2}$$

$$\text{and } n_2 = \frac{0.5 \times 10^3}{10^5} = 5 \times 10^{-3}$$

So

$$\bar{M}_n = \frac{(5 \times 10^{-2} \times 10^4) + (5 \times 10^{-3} \times 10^5)}{5 \times 10^{-2} + 5 \times 10^{-3}}$$

$$= \frac{1000}{0.055} \approx 18\,000$$

This average is weighted in favour of the constituent with the lower molecular mass, expressing the fact that there are many more low-mass molecules than heavy ones.

A useful measure of the spread of molecular mass is the **dispersion**, defined as

$$\text{dispersion} = \frac{\bar{M}_w}{\bar{M}_n}$$

For Figure 7.25 dispersion is $55\,000/18\,000$ which is about 3.

SAQ 7.11 (Objective 7.8)
Determine the dispersion in 1 kg of a mixture containing 25% by weight of species with constituent molecular masses 10 000, 20 000, 30 000 and 40 000.

The above derivations can be extended to cover polymers such as that of Figure 7.24 where there is a continuous distribution of chains of different length. In smooth distributions such as this the peak is straddled by $\bar{M}_n$ (below) and $\bar{M}_w$ (above).

Polymer properties are germane not just to how the polymer behaves in use, but also to how it may be processed. We need to look therefore at the effect of polymer grade on service and processing properties.

# ▼GPC▲

Finding the distribution of molecular masses in a polymer requires the molecules to be separated by size. **Gel permeation chromatography** (GPC) does this. First the polymer has to be dissolved in a suitable solvent. The solution, at chosen concentration and temperature, is presented at the top of a column containing a gel formed of tiny spherical particles (Figure 7.26). As the solution seeps through the gel, the polymer molecules repeatedly get trapped in and escape from the spaces between the gel particles. The longer molecules take longer on average to escape, so the smaller polymer molecules get ahead of the larger ones and by the time the whole column has been traversed the separation is sufficient for an adequate measurement of the distribution of molecular masses.

The solution coming from the foot of the column is collected in successive samples of about $1\,cm^3$ and the weight fraction of polymer solute in each is estimated from precision measurements its refractive index.

To be able to convert these data into a distribution of *molecular* masses, the column must be calibrated by recording the times for standard solutions of polymer of known molecular mass to traverse the column. In commercial instruments, temperature control, refractive index measurements and interpretation all take place within the instrument; you pour the solution in and get a chart out.

GPC is a useful method for comparing grades of a single polymer. However, the behaviour of the column can drift and in practice samples should be run in close succession and frequent calibration checks should be made.

Figures 7.24 and 7.35 are taken from the output of a GPC machine. The machine automatically decides how to smooth the histogram produced by sampling.

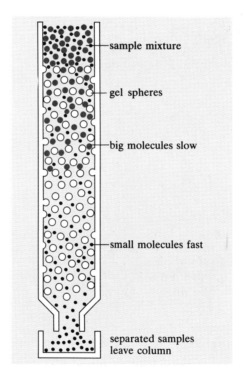

sample mixture

gel spheres

big molecules slow

small molecules fast

separated samples leave column

Figure 7.26 GPC column

## Service properties

Below a threshold known as the **critical entanglement mass**, $M_c$, polymers are too weak to be useful. Beyond $M_c$ the service mechanical properties, particularly long-term creep and strength, are sensitive to molecular mass. Figure 7.27 shows considerable benefits gained by increasing molecular mass up to a plateau region (normally at a molecular mass of the order of several million). In this region, further improvements in strength with increasing molecular mass become smaller — although for some critical applications like HDPE hip-joint cups they may still be worth achieving.

'Entanglement' calls up images of long molecules sufficiently entangled to be inextricable merely by pulling. This is a good model to carry in mind, but realize that the inherent stiffness of the molecules and the strengths of secondary bonds holding them together must influence their ability to tangle and their response to tension. Experiments with monodisperse samples show the critical lengths to vary from about 290 chain atoms for linear polyethylene to 675 for polystyrene.

You know how much easier it is to disentangle string if you cut a few critical points where it is knotted together. If the average molecular mass of a polymer is near the entanglement value, a few cuts can quickly break the cohesion. In crystalline material, cuts in the interlamellar (amorphous) regions soon reduce the material's strength. So the degradation properties of polymers in service are significantly affected by the molecular mass distribution. A healthy proportion of ultra-long molecules which link lamellae will offer protection against cracking under ultraviolet light or heat.

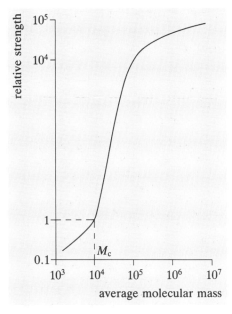

Figure 7.27 Polymer strength as function of molecular mass

## Processing properties

Both injection moulding and extrusion forming use molten polymer, so how molten polymers flow is a most important process property. Basically, low-molecular-mass polymers flow more easily than high-molecular-mass grades, so low-molecular-mass material can be shaped with less force or at lower temperature (or both). Capital and running costs of machinery are therefore, potentially, lower if the product can be made from a grade with low molecular mass. Manufacturers might nevertheless still choose to use more powerful machinery and higher temperatures than the material actually warrants so as to keep the production rate high. The unit cost is inversely related to the rate of production.

In molten polymer, resistance to flow arises mainly from physical entanglement of polymer molecules. The longest molecules are most significant here so $\bar{M}_w$ is the measure to use to relate flow to the molecular mass. If $\bar{M}_w$ is small, below the critical entanglement threshold, it and viscosity (at low shear rates and at a given temperature) are proportional ($\eta \propto \bar{M}_w$). Once entanglement is involved, as it must be for a useful polymer, a much stronger relation develops with

$$\eta \propto \bar{M}_w^{3.5}$$

Indeed the jump from one relation to the other is a way of estimating the critical entanglement molecular mass.

These laboratory observations of viscosity do not, however, adequately describe behaviour in processing machines, where high shear strain rates are necessary. Then the looped molecules tend to pull each other out straight and the flow rate goes up. Figure 7.28 indicates how, because of this effect, large changes in shear strain rate can be induced by modest increases of driving force. But this is a 'viscoelastic' effect; when the shearing forces are removed, random configurations are re-established. ▼**Two polymer products**▲ looks at the benefits and problems of these effects.

## 7.4.2 Building polymer molecules and controlling MMD

There are two mechanisms by which monomers may be polymerized, namely **step polymerization** and **chain polymerization**. (These used to be known as 'condensation' and 'addition' respectively.)

Step polymerization has the monomer molecules gradually 'condensing' to bigger things. First, most monomers make pairs, then the pairs join, then these bigger fragments join end-to-end, and so on until there are so few 'ends' left that they cannot easily find each other. Chain growth involves repeated addition of a monomer molecule to one end of the growing polymer molecule, like adding extra links one at a time to make a chain.

The two methods are appropriate to different classes of monomer, but the important distinction here is that they yield quite different molecular mass distributions. Chain reactions can achieve much longer molecules because small, mobile monomer molecules can find the chain ends. But in a tangled mass of medium-length molecules formed by step reactions, it becomes increasingly unlikely that free ends will come together to enable the reaction to go to completion.

### Step-growth polymerization

The main products produced this way are the polyesters, such as Terylene (PET), and the polyamides, for example the various types of nylon. These groups: are made by the reaction of an organic acid —COOH either with an —OH (Figure 7.29) or with an $NH_2$ (which also liberates $H_2O$). A useful ploy for getting high molecular masses is to use monomers which are already quite long. Thus nylon 6,6 is made from monomers containing six carbon atoms in a line.

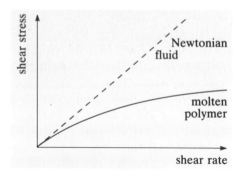

Figure 7.28 Large changes of shear strain rate from modest change of shear stress

# ▼Two polymer product

The dependence of flow rate on stress makes injection moulding of very thin sections possible. As molten polymer is driven into a narrow channel, it flows more easily; so it can get through to the end of a mould. Items such as cheap food pots, with a wall thickness about 0.5 mm, would be impossible to mould otherwise, even with a relatively low-molecular-mass, low-strength polymer.

At the other end of the scale, consider making heavy-gauge gas piping by extrusion. As the pipe leaves the die, the driving forces are suddenly removed and aligned molecules relax into randomness. Long thin shapes tend to become shorter and fatter, so the pipe coming out of the extruder swells laterally slightly ('die swell'). The die shape must allow for this swelling to give the final specified dimensions in the product. Because an extrusion has to be self supporting as soon as it leaves the die, grades of polymer with higher molecular mass are better suited to this process route. Such polymers exacerbate the die-swell problem. Forced cooling of extrusions by air jets or water baths can be used to get the right compromise between extrudability, immediate rigidity and dimensional accuracy. Medium density polyethylene is chosen for gas pipes, giving adequate service properties without excessive processing costs.

The monomer molecules have two sites which react wherever they meet, and to begin with that happens everywhere. When half of the possible reactions have happened the average length of the chain must be about two monomer units, two thirds of all sites reacted makes the average about 3 units, and so on. Even when 90% have gone, the chains are still only around 10 units long, though by then statistics are beginning to get a grip and a wider range of molecular lengths is developing. Critical entanglement for nylon 6,6 requires at least 50 monomer molecules to have joined. A theoretical model gives the graphs of Figure 7.30 which shows that to get high molecular masses the reaction must go very close to completion. This means tailoring reactive conditions to bring those last few ends together. The model also predicts a spread of molecular masses with $\bar{M}_w/\bar{M}_n$ exactly 2. Values only slightly different from this are found in commercial step-grown polymers. To achieve high molecular mass, exactly equal numbers of the two types of reactive ends must be provided, otherwise all ends will soon become identical and further reaction will stop. In nylon 6,6 manufacture this is neatly achieved by reacting the acid and amine in water when 'nylon salt', the pure 'dimer', is precipitated out of solution. This is then the starting material for polymerization. It is dispersed in a non-reacting diluent so that as the reaction proceeds and the longer molecules start to entangle some mobility is preserved.

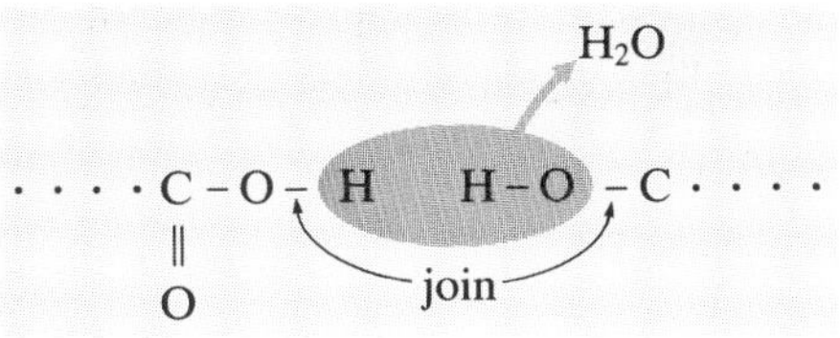

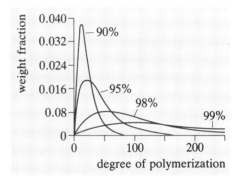

Figure 7.29

Figure 7.30 Change of MMD at high conversion of monomer to polymer

## Chain-growth polymerization

Chain polymerization is the mode of growth for polymers whose monomers contain C=C bonds, for example ethylene ($CH_2$=$CH_2$), vinyl chloride ($CH_2$=CHCl) and many others. Polymerization depends on disrupting the $sp^2$ hybrid bond and establishing $sp^3$ configuration. Then long chains —C—C—C—C $\cdots$ can build. There are several ways of doing this, and there are various catalysts which govern the tacticity of the polymer. I will look at one relatively simple mechanism.

Many unstable symmetrical molecules have the property of breaking in half under thermal vibration (or by absorbing energetic ultraviolet radiation), the two halves becoming free radicals. These are incomplete molecules with an unpaired electron, looking for an opportunity to bond to something. An example is benzoyl peroxide $C_6H_5$—O—O—$C_6H_5$ which is easily cracked to make two bits of $C_6H_5$—O• with the dot representing an electron needing a mate. When the orbital of such an electron overlaps with the $\pi$ bond orbital of a C=C bond a reaction starts.

$$C_6H_5-O\bullet + CH_2=CHCl \longrightarrow C_6H_5-O-CH_2-\overset{\bullet}{C}HCl$$

The original free radical has satisfied itself but now the monomer molecule has a dangling bond. At the next meeting with a double bonded monomer the process repeats and you can see the chain is on its way.

Continued addition of monomer units is the most likely further reaction and chain growth is extremely rapid for most species. But a **chain transfer reaction** can occur which terminates the polymer chain:

$$\cdots H_2C\text{—}\overset{\bullet}{C}HCl + CH_2\text{=}CHCl \longrightarrow \cdots CH\text{=}CHCl + CH_3\text{—}\overset{\bullet}{C}HCl$$

| growing polymer chain | monomer | dead polymer chain | activated monomer |

The growing polymer chain is stopped and a new one started. This chain transfer reaction has a high activation energy and so increases in importance with increasing temperature.

Two other termination mechanisms act by free radical ends meeting each other. Obviously they may just join up (**coupling**):

$$\cdots CH_2\text{—}\overset{\bullet}{C}HCl + ClH\overset{\bullet}{C}\text{—}CH_2 \cdots \rightarrow \cdots CH_2\text{—}CHCl\text{—}ClHC\text{—}CH_2 \cdots$$

coupled chain

or, by a hydrogen jumping chains (**disproportionation**):

$$\cdots H_2C\text{—}\overset{\bullet}{C}HCl + ClH\overset{\bullet}{C}\text{—}CH_2 \cdots \rightarrow \cdots CH\text{=}CHCl + ClCH_2\text{—}CH_2 \cdots$$

| active chain end | active chain end | dead chain | dead chain |

and neither of these result in a new chain starting up.

EXERCISE 7.5 What are the molecular mass effects of the three different kinds of termination? What chain structural differences arise from these termination reactions?

Polymer fabrication thus comprises initiation, growth and termination reactions, and the relative rates of these reactions determine the extent to which polymer chains develop from a mass of monomer. For example, if the initiation step is slow, the growth very fast and termination non-existent, a small number of very long molecules forms until all the monomer is consumed. Conversely if termination reactions are fast, only short polymer molecules result.

All the rates can be adjusted by choice of initiator, by controlling monomer dispersion and by adjusting temperature. In these ways, MMD of a polymer is controlled. The diversity of systems for implementing this control is now great, but one factor is immutable. Polymerization reactions are strongly exothermic, so removing heat to keep the reaction down to a safe rate is vital. Dispersing the monomer in a liquid greatly increases the thermal capacity of the reacting mass and makes heat transfer more efficient. A typical process for propylene has the gaseous monomer dissolved in heptane ($C_7H_{16}$, a liquid hydrocarbon) at a few atmospheres pressure (to increase solubility) and brought into contact with the catalyst in a closed vessel purged of air. The catalyst is in the form of a slurry of fine particles. Heat extraction

# ▼ Dental fillings ▲

This application requires a cement which is initially pasty, so that it can be worked into the small, drilled-out cavity, and which then hardens to an abrasion-resistant, thermally insulating, non-toxic material which is unaffected by saliva and matches the colour of the living tooth. Polymers are superior in many ways to the older amalgams.

The dentist keeps several pastes of different tints, consisting of a basic white filler ($SiO_2$) with added pigment. The liquid part is polymer of low molecular mass (below its entanglement threshold) and a light-sensitive hardener. The polymer is designed to be further polymerized in the tooth. When the cavity has been packed, the hardener is stimulated by radiation from a violet lamp and the whole mass polymerizes to a three-dimensional cross-linked piece.

The polymer system exhibits both step and chain polymerization. The low-molecular-mass fraction of the paste is a polyurethane made by step reaction between a di-isocyanate and di-ol:

$$O\text{=}C\text{=}N\text{—}\boxed{\phantom{xx}}\text{—}N\text{=}C\text{=}O$$
$$+$$
$$H\text{—}O\text{—}\boxed{\phantom{xx}}\text{—}O\text{–}H$$
$$\downarrow$$
$$O\text{=}C\text{=}N\text{—}\boxed{\phantom{xx}}\text{—}N\text{–}CO$$
$$\qquad\quad\; H \quad O\text{—}\boxed{\phantom{xx}}\text{—}O\text{–}H$$
urethane link

To keep the viscosity low, the molecular mass must be kept below the critical entanglement threshold ($M_c$), so an excess of the di-isocynate is used. Then, before many steps of reaction have occurred, all the polymer molecules have isocyanate

restricts the temperature rise to about 60 °C. In this case initiation of polymer chains (on the particle surfaces) occurs throughout the time allowed. The products include molecules which have been growing for a long time as well as those grown for shorter times so a broad MMD with $\bar{M}_w/\bar{M}_n$ in the range 8 to 12 is usual. The solid polymer and catalyst are separated by filtration and the solvent recycled to the start.

An interesting example of use of both polymerization methods is given in ▼Dental fillings▲, and the way competing reaction rates govern copolymer structure is discussed in ▼Copolymer structures▲.

# ▼Copolymer structures▲

ends so the reaction stops. To tailor these molecules for the *in situ* polymerization the excess isocyanate monomer is separated and a short chain alcohol containing double bonds is introduced. This makes urethane links with the terminal groups of the polymer,

$$\boxed{\phantom{}}\!-\!N\!=\!C\!=\!O$$
$$+$$
$$H\!-\!O\!-\!CH_2\!-\!CH\!=\!CH_2$$
$$\downarrow$$
$$\boxed{\phantom{}}\!-\!\underset{\underset{H}{|}}{N}\text{-}\underset{\underset{O\text{-}CH_2CH=CH_2}{|}}{CO}$$

and the final double bonds are now available for the final polymerization in the tooth by chain reactions. The hardener splits into reactive free radicals which initiate polymerization by attacking the double C=C bonds. Because there is a double bond on the ends of each polyurethane molecule there are sufficient sites to ensure three-dimensional cross linking of all the polymer molecules to build a rigid block.

*In situ* polymerization is not limited to the dentist's chair nor to this formulation. The principle of using low-molecular-mass polymers for their easy flow properties and then using further polymerization to make them solid has wide application. Paints and many glues are designed to work like this, as are glass fibre reinforced resins for repairing car bodies. **Reaction injection moulding** is an increasingly competitive production route which uses very fluid polymers to give access to the mould and rapid chain-growth polymerization for solidification.

How can the different copolymers structures in Figure 5.61 be realised? The idea of competing reaction rates in chain polymerization gives us some ideas. Imagine polymerization going on in a mixture of two monomers A and B. At some stage a chain has an end which is 'activated A', that is $\cdots A^{\bullet}$. Is it more likely to react with another A molecule or with a B? Similarly, which of A or B is more likely to join a chain $\cdots B^{\bullet}$? Let's define these 'reactivity ratios':

$$r_1 = \frac{\text{rate } A^{\bullet} \text{ reacts with } A}{\text{rate } A^{\bullet} \text{ reacts with } B}$$

$$r_2 = \frac{\text{rate } B^{\bullet} \text{ reacts with } B}{\text{rate } B^{\bullet} \text{ reacts with } A}$$

These rates are measurable by experiment.

Clearly if both $r_1$ and $r_2$ are approximately unity, the active ends don't mind which species their next partner is and random copolymers result, Figure 5.61(a).

If both $r_1$ and $r_2$ are much less than 1, $A^{\bullet}$ is more likely to react with B and $B^{\bullet}$ with A giving an alternating copolymer. To make block copolymers $r_1$ and $r_2$ should be much larger than unity so that the preferences are A—A—A and B—B—B, with occasional switches. In practice this method isn't used because the reactivity ratios are rarely convenient. Instead the different monomers are introduced to the reaction in turn. Table 7.6 quotes some values for reactivity ratios.

SAQ 7.12 (Objective 7.9)
What kind of copolymer will be formed by free-radical polymerization of the following monomer mixtures:

(a) styrene + acrylonitrile
(b) styrene + butadiene
(c) vinyl chloride + vinyl acetate
(d) acrylonitrile + butadiene
(e) styrene + vinyl acetate.

Table 7.6  Typical monomer reactivity ratios

| Monomer 1 | Monomer 2 | $r_1$ | $r_2$ |
|---|---|---|---|
| acrylonitrile | butadiene | 0.02 | 0.3 |
| | methyl methacrylate | 0.15 | 1.22 |
| | styrene | 0.04 | 0.40 |
| | vinyl acetate | 4.2 | 0.05 |
| | vinyl chloride | 2.7 | 0.04 |
| butadiene | methyl methacrylate | 0.75 | 0.25 |
| | styrene | 1.35 | 0.78 |
| | vinyl chloride | 8.8 | 0.035 |
| methyl methacrylate | styrene | 0.46 | 0.52 |
| | vinyl acetate | 20 | 0.015 |
| | vinyl chloride | 10 | 0.1 |
| styrene | vinyl acetate | 55 | 0.01 |
| | vinyl chloride | 17 | 0.02 |
| vinyl acetate | vinyl chloride | 0.23 | 1.68 |

## 7.4.3 Degradation by oxygen

Fire is the ultimate service hazard for all organic materials. It returns them by oxidation to the thermodynamic stability of the oxides $CO_2$ and $H_2O$ whence photosynthesis originally brought them. Not only is $\Delta H$ negative for combustion (these oxides are strongly bonded) but $\Delta S$ is very positive when so many small molecules are formed. Combining these makes $\Delta G$ negative at all temperatures. The apparent chemical stability of polymers must therefore be ascribed to the high activation energy needed to set these reactions going. But much less drastic intervention by oxygen can render polymers unserviceable.

Almost every plastic artefact suffers two opportunities for degradation. Firstly, in the processes of forming the polymer into an object, high temperatures are used. Secondly, in use the object will be exposed to sunlight. The ultraviolet component of sunlight comprises energetic photons capable of disrupting chemical bonds. As in polymerization, free-radical chemistry plays a large part in the processes of degradation. Heat and ultraviolet light initially damage polymer molecules by establishing short-lived free radicals. Many of these recombine without damage, but if oxygen joins the act by reacting with these free radicals the consequences are varied. Chains may be broken, reducing the molecular mass and weakening the plastic by disentangling the vital bridge molecules linking crystal lamellae. Alternatively, cross-linking may be provoked, increasing the molecular mass and reducing flexibility. The plastic becomes brittle and cracks. Figure 7.31 shows cracked and embrittled PVC garden furniture in which exposure to ultraviolet light has been exacerbated by high surface temperatures caused by sunlight. Higher temperatures increase oxygen diffusion and oxidation reaction rates.

The susceptibility of a polymer to degradation is determined primarily by its 'weak links', which are the first to be disrupted. Free-radical-initiated polymers will contain residues of initiator which are obvious

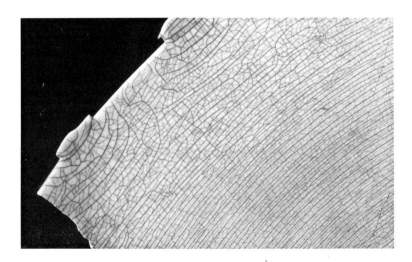

Figure 7.31 Ultraviolet degradation in PVC

starting points. Chain irregularities, such as 'head-to-head' groups resulting from termination by coupling, are also vulnerable. Other weak points are chain ends, hence low-molecular-mass polymer tends to be more susceptible to degradation. Less obvious, but more pernicious, are the properties of the repeat unit itself. Some structures give longer-lived free radicals, which can wait for an oxygen molecule to arrive. An example is a C—H bond, where the C is joined to three other carbon atoms or is adjacent to a double C=C bond. Polypropylene and butadiene rubbers are examples containing such hazards all along their molecules. See Figure 7.32.

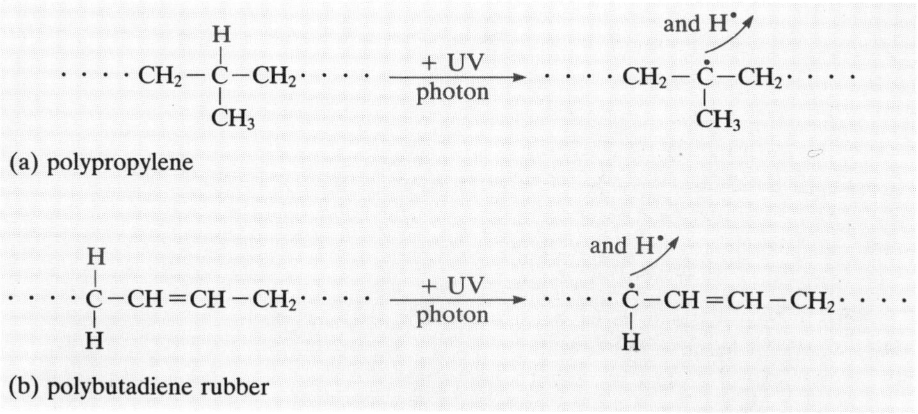

(a) polypropylene

(b) polybutadiene rubber

Figure 7.32 Degradation of polypropylene and polybutadiene

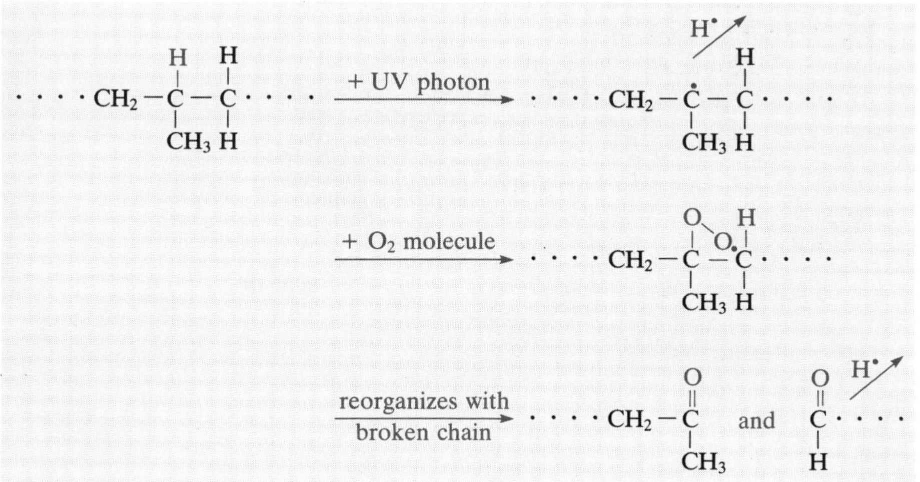

Figure 7.33 Chain cleavage

The C=O (carbonyl) group is itself vulnerable to ultraviolet radiation, so polymers such as polyamides, polyesters and polycarbonates, which have this group, may be prone to attack. A typical reaction scheme leading to chain cleavage is outlined in Figure 7.33.

Protection against chemical degradation by oxygen consists either of absorbing the ultraviolet light in a pigment, such as carbon black, or using anti-oxidants, which latch on to the oxygen free-radicals and can suppress the later reactions. ▼A forensic study▲ describes a practical instance of these problems.

# ▼ A forensic study ▲

Polypropylene is much used for the casings of vehicle batteries. It is mechanically robust, chemically inert to battery acid and easily formed by injection moulding. Some large, expensive, heavy-duty batteries with polypropylene casings were sold to the Israeli army. In service on fork-lift trucks, the tops of these batteries were open to very sunny skies. Several battery cases deteriorated very badly (Figure 7.34). Oxidation from excessive exposure to ultraviolet light had degraded the plastic. The manufacturer claimed he was blameless, since the original order had not specified that protective additives against ultraviolet light would be required. The customer had misused the product. But the customer wondered why some cases had cracked, whereas others had not. Could the

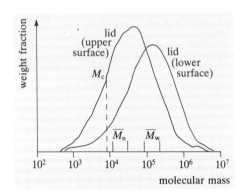

Figure 7.35 Change in MMD for failed batteries

material be defective in some way which enhanced its susceptibility to ultraviolet light? The argument had to be settled in order to attribute liability. The cost of replacements was not trivial and the Israeli army was a big customer.

Infra-red spectroscopy gave the answer. Carbonyl groups, $C{=}O$, are indicators of degradation and can be found by looking in the infra-red spectrum for absorption lines corresponding to vibrations of $C{=}O$ pairs. Specimens for the spectrometer were thin films of plastic made by dissolving some shavings from the battery cases in a pure solvent known not to contain $C{=}O$, and then evaporating the solution spread out on a polished plate. There had to be no risk of oxidizing the molecules in this step. Evaporation under vacuum was used since it excludes oxygen and avoids heating.

Results from the cracked parts confirmed the obvious severe oxidation. GPC revealed the consequent shift downwards of the molecular mass distribution (Figure 7.35). The loss of ultra-long molecules needed to tie lamellae together allowed residual stresses in the mouldings to crack the material. Note that in this degraded material from the upper surface, a much larger fraction of the material now falls below $M_c$, the critical entanglement mass of polypropylene. Samples from cases which had stayed sound in service showed very little carbonyl in the tops, despite long ultraviolet exposure. The critical test was for carbonyl in the unexposed parts of the cases. Specimens from the bottoms of sound cases showed no carbonyl at all, whereas those from damaged cases did show evidence of slight oxidation. This was a crucial observation, since the oxidation reactions are to some extent 'auto-catalytic', that is the products encourage further reaction. The seeds of the degradation had indeed been sown before service.

How did the cases get oxidized before exposure? The most common cause is allowing air to get at the hot polymer in the mould, perhaps in conjunction with using too high a melt temperature to get low viscosity for accurate moulding. The outcome was that the manufacturer had to accept liability for the damage and take firmer control of the injection and welding procedures. In view of the harsh service environment, ultraviolet inhibitor was specified for subsequent orders.

Figure 7.34 Left, four battery lids from old stock. Top right and half-section at bottom right show 'chalking' at weld zone

Table 7.7 is an empirical table of both short-term (200 hour) and long term (1000 day) stability in air. It takes account of both depolymerization and resistance to oxidation. However, it does not allow for ultraviolet resistance; some heat-stable polymers, like Kevlar for example are very sensitive to ultraviolet owing to their chain structure of coupled rings and $C{=}O$ groups.

At the lower end of the table are the familiar commodity thermoplastics. At the higher end are the new generation of plastics for high-temperature applications. These get their chemical and physical stability from closely bound molecular structures with a preponderance of closed rings. Polyimide (see Table 4.5) was developed mainly for thin-film application in electronic products where resistance to soldering is needed for mass-produced circuits.

Table 7.7  Heat stability of polymers

| Polymer | Ultimate end-use temperature range in 200 h/°C | Ultimate end-use temperature range in 1000 days/°C |
|---|---|---|
| polybenzimidazole | 350–400 | 250–300 |
| polyimides | 300–350 | 180–250 |
| aromatic polyamides (e.g. Kevlar) | 250–300 | 180–230 |
| polyfluorocarbons (e.g. PTFE) | 230–300 | 150–220 |
| silicone elastomers | 200–280 | 130–180 |
| fluor elastomers (e.g. Viton rubber) | 200–260 | 130–170 |
| polysulphone | 160–180 | 130–150 |
| poly(phenylene oxide) (PPO) | 160–180 | 130–150 |
| cross-linked aromatic polyester | 180–250 | 120–150 |
| linear polyester (PET) | 140–200 | 100–135 |
| polycarbonate | 120 | 100–135 |
| cross-linked polyurethanes | 150–250 | 100–130 |
| epoxy resins | 140–250 | 80–130 |
| linear polyurethanes | 130–180 | 70–110 |
| polyamides | 100–150 | 80–100 |
| polyolefins | 70–100 | 60–90 |
| poly(methacrylates) e.g. PMMA | 70–100 | 60–80 |
| polyisoprene e.g. natural rubber | 60–90 | 60–80 |
| polystyrene | 60–90 | 60 |
| poly(vinyl chloride) | 60–90 | 50 |

## Summary

Properties of polymers are related to their molecular mass distributions. A weight average and a number average molecular mass are defined. Their ratio ('dispersion') is used to express the spread of molecular masses in a specimen.

Polymerizations are either step or chain reactions. Chain reactions more readily provide very high molecular masses. Polymers degrade by the combined action of oxygen and either heat or ultraviolet radiation. Together they stimulate free radicals and then reaction with oxygen.

EXERCISE 7.6  What polymers would you *not* select for use under the bonnet of a car, where temperatures in excess of 70 °C at the engine manifold can normally be expected? Explain your choice.

SAQ 7.13  (Objective 7.10)
Explain why low-molecular-mass LDPE is more susceptible to photo-oxidation than high-molecular-mass HDPE.

SAQ 7.14   (Objectives 7.7, 7.8, 7.9 and 7.10)
Select, from the key, pairs of characteristics which suggest that a polymeric materials would be:
(a) made by chain growth
(b) made by step growth
(c) insensitive to ultraviolet degradation
(d) sensitive to ultraviolet degradation
(e) suitable for injection moulding.

Key

The polymer:

(i)   has very high molecular mass
(ii)  has molecular mass not much above the critical entanglement value
(iii) has $\bar{M}_w/\bar{M}_n \approx 2$
(iv)  contains carbon black
(v)   contains head-to-head terminations
(vi)  contains —C—N— groups
              ‖  |
              O  H

(vii) has low melting temperature.

## Objectives for Chapter 7

You should now be able to do the following.

7.1 Sketch an Ellingham diagram to express a simple free energy model of how temperature governs the thermodynamic possibility of a chemical reaction. (SAQ 7.2, 7.3)

7.2 Use that model to explain the thermal stability of refractory oxides, how metals may be won from their ores, and how polymers may be formed and may decay. (SAQ 7.4, 7.5)

7.3 Describe the sequences of events as pottery is fired, as cement sets, and as iron is formed in a blast furnace.

7.4 Describe the essential features of any electrochemical system and within that generality explain: anodic and cathodic reactions; how to measure a standard electrode potential; the dual 'permissive' and 'activation' roles of the contact potentials. (SAQ 7.6, 7.7, 7.8, 7.9, 7.10)

7.5 Given an appropriate Tafel diagram comment on the feasibility of a proposed electrochemical reaction.

7.6 Describe qualitatively the factors influencing dry and wet corrosion of metals. (SAQ 7.8, 7.9, 7.10)

7.7 Discuss the implications of polymer molecular mass for processing and service. (SAQ 7.14)

7.8 Calculate the weight and number average molecular masses of polymers from simple data and say why the ratio of these quantities is important. (SAQ 7.11, 7.14)

7.9 Distinguish between chain and step polymerization and give examples of each type; outline the factors that affect whether copolymers are block, random or alternating. (SAQ 7.12, 7.14)

7.10 Outline degradative reactions of polymers with oxygen when heated or exposed to ultraviolet radiation. (SAQ 7.13, 7.14)

7.11 Define and use the following terms and concepts.

| | |
|---|---|
| anode | gel water |
| cathode | number average |
| chain polymerization | overpotential |
| chain transfer reaction | oxidation |
| contact potential | Pilling–Bedworth ratio |
| coupling | polarization diagrams |
| critical entanglement mass | reaction injection moulding |
| dispersion | redox |
| disproportionation | reduction |
| electrode potential | standard electrode potential |
| electroplating | step polymerization |
| Ellingham diagram | Tafel plot |
| exchange current | weight average |
| gel permeation chromatography | weight fraction |

# Answers to Exercises

**EXERCISE 7.1** Avogadro's number is $6 \times 10^{23}$, so $\frac{4}{3}$ of a mole is

$$\frac{4}{3} \times 6 \times 10^{23} \text{ atoms}$$
$$= 8 \times 10^{23} \text{ atoms}$$

**EXERCISE 7.2** Figure 5.30 shows the mixture should become mullite $(3Al_2O_3.2SiO_2)$ and free silica. Nothing further should happen until 1870 K when a eutectic melt appears. In Figure 7.36 silica has been removed leaving just mullite crystals.

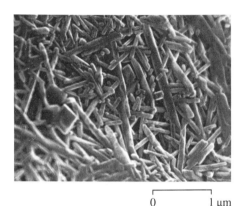

Figure 7.36 Mullite crystals

**EXERCISE 7.3** With the reaction 'just switched on', the current will be approximately $i_e$, which Table 7.4 gives as $10^{-1}$ A m$^{-2}$ for hydrogen discharge on nickel. So with 700 m$^2$ of electrode used for 1 day (86 400 s) the charge passed will be

$$Q = \frac{10^{-1} \times 700 \times 86\,400}{10^5} \text{ faradays}$$

$$\approx 60 \text{ faradays}$$

Each faraday discharges 1 g of H$^+$ ions ($10^{-6}$ tonnes), so the production rate is $60 \times 10^{-6}$ tonnes per day. This is hardly an economic rate. Sufficient overpotential must be applied to increase the current a million fold. A volt or so would do.

**EXERCISE 7.4** Figure 7.17 shows the current of cadmium ions at $\frac{1}{4}$ volt overpotential to be $10^5$ A m$^{-2}$, or $10^5$ C s$^{-1}$ m$^{-2}$. Depositing 1 mole of cadmium requires 2 faradays of charge, so the deposition rate is 0.5 mole s$^{-1}$ m$^{-2}$, or 56 g s$^{-1}$ m$^{-2}$. After a minute 3.36 kg of metal will have been laid down over 1 m$^2$. The thickness is therefore (3.36/8650) m or 0.39 mm.

**EXERCISE 7.5** Transfer and disproportionation reactions lead to *lower* molecular mass polymer whereas coupling leads to *higher* molecular masses. However, one of the two chain ends created in the former reactions possess reactive double bonds, so further activation can occur with free radicals. Transfer to polymer chains (rather than monomer) leads to branched polymer and you will notice that the final joint in the coupled chain has two chain-atoms next to one another (head-to-head joint). This joint is unstable compared to a normal joint, where chlorine atoms are separated by a methylene ($-CH_2-$) group.

**EXERCISE 7.6** Clearly from Table 7.7, you might expect to exclude polystyrene and PVC since their 1000 day thermal lives are 60 °C and 50 °C respectively. Polyolefins, PMMA and polyisoprene are more difficult to assess since 70 °C falls in the 1000 day range. Although the engine manifold is at 70 °C, that temperature will fall off rapidly with distance of the component from the engine. Since the commodity polymers are also the cheapest, one could not exclude polyolefins or natural rubber from the selection. Polypropylene is in fact widely used for battery cases and natural rubber for engine mountings.

# Answers to self-assessment questions

**SAQ 7.1** 'Permission' for a reaction is granted by the free-energy change which would accompany the reaction *if* it occurred. The free energy $G$ must fall, that is $\Delta G$ must be negative. As Chapter 4 showed, whether $\Delta G$ is negative depends on the balance of signs and terms on the right-hand side of

$$\Delta G = \Delta H - T\Delta S$$

'Critical temperature' events are those where the 'sticking' term $\Delta H$ and the 'rattle' term $T\Delta S$ act in opposition. In Figure 4.70 this happens in the two quadrants where temperature governs permission. If $\Delta H$ and $\Delta S$ are both positive, $\Delta G < 0$ at some sufficiently high temperature, whereas if both are negative, $\Delta G < 0$ only if $T$ is low enough.

The rate of permitted change is governed by activation energies. The Arrhenius equation shows that the chance of individual molecules being sufficiently energetic to react rises steeply with temperature. When diffusion is the mechanism for bringing potential reactants together, activation of diffusion may limit the rate of the reaction.

**SAQ 7.2** The entropy decreases as the oxygen gas is incorporated into the solid oxide phase. Therefore $\Delta S$ is negative and the gradient $-\Delta S$ is positive.

**SAQ 7.3**

$$2C + O_2(g) \longrightarrow 2CO(g)$$

says that the oxidation of carbon by 1 mole of gaseous oxygen produces 2 moles of gaseous CO. Therefore the entropy increases as the oxide is formed. This is the opposite of what happens in metal oxidation, $M(s) + O_2(g) \longrightarrow MO_2(s)$ in which the amount of gas decreases. Hence the lines slope oppositely.

**SAQ 7.4** In the lowest parts of the furnace

$$SiO_2 + 2C \longrightarrow Si + 2CO$$

becomes possible at 1950 K. Thus small amounts of silicon (about 1%) appear as an impurity in the molten iron. Reduction of $Al_2O_3$ and CaO requires temperatures in excess of about 2500 K, so these oxides remain in the slag.

Table 7.8

| Reaction | $\Delta G$ at 500 °C/(kJ mole$^{-1}$) | $\Delta G$ at 1000 °C/(kJ mole$^{-1}$) |
|---|---|---|
| $PbS \rightarrow Pb + S$ | $+95$ | $+60$ |
| $2PbO \rightarrow 2Pb + O_2$ | $+280$ | $+180$ |
| $S + O_2 \rightarrow SO_2$ | $-300$ | $-265$ |
| $PbS + 2PbO \rightarrow 3Pb + SO_2$ | $+75$ | $-25$ |

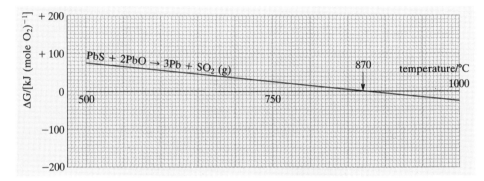

Figure 7.37 Ellingham diagram for lead reduction

SAQ 7.5 The reactions to be added are given in the first three lines of Table 7.8. The last line gives the overall reaction, with $\Delta G$ found by totalling the values in the lines above. $\Delta G$ plotted against temperature for the last reaction gives a straight line crossing $\Delta G = 0$ at 870 °C (Figure 7.37), which is a fairly easily achieved temperature in a wood-fired furnace with a good draught.

SAQ 7.6 The SEP of zinc is $-0.76$ V. The SEP of copper is $+0.34$ V. The relative potential between them is $+0.34$ V $- (-0.76$ V$) = 1.1$ V. Zinc's SEP is more negative (electrons are shed more easily), so the zinc will go into solution and copper will be plated out.

SAQ 7.7 Changing from a copper electrode to one of titanium would apparently reduce hydrogen production by a factor of 100. But as soon as the cathode has a layer of copper it is a copper electrode again!

Provided the copper solution is of reasonable concentration, the exchange current for copper deposition is likely to exceed that for hydrogen production by at least a factor of 1000. That copper was not deposited implies that copper ions were not diffusing to the cathode fast enough to supply the current. Stirring cured the problem.

SAQ 7.8 Not because the groove traps moisture, but because having done so the groove is starved of oxygen. This cannot be the cathode, so it is a potential anode — hence the oxidation of iron.

SAQ 7.9 Bronze, a copper alloy, will form the cathode in a cell with steel so the steel (anode) will corrode. Magnesium has a much more negative standard electrode potential than iron so, even in non standard conditions, may be expected to become the anode of any corrosion cells. The magnesium blocks corrode away and have to be replaced periodically. This is called 'sacrificial protection'.

SAQ 7.10 The pipeline must be made the cathode, so it should be connected to the negative side. The scrap iron should be connected to the positive. The scrap-iron anode will then corrode and harmless cathodic reactions will occur at the pipe.

(a) Saline groundwater will be more conductive, so the current will increase.

(b) Without oxygen, the cathodic reaction

$$H_2O + \tfrac{1}{2}O_2 + 2e^- \longrightarrow 2OH^-$$

cannot happen. The alternative cathodic reaction $2H^+ + 2e^- \rightarrow H_2$ polarizes to a lower current.

(c) To be economic, the current must be low, so the area of cathode is kept small. An uncoated join would be extravagent of current.

SAQ 7.11 To calculate the dispersion, we clearly must calculate both $\bar{M}_w$ and $\bar{M}_n$ for the mixture. The weight fraction of each constituent is 0.25, so

$$\begin{aligned}
\bar{M}_w =\ & (0.25 \times 10\,000) \\
& + (0.25 \times 20\,000) \\
& + (0.25 \times 30\,000) \\
& + (0.25 \times 40\,000) \\
=\ & 2500 + 5000 + 7500 \\
& + 10\,000 \\
=\ & 2.5 \times 10^4
\end{aligned}$$

To find $\bar{M}_n$, we calculate separately the mole fractions. Thus

$$n_1 = \frac{0.25 \times 10^3}{10\,000} = 2.5 \times 10^{-2}$$

and therefore
$n_1 M_1 = 2.5 \times 10^{-2} \times 10^4 = 2.5 \times 10^2$.

Similarly,

$$n_2 = 1.25 \times 10^{-2}$$
$$\text{and } n_2 M_2 = 2.5 \times 10^2$$
$$n_3 = 0.833 \times 10^{-2}$$
$$\text{and } n_3 M_3 = 2.5 \times 10^2$$
$$n_4 = 0.625 \times 10^{-2}$$
$$\text{and } n_4 M_4 = 2.5 \times 10^2$$

$$\begin{aligned}
\bar{M}_n &= \frac{(2.5 + 2.5 + 2.5 + 2.5)10^2}{(2.5 + 1.25 + 0.833 + 0.625)10^{-2}} \\
&= \frac{10}{5.208} 10^4 \\
&= 1.92 \times 10^4
\end{aligned}$$

So

$$\frac{\bar{M}_w}{\bar{M}_n} = \frac{2.5 \times 10^4}{1.92 \times 10^4} = 1.3$$

SAQ 7.12

(a) From Table 7.6, the reactivity ratios of acrylonitrile and styrene reacting together are $r_1 = 0.04$ and $r_2 = 0.40$. Both are less than one, so there will be a tendency to form an alternating copolymer structure.

(b) The reactivity ratios are $r_1 = 1.35$ and $r_2 = 0.78$. So $r_1 \approx r_2 \approx 1$, and the copolymer will tend to be random.

(c) Vinyl chloride–vinyl acetate copolymers will tend to be random but with longer blocks of vinyl chloride repeat units than vinyl acetate blocks.

(d) Acrylonitrile–butadiene polymers will tend to alternate since $r_1$ and $r_2$ are both less than zero (0.02 and 0.3 respectively).

(e) Polymerization of styrene + vinyl acetate mixtures will be complex since $r_1 = 55$ and $r_2 = 0.01$. The reactivity ratios imply that styrene much prefers to react with itself than with vinyl acetate. By contrast, vinyl acetate much prefers to react with styrene than with itself, so the structure of the polymer will consist of long blocks of styrene interspersed with single or short blocks of repeat units of vinyl acetate.

SAQ 7.13 LDPE has numerous side branches, both long and short, produced by chain transfer to polymer chains. These points where the branches meet the main chain are just like the group adjacent to the methyl groups in PP so they make LDPE more susceptible to degradation than HDPE. In addition, low-mass polymer has end groups with double bonds, which are produced by termination by disproportionation or by chain transfer to monomer. They too make LDPE more liable to degradation.

SAQ 7.14

(a) (i), (v)
(b) (iii), (vi)
(c) (i), (iv)
(d) (ii), (vi)
(e) (ii), (vii)

# Appendix 1  Periodic table

| Period | Ia | IIa |
|---|---|---|
| 1 | 1 **H** hydrogen $1s^1$ 1.00 | |

| 2 | 3 **Li** lithium $2s^1$ $1s^2$ 6.94 b | 4 **Be** beryllium $2s^2$ $1s^2$ 9.01 c |
|---|---|---|

| 3 | 11 **Na** sodium $3s^1$ 23.0 | 12 **Mg** magnesium $3s^2$ 24.3 h |
|---|---|---|

| 4 | 19 **K** potassium $4s^1$ 39.1 | 20 **Ca** calcium $4s^2$ 40.1 f |
|---|---|---|

| 5 | 37 **Rb** rubidium $5s^1$ 85.5 | 38 **Sr** strontium $5s^2$ 87.6 |
|---|---|---|

Period 6 (group IIIb series):

| 55 **Cs** caesium | 56 **Ba** barium | 57 **La** lanthanum | 58 **Ce** cerium | 59 **Nd** praseo-dymium | 60 **Nd** neodymium | 61 **Pm** promethium | 62 **Sm** samarium | 63 **Eu** europium | 64 **Gd** gadolinium | 65 **Tb** terbium | 66 **Dy** dysprosium | 67 **Ho** holmium | 68 **Er** erbium | 69 **Tm** thulium | 70 **Yb** ytterbium |
|---|---|---|---|---|---|---|---|---|---|---|---|---|---|---|---|
| $6s^1$ | $6s^2$ | $6s^2$ $5d^1$ | $6s^2$ $5d^0$ $4f^2$ | $6s^2$ $5d^0$ $4f^3$ | $6s^2$ $5d^0$ $4f^4$ | $6s^2$ $5d^0$ $4f^5$ | $6s^2$ $5d^0$ $4f^6$ | $6s^2$ $5d^0$ $4f^7$ | $6s^2$ $5d^1$ $4f^7$ | $6s^2$ $5d^0$ $4f^9$ | $6s^2$ $5d^0$ $4f^{10}$ | $6s^2$ $5d^0$ $4f^{11}$ | $6s^2$ $5d^0$ $4f^{12}$ | $6s^2$ $5d^0$ $4f^{13}$ | $6s^2$ $5d^0$ $4f^{14}$ |
| 132 b | 137 b | 139 | 140 | 141 | 144 | 145 | 150 | 152 | 157 h | 159 | 162 | 165 | 167 | 169 | 173 |

Period 7:

| 87 **Fr** francium | 88 **Ra** radium | 89 **Ac** actinium | 90 **Th** thorium | 91 **Pa** protact-inium | 92 **U** uranium | 93 **Np** neptunium | 94 **Pu** plutonium | 95 **Am** americium | 96 **Cm** curium | 97 **Bk** berkelium | 98 **Cf** californium | 99 **Es** einsteinium | 100 **Fm** fermium | 101 **Md** mendelevium | 102 **No** nobelium |
|---|---|---|---|---|---|---|---|---|---|---|---|---|---|---|---|
| $7s^1$ | $7s^2$ | $7s^2$ $6d^1$ | $7s^2$ $6d^2$ $5f^0$ | $7s^2$ $6d^1$ $5f^2$ | $7s^2$ $6d^1$ $5f^3$ | $7s^2$ $6d^1$ $5f^4$ | $7s^2$ $6d^0$ $5f^6$ | $7s^2$ $6d^0$ $5f^7$ | $7s^2$ $6d^1$ $5f^7$ | $7s^2$ $6d^0$ $5f^9$ | $7s^2$ $6d^0$ $5f^{10}$ | $7s^2$ $6d^0$ $5f^{11}$ | $7s^2$ $6d^0$ $5f^{12}$ | $7s^2$ $6d^0$ $5f^{13}$ | $7s^2$ $6d^0$ $5f^{14}$ |
| 223 | 226 | 227 | 232 | 231 | 238 | | | | | | | | | | |

group    Ia      IIa  ——————————————————————————— IIIb ———————————————————

The bottom line in each entry shows atomic mass and, for selected elements, crystal structure at room temperature.
b = body centred cubic    f = face centred cubic    d = diamond    h = hexagonal close packed

| | | | | | | | | | | | | | | | | | 2 **He** |
|---|---|---|---|---|---|---|---|---|---|---|---|---|---|---|---|---|---|
| | | | | | | | | | | | | | | | | | helium |
| | | | | | | | | | | | | | | | | | $1s^2$ |
| | | | | | | | | | | | | | | | | | 4.00 |

| 5 **B** | 6 **C** | 7 **N** | 8 **O** | 9 **F** | 10 **Ne** |
|---|---|---|---|---|---|
| boron | carbon | nitrogen | oxygen | fluorine | neon |
| $2p^1$ | $2p^2$ | $2p^3$ | $2p^4$ | $2p^5$ | $2p^6$ |
| $2s^2$ | $2s^2$ | $2s^2$ | $2s^2$ | $2s^2$ | $2s^2$ |
| $1s^2$ | $1s^2$ | $1s^2$ | $1s^2$ | $1s^2$ | $1s^2$ |
| 10.8 | 12.0 d* | 14.0 | 16.0 | 19.0 | 20.2 |

| 13 **Al** | 14 **Si** | 15 **P** | 16 **S** | 17 **Cl** | 18 **Ar** |
|---|---|---|---|---|---|
| aluminium | silicon | phosphorus | sulphur | chlorine | argon |
| $3p^1$ | $3p^2$ | $3p^3$ | $3p^4$ | $3p^5$ | $3p^6$ |
| $3s^2$ | $3s^2$ | $2s^2$ | $3s^2$ | $3s^2$ | $3s^2$ |
| 26.9 f | 28.1 d | 31.0 | 32.1 | 35.5 | 39.9 |

| 21 **Sc** | 22 **Ti** | 23 **V** | 24 **Cr** | 25 **Mn** | 26 **Fe** | 27 **Co** | 28 **Ni** | 29 **Cu** | 30 **Zn** | 31 **Ga** | 32 **Ge** | 33 **As** | 34 **Se** | 35 **Br** | 36 **Kr** |
|---|---|---|---|---|---|---|---|---|---|---|---|---|---|---|---|
| scandium | titanium | vanadium | chromium | manganese | iron | cobalt | nickel | copper | zinc | gallium | germanium | arsenic | selenium | bromine | krypton |
| | | | | | | | | | | $4p^1$ | $4p^2$ | $4p^3$ | $4p^4$ | $4p^5$ | $4p^6$ |
| $4s^2$ | $4s^2$ | $4s^2$ | $4s^1$ | $4s^2$ | $4s^2$ | $4s^2$ | $4s^2$ | $4s^1$ | $4s^2$ | $4s^2$ | $4s^2$ | $4s^2$ | $4s^2$ | $4s^2$ | $4s^2$ |
| $3d^1$ | $3d^2$ | $3d^3$ | $3d^5$ | $3d^5$ | $3d^6$ | $3d^7$ | $3d^8$ | $3d^{10}$ | $3d^{10}$ | $3d^{10}$ | $3d^{10}$ | $3d^{10}$ | $3d^{10}$ | $3d^{10}$ | $3d^{10}$ |
| 50.0 h | 47.9 h | 50.9 b | 52.0 b | 54.9 | 55.9 b | 58.9 h | 58.7 f | 63.6 f | 65.4 h | 69.7 f | 72.6 d | 74.9 | 79.0 h | 79.9 | 83.8 |

| 39 **Y** | 40 **Zr** | 41 **Nb** | 42 **Mo** | 43 **Tc** | 44 **Ru** | 45 **Rh** | 46 **Pd** | 47 **Ag** | 48 **Cd** | 49 **In** | 50 **Sn** | 51 **Sb** | 52 **Te** | 53 **I** | 54 **Xe** |
|---|---|---|---|---|---|---|---|---|---|---|---|---|---|---|---|
| yttrium | zirconium | niobium | molybdenum | technetium | ruthenium | rhodium | palladium | silver | cadmium | indium | tin | antimony | tellurium | iodine | xenon |
| | | | | | | | | | | $5p^1$ | $5p^2$ | $5p^3$ | $5p^4$ | $5p^5$ | $5p^6$ |
| $5s^2$ | $5s^2$ | $5s^1$ | $5s^1$ | $5s^2$ | $5s^1$ | $5s^1$ | $5s^0$ | $5s^1$ | $5s^2$ | $5s^2$ | $5s^2$ | $5s^2$ | $5s^2$ | $5s^2$ | $5s^2$ |
| $4d^1$ | $4d^2$ | $4d^4$ | $4d^5$ | $4d^5$ | $4d^7$ | $4d^8$ | $4d^{10}$ | $4d^{10}$ | $4d^{10}$ | $4d^{10}$ | $4d^{10}$ | $4d^{10}$ | $4d^{10}$ | $4d^{10}$ | $4d^{10}$ |
| 88.9 | 91.2 h | 92.9 b | 95.9 b | 98.9 | 101 | 103 f | 106 f | 108 f | 112 h | 115 | 119 d | 121 | 128 h | 127 | 131 f |

| 71 **Lu** | 72 **Hf** | 73 **Ta** | 74 **W** | 75 **Re** | 76 **Os** | 77 **Ir** | 78 **Pt** | 79 **Au** | 80 **Hg** | 81 **Tl** | 82 **Pb** | 83 **Bi** | 84 **Po** | 85 **At** | 86 **Rn** |
|---|---|---|---|---|---|---|---|---|---|---|---|---|---|---|---|
| lutetium | hafnium | tantalum | tungsten | rhenium | osmium | iridium | platinum | gold | mercury | thallium | lead | bismuth | polonium | astatine | radon |
| | | | | | | | | | | $6p^1$ | $6p^2$ | $6p^3$ | $6p^4$ | $6p^5$ | $6p^6$ |
| $6s^2$ | $6s^2$ | $6s^2$ | $6s^2$ | $6s^2$ | $6s^2$ | $6s^2$ | $6s^1$ | $6s^1$ | $6s^2$ | $6s^2$ | $6s^2$ | $6s^2$ | $6s^2$ | $6s^2$ | $6s^2$ |
| $5d^1$ | $5d^2$ | $5d^3$ | $5d^4$ | $5d^5$ | $5d^6$ | $5d^7$ | $5d^9$ | $5d^{10}$ | $5d^{10}$ | $5d^{10}$ | $5d^{10}$ | $5d^{10}$ | $5d^{10}$ | $5d^{10}$ | $5d^{10}$ |
| $4f^{14}$ | $4f^{14}$ | $4f^{14}$ | $4f^{14}$ | $4f^{14}$ | $4f^{14}$ | $4f^{14}$ | $4f^{14}$ | $4f^{14}$ | $4f^{14}$ | $4f^{14}$ | $4f^{14}$ | $4f^{14}$ | $4f^{14}$ | $4f^{14}$ | $4f^{14}$ |
| 175 | 178 | 181 b | 184 b | 186 | 190 | 192 f | 195 f | 197 f | 201 | 204 | 207 f | 209 | 209 | 210 | 222 |

| 103 **Lr** | 104 **Rf** | 105 **Ha** | 106 |
|---|---|---|---|
| lawrencium | rutherfordium | hahnium | |
| $7s^2$ | | | |
| $6d^1$ | | | |
| $5f^{14}$ | | | |

| | IVb | Vb | VIb | VIIb | | VIII | | Ib | IIb | IIIa | IVa | Va | VIa | VIIa | 0 |
|---|---|---|---|---|---|---|---|---|---|---|---|---|---|---|---|

*Carbon also exists as graphite, which has a close packed hexagonal structure.

# Appendix 2 Names, abbreviations and descriptions of common polymers

| Abbreviation | Chemical or technical name | Repeat unit | Comments |
|---|---|---|---|
| *Thermoplastics* | | | |
| HDPE | High density polyethylene | See PE | Characterized by side branches |
| LDPE | Low density polyethylene | See PE | |
| MDPE | Medium density polyethylene | See PE | |
| PA6 | Nylon 6 | $-N(H)-(CH_2)_5-C(=O)-$ | The chemical name for nylons is polyamide. 6 carbon atoms in the repeat unit |
| PA6, 6 | Nylon 6, 6 | $-N(H)-(CH_2)_6-N(H)-C(=O)-(CH_2)_4-C(=O)-$ | $-C(=O)-N(H)-$ is the *amide* group. 6 carbon atoms in each part of the repeat unit. |
| PAN | Polyacrylonitrile | $-CH_2-CH(CN)-$ | Normally atactic |
| PC | Polycarbonate | $-O-C_6H_4-C(CH_3)_2-C_6H_4-O-C(=O)-$ | $-O-C(=O)-$ is the *ester* group. See also PET |
| PE | Polyethylene | $-CH_2-CH_2-$ | |
| PET | Poly(ethylene terephthalate) | $-O-CH_2-CH_2-O-C(=O)-C_6H_4-C(=O)-$ | |
| PMMA | Poly(methyl methacrylate) | $-CH_2-C(CH_3)(COOCH_3)-$ | Normally atactic |
| POM | Polyoxymethylene | $-CH_2-O-$ | Also known as acetal resin |
| PP | Polypropylene | $-CH_2-CH(CH_3)-$ | Normally isotactic |
| PPO | Polyphenylene oxide | $-C_6H_2(CH_3)_2-O-$ | Often blended with PS |
| PS | Polystyrene | $-CH_2-CH(C_6H_5)-$ | Normally atactic |

| Abbreviation | Chemical or technical name | Repeat unit | Comments |
|---|---|---|---|
| *Thermoplastics* | | | |
| PTFE | Poly(tetrafluoroethylene) | $-CF_2-CF_2$ | |
| PVC | Poly(vinyl chloride) | $-CH_2-CH-$ with $Cl$ | Atactic and often heavily plasticized |
| PVAC | Poly(vinyl acetate) | $-CH_2-CH-$ with $O$, $O=C-CH_3$ | |
| PVA(L) | Poly(vinyl alcohol) | $-CH_2-CH-$ with $OH$ | |
| PI | Polyimide | | |
| PU | Polyyurethane | $-O-\overset{O}{C}-\overset{H}{N}-R-\overset{H}{N}-\overset{O}{C}-O-R'-$ | $-O-\overset{O}{C}-\overset{H}{N}-$ is the *urethane* group. Various groups at R and R′ give wide range of urethanes |
| UPVC | Unplasticized PVC | See PVC | |
| *Elastomers* | | | |
| BR | Polybutadiene rubber | $-CH_2-C=CH-CH_2$ with $H$ | |
| CR | Polychloroprene rubber/Neoprene | $-CH_2-C=CH-CH_2-$ with $Cl$ | Note the similarity of the molecular structures |
| IR | Polyisoprene rubber | $-CH_2-C=C-CH_2-$ with $CH_3$ $H$ | |
| NR | Natural rubber | $-C=CH-CH_2-$ with $CH_3$ | |

| Abbreviation | Chemical or technical name | Repeat unit | Comments |
|---|---|---|---|
| *Thermoplastics* | | | |
| PIB | Polyisobutylene/Butylrubber | $-CH_2-\underset{\underset{CH_3}{|}}{\overset{\overset{CH_3}{|}}{C}}-$ | |
| VMQ | Silicone rubber | $-\underset{\underset{CH_3}{|}}{\overset{\overset{R}{|}}{Si}}-O-$ | $R = H$ or $CH_3$ or $C_6H_5$ |
| *Copolymers* | | | |
| ABS | Acrylonitrile–butadiene–styrene | | SAN with BR |
| HIPS | High impact polystyrene | | PS with BR |
| NBR | Nitrile rubber | | Butadiene with acrylonitrile |
| PVCA | Vinyl chloride-vinyl acetate | | |
| SAN | Styrene–acrylonitrile | | |
| SBR | Styrene–butadiene rubber | | random structure |
| SBS | Styrene-butadiene-styrene rubber | | block structure |

| Abbreviation | Chemical or technical name | Reactive groups | Comments |
|---|---|---|---|
| *Thermosetting plastics and composites** | | | |
| CFRP | Carbon (fibre) reinforced plastic | See UP and EP | Plastic is usually either UP or EP |
| EP | Epoxy resin | $-CH-CH-$ (epoxide, O bridging) | |
| GFRP | Glass (fibre) reinforced plastic | see UP and EP | As above |
| MF | Melamine formaldehyde | $-N(CH_2OH)_2-$ | Cross-link on heating |
| PF | Phenol formaldehyde | (phenol ring with OH, $CH_2$, $CH_2$ groups) | Cross-link on heating |
| PU | polyurethane | $-N=C=O-$ | See also thermoplastics |
| UP | Unsaturated polyester | $-\overset{\overset{O}{\|}}{C}-CH=CH-\overset{\overset{O}{\|}}{C}-$ | Cross-links are produced via *unsaturated* (C=C) bonds in linear polyesters and polyurethanes |

*Since the chemical composition of cross-linked plastics cannot be specified accurately, this table includes only the reactive groups, not the molecular products.

# Acknowledgements

Grateful acknowledgement is made to the following for the figures used in this book:

Figure 2.8(b), N. Stoloff, form *Physical Bases of Yield Fracture*; Conference Proceedings © 1966 Institute of Physics. Figure 2.23, B. G. Butterfield and M. A. Meylan *Three Dimensional Structure of Wood* 2nd ed. Chapman and Hall 1980. Figure 2.29(a), C. Hall *Polymer Materials* Macmillan © 1981 C. Hall. Figure 2.35(a), H. M. Finniston *Structure Characteristics of Materials* © 1971 Elsevier. Figure 2.35(b), R. M. Rose, L. A. Shepard and J. Wulf *The Structure and Properties of Materials* Vol IV 'Electronic properties' © 1966 John Wiley. Figure 2.38(a), R. Carey and D. Isaac *Magnetic Domains and Techniques for their Observation* English Universities Press © 1985 R. Carey and D. Isaac. Figure 2.39(a), © Dr R. P. Kambour. Figure 2.56, W. T. Read Jnr *Dislocations in Crystals* © 1953 McGraw-Hill. Figure 2.58(a), R. B. Nicholson, from A. H. Cottrell *An Introduction to Metallurgy* Edward Arnold © 1967 A. H. Cottrell. Figure 4.12, © The Welding Institute. Figure 4.33, J. Brett and L. Seigle 'The role of diffusion versus plastic flow in the sintering of model compacts' *Acta Metallurgica* Vol 14 © 1966 Pergamon Press. Figure 5.2, J. H. Debussy *Materials and Technology* Vol III 'Metals and Ores' © 1970 Longman-J. H. Debussy. Figure 5.18, Dr A. G. Cullis *Philosophy Magazine* Vol. 30 1974 Taylor & Francis Ltd. Figure 5.35(b), K. Reynolds. Figure 5.45, M. F. Ashby 'The deformation of plastically non-homogeneous alloys' in A. H. Kelly and R. B. Nicholson (eds.) *Strengthening Methods in Crystals* Elsevier 1971 © M. F. Ashby. Figure 6.49(b), R. Hermann. Figure 7.5(a), S. Lewis. Figures 2.1, 2.5, 2.6, 2.9, 2.16, 2.17, 2.18, 2.30, 2.34, 2.36, 4.25, 6.42, Richard Black. Figures 1.37, 2.8, 2.26, 2.40(a), 2.59, 2.60, 2.61(a), 5.22, 6.41, 7.4(b), 7.5(b), 7.36, Naomi Williams.

# Index

Page number prefixed by ▼ refer to red text